PHYSICAL STRUCTURE

in

SYSTEMS THEORY

PHYSICAL STRUCTURE
in
SYSTEMS THEORY

Network Approaches to Engineering and Economics

edited by

J. J. van DIXHOORN

Department of Electrical Engineering
Twente University of Technology
Enschede, The Netherlands

and

F. J. EVANS

Department of Electrical and Electronic Engineering
Queen Mary College
University of London, England

1974

ACADEMIC PRESS
LONDON NEW YORK SAN FRANCISCO
A Subsidiary of Harcourt Brace Jovanovich, Publishers

ACADEMIC PRESS INC. (LONDON) LTD.
24–28 Oval Road,
London NW1

United States edition published by
ACADEMIC PRESS INC.
111 Fifth Avenue
New York, New York 10003

Library of Congress Catalog Card Number: 74-5658
ISBN: 0 12 712450 0

Text set in 10/11 pt. IBM Press Roman, printed by photolithography, and bound in Great Britain at The Pitman Press, Bath

LIST OF CONTRIBUTORS

K. M. ADAMS; *Department of Electrical Engineering, Delft University of Technology, Delft, The Netherlands*

J. J. VAN DIXHOORN; *Department of Electrical Engineering, Twente University of Technology, Enschede, The Netherlands*

J. M. ENGLISH; *School of Engineering and Applied Science, U.C.L.A., Los Angeles, California, U.S.A.*

F. J. EVANS; *Department of Electrical and Electronic Engineering, Queen Mary College, University of London, England*

O. I. FRANKSEN; *Electric Power Engineering Department, Technical University of Denmark, Lyngby, Denmark*

D. J. HOLDING; *Department of Electrical and Electronic Engineering, Queen Mary College, University of London, England*

D. L. JONES; *A.W.R.E., Aldermaston, Berkshire, England*

J. KORN; *Faculty of Engineering, Science and Mathematics, Middlesex Polytechnic, Enfield, Middlesex, England*

J. W. LYNN; *Department of Electrical Engineering and Electronics, University of Liverpool, England*

J. W. POLDER; *Department of Mechanical Engineering, Eindhoven University of Technology, Eindhoven, The Netherlands*

J. RIETMAN; *Royal Naval College, Den Helder, The Netherlands*

R. A. RUSSELL; *Department of Electrical Engineering and Electronics, University of Liverpool, England*

M. J. SEWELL; *Department of Mathematics, University of Reading, Berkshire, England*

J. U. THOMA; *Consulting Studio, Bellevue 23, Zug, Switzerland*

J. C. WILLEMS; *Mathematical Institute, University of Groningen, The Netherlands*

It is surprising how few ultimate types of elements there are that form the building blocks of the great variety of engineering structures. . . . . . The great variety of structures differ only by the manner of interconnections . . . . . and the variety of theories only by the type of hypothetical reference frame assumed.

Gabriel Kron
'Tensors for circuits'–1942

PREFACE

In recent years there have been considerable developments in the area of generalised dynamic systems analysis. It is a field that embraces a variety of disciplines, but is substantially based on an integration of thermodynamic, network, field and variational concepts. It represents an attempt to strengthen the *physical* basis of the subject, and this has already resulted in new unifying concepts related to the classification of physical variables, the generalisation of electrical network theory to other domains, and the powerful, but not yet well known, methods of bond graphs.

In February 1973, following an initiative by F. J. Evans of Queen Mary College, London, a three-day colloquium was held at Twente University of Technology, bringing together, by invitation, fifteen active workers in this field. Amongst them were O. I. Franksen (Denmark) and J. M. English (U.S.A.) whose common interests extended the scope of the meeting to include economic networks.

It was fortunate that so many enthusiasts were able to come together, and even more so that they were willing to produce these papers in a form which reflects the many stimulating discussions that were held.

The contributions in this volume are preceded by a chapter of historical and introductory nature. Five chapters are concerned with network and bondgraph modelling in general, and their applications in thermal and mechanical systems particularly. The following five chapters explore the possibilities of a unified theory of physical systems arising from the common ground between generalised network theory, irreversible thermodynamics and variational analysis. Both of the final two chapters pursue possible applications in the modelling and theory of economic systems.

The volume addresses itself mainly to those in higher engineering education interested in a unified systems approach on a physical basis, and it is hoped that they will be found throughout departments of mechanical, electrical, control and systems engineering, applied mathematics and business economics.

J. J. VAN DIXHOORN

CONTENTS

TOWARDS MORE PHYSICAL STRUCTURE IN SYSTEMS THEORY

F. J. Evans *and* J. J. van Dixhoorn
Queen Mary College, University of London, England
Twente University of Technology, Enschede, The Netherlands

1. PHYSICS AND MATHEMATICS

Mathematics is the language of the physicist and engineer and, broadly speaking, this volume is concerned with the way in which mathematics and physics impinge on one another. This is done by a number of specific studies, most of which are firmly based in the real physical world. The mathematicians' major criterion is that of consistency, and this must be satisfied by any "system" that he produces. There are no demands that his systems be recognizable in the physical world, for it is well known that a large body of mathematical knowledge came originally from purely abstract notions devoid of any physical connotations. But constantly both the physicist and engineer call on this storehouse of mathematical structures and fit some of the contents to their prevailing problems.

Many examples can be given of a striking correlation between forms that have been derived from opposite ends of the mathematical-physical spectrum. This is particularly true of network theory and associated topics to which a large part of this book is devoted. In this middle ground there is much scope for both the intuitive and more formalistic approaches to meet productively, and we hope that, to some extent, this has occurred in this collection of papers.

This field is one that generates considerable feeling and difference of opinion, and it is still too early in its development to ask that all subscribe to a well-defined common view. This is apparent from the contributions, but we prefer to accept this as a promise of future evolution and not as a sign of present disarray. In fact, it is remarkable that so much common ground exists between such a diverse group of people brought together for the first time. The results of the meeting only confirm our previously held conviction that in recent systems theory insufficient attention has been given to that which could be called "physical structure" in its own right, as opposed to structure that arises purely from mathematical considerations.

The whole subject of linear systems analysis for example illustrates how the mathematical description of physical systems is often modified to provide ease of solution, but how, during this process, major physical characteristics can be lost. Implicit in the transfer function approach is the assumption that no element in a system can affect any other element that occurs "earlier" in the causal chain, except by the addition of some "feedback" path which is often neglected. Such a mechanism is completely contrary to that which exists universally in the physical world, where no system element can be connected to another without the one reacting on the other. In engineering, considerable effort is devoted to designing non-interactive elements, and, although this has achieved a high degree of success, it does not necessarily justify our ignoring the real nature of acquiring physical measurements, that is that the insertion of any measuring device must affect the system. To simplify the description of a physical system as a mathematical expedient (i.e. for reasons of solution) can, in some cases, prevent the behaviour of the system from being described in terms of particularly physical concepts (e.g. in terms of some potential functions), and reduce the analysis to numerical techniques.

It would seem to be indispensable to a full understanding of how systems actually "work", that both physical and mathematical concepts be integrated to a far greater degree than at present. This thought is not new, of course, but it is probably more important now than ever before if we are to remove some of the dividing barriers, classifying systems for mathematical reasons, that have been erected in the past, but that are now restricting our thinking.

2. HISTORICAL DEVELOPMENT OF NETWORK MODELLING

Electrical networks

All equations describing physical systems are concerned with statements of conservation. In electrical network theory these statements are also easily related to the observable structure of the system. The linear graph, used to describe a network, has the same topological properties as the real system itself, and this constitutes a unique advantage. Furthermore electrical network theory possesses intimate associations with a large body of electromagnetic field theory. Finally at an early stage of technical development electrical network theory had reached a useful level, offering many possibilities for calculation. So it is not surprising that it is electrical networks that have become the accepted framework on which to build analogies between different types of physical system, once an adequate classification of variables has been established.

As early as 1847 Kirchhoff formulated his famous topological theorems for electrical networks by which the foundation was laid for the later graph-theoretic treatment. It was nearly 100 years before the great generality of these theorems was realized and related to classical mechanics. Kirchhoff's nodal, or current, law which states that charge can neither be stored or lost at a node, is now seen as analogous to that made by D'Alembert, nearly 100 years earlier when he restated Newton's second law to show that the sum of all the forces acting on a body was zero (cf. the algebraic sum of all currents at a node is zero). The correspondence of this law and its dual, that the algebraic sum of the voltage differences in any closed circuit is also zero, to classical mechanics is discussed by Adams and also by a number of other authors.

The elements that compose an electric circuit have the added advantage that they can be well defined in terms of a characteristic relationship between the measurable variables of current and voltage. By the fact that these relationships are virtually linear a considerable ease of solution of the associated differential equations results for the steady state sinusoidal case, greatly enhanced by the use of complex variables and the introduction of the impedance concept by Steinmetz in 1893.

Also in 1893 Oliver Heaviside first provided a solution to the non-periodic transient behaviour of switched circuits by his operational calculus. The mathematical rigour of this approach was increased both by Van der Pol in 1922 and later by Doetsch in 1937 who reintroduced the Laplace transform which had been first proposed in 1779. It was not until as late as 1942 that in the very related field of automatic control the Fourier and Laplace transforms were really adopted as the means to solve linear differential equations, and so the door was opened to transfer function analysis and all related techniques.

Modern developments in mathematical graph theory also occupy an important place in the development of general electrical network theory, and are likely to do so to an increasing extent. Whitney's proof (1932) [1, 2] of the non-existence of duals for non-planar graphs and its connection with the validity of the force-current analogy, as opposed to the force-voltage analogy, is an interesting case.

It is also important to note that in graph theory there exists a systematic method of selecting sets of independent variables with which to describe a system. The partitioning of a graph into two interconnected sub-graphs, the tree and the set of links, provides this facility.

Added impetus to the application of graph theory undoubtedly comes from the associated use of matrix algebra, not only as a descriptive notation, but also as the ideal symbolism for computer-aided methods (Veblen, 1931, [3]).

The non-electrically oriented reader can find suitable introductions to such methods in many standard text books, see references [6], [7] and [8], which also provide examples of their use in an inter-disciplinary manner. Fundamental treatments are given in Belevitch [4] and Seshu and Reed [5].

Analogies

In 1926 *The Engineer* published a series of articles on "Models and analogies for demonstrating electrical principles" [9], showing that an early impulse in the search for electrical analogies was to find visible systems which could demonstrate the hidden phenomena in electrical networks. It is interesting to reflect that the newly acquired ability to actually observe the variation of fast transients in electrical circuits by the newly invented instruments, such as the oscillograph and cathode ray oscilloscopes, reversed this line of thought. Actually it was one year earlier, in 1925, that Nickle [10] had written "Oscillographic solution of electromechanical systems". He started a new era interested in electrical analogues of *non*-electrical systems.

Network-theoretic concepts, like mechanical and acoustical impedances were successfully introduced for the development of electro-acoustic devices, like microphones and telephone-receivers. Olson's "Dynamical Analogies", 1943, [11], remains a comprehensive and valuable work in this field, where the linear description is valid and mechanical movements are restricted to one dimensional translation.

The derivation of electrical analogies for non-electroacoustical systems has been much less straightforward. An excellent literature survey on "Engineering Analogies" was published in 1963 by Murphy *et al.* [12] and in this survey it is shown that useful analogies can arise from isomorphic mathematical structures. But the models often lack consistency in terms of variable correspondence and do not show a systematic physical structure.

Across- and through-variables

From the mentioned electro-mechanical-acoustical field, however, very fundamental results in the search for structure can be traced. The original analogy between mechanical and electrical systems was formulated by making force analogous to voltage. Maxwell has been the first to state this in [13], referring to the identical form of Lagrange's equations in mechanics. The word "resistance" is traditionally used in this sense: a high resistance implies either a large mechanical force or a large voltage. Mathematically the force-current analogy is equally valid. It was not until the 1930s that the practical and theoretical advantages of this "mobility analog" were formulated independently of each other by Darrieus in 1929 [14], Hähnle in 1932 [15] and Firestone in 1933 [16]. Firestone introduced the distinction between two types of physical variables: across-variables and through-variables. Across-variables, like velocity, voltage, pressure, temperature, and concentration-(chemical potential) are expressed as differences and are measured across two spatially different points. The corresponding through-variables, like force, current, fluid flow, heat flow, and mass flow are measurable at a single point. The product of each variable pair has the dimension of power.

The operational aspects of the classification are briefly discussed by Franksen. As Jones and his co-authors show, this classification carries with it certain implications of quite a fundamental nature, and this is also confirmed by Korn *et al.* One cannot dissociate the topology from the "through" and "across" concepts.

From a philosophical point of view, it can be seen that the existence of two basic types of variable, those we have termed "through" and "across", provides the simplest means of generating invariant functions such as energy and power. If there exist two basic types of variable with which to describe the physical world, (viz. co-variant, contra variant) and if there is to be any consensus of agreement by two observers of any particular event when viewed from their respective frames of reference, then a product of a co-variant and a contra-variant can generate the required invariant (e.g. voltage times current gives power).

The work of Trent (1953) [17]; "Isomorphisms between oriented linear graphs and lumped physical systems" provided a more rigorous basis for Firestone's "complete" analogies as well as algorithmic rules for their construction, by the use of the linear graph and its associated matrix algebra.

Network graphs in physical systems engineering

Trent's work formed the basis from which several other authors developed a unified approach to dynamic system analysis. Koenig and Blackwell (1961) [13] could probably be considered as providing the most outstanding contribution at that time with their "Electro-mechanical system theory". In this they ask "Is there an engineering discipline which truly encompasses the wide variety of components

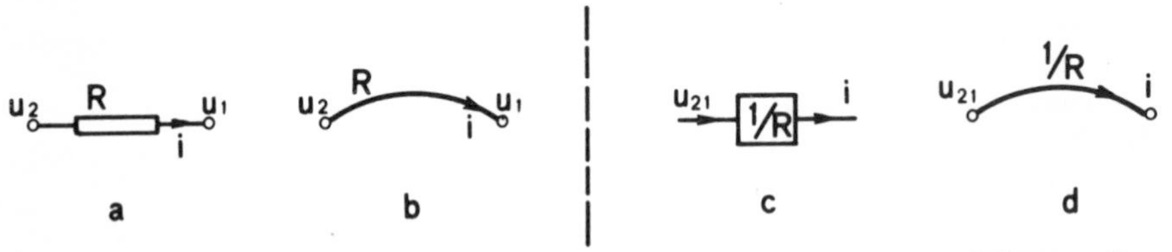

Fig. 1. (a) network component; (b) network graph of (a); (c) block diagram of the operation $i = 1/R \,.\, u_{21}$; (d) signal flow graph of the same operation.

or hardware found in the area referred to as systems?". Maxwell had, in fact, asked the same question when reflecting on the form of Lagrange's equation, which is generally applicable to a wide class of system, and introduced for the first time the notion of the generalized coordinate and the generalized force.

Koenig and Blackwell decided that if the key to such a discipline did exist it was in the concept of the linear graph, and not essentially in the notions of analogies. It could be said that they were responsible for the establishment of the field of "physical systems engineering" by their thorough treatment of a wide variety of engineering devices, using the full power of matrix methods. They did, however, leave several outstanding questions concerning motion in two or three dimensions, and others related to nonlinearity.

All of these developments should be distinguished from those that came under the heading of signal flow graphs. As their title suggests, these linear graphs represent a system as a signal processing device, from input to output, and are directly related to the transfer function models for linear systems and to the block diagrams for general systems (Fig. 1). To prevent confusion, a linear graph, representing the topology of a network, will be called a network graph.

Network graphs in dynamic systems education

In September 1961 a workshop was organized at Massachusetts Institute of Technology, Cambridge, USA to bring together 20 key people from different departments, having a common interest in dynamic analysis of lumped parameter linear systems [19]. It was generally recognized that many of the offerings of various departments were very fundamental and had much in common with those of other departments, but that subjects are taught with a specialized terminology and oriented toward only one class of application. It was believed that there would be considerable value in the development of certain of these common areas on an interdisciplinary basis. A proposed new interdisciplinary undergraduate subject, "Dynamic Systems", was considered to be ideally suited for such a role.

Such a course was started in 1962. Seven years later the class-room tested and reshaped material was published by Shearer *et al.* in the form of an excellent book [20], in which the main step in the process of modelling the real physical world with its complications to mathematical formulas is taken by using network graphs together with across- and through-variables. Many comparable texts on an introductory level using these methods have appeared [21–25].

The non-electric network models in such courses are always restricted to simple, one-dimensional mechanical systems, to systems having rotations about fixed parallel shafts, to fluid systems with constant density (hydraulical or acoustical) or to conductive or radiative thermal systems.

Energy conversion between the different kinds of energy is treated, but thermal energy conversion is always excluded.

As far as the second author is aware there has neither been any progress on an advanced level beyond these restrictions or those reached by Koenig and Blackwell. This state of affairs clearly prohibits the use of network graphs for most practical problems, neither can it be expected that graduate teachers are very motivated to integrate such a network approach in their own specialistic courses. A breakthrough is needed.

Genesis of bond graphs

Henry Paynter's bond graphs are the breakthrough. The original concept that is central to the bond graph approach is that of the multi-energy port representation, associated with a reticulation (i.e. the making of a network, Latin: reti = fishnet) of the system. Paynter has said "The rational process of endowing a system with structure we call reticulation". The bond graph embraces a greater number of the distinct concepts outlined above than any other single formulation to date. In his class notes (1960–1) [27] Paynter presents the broad background leading to his comprehensive treatment, which he first started to present to his students at M.I.T. in 1955. He broke away from the dominating overtones of electrical networks, although always acknowledging the fact that his generalizations were firmly rooted both in Kirchhoff's original network theorems and the sweeping field concepts of Maxwell. It was Kirchhoff who had first shown the relation between the field theorems of Stokes and Gauss and the network voltage and current laws respectively.

Paynter adopted the notion of the energy port that had been previously introduced into network theory in 1949 by Wheeler [26]. The term was a general description of an entrance or exit of a network, or of an electromagnetic waveguide and it was argued that energy transfer is not restricted to a circuit with wires, conduction and two terminals. Paynter considered that other physical systems could be represented by a multi-port reticulation in a given region of space, and the limitations of the conventional form of diagram and symbols had to be removed by resort to a new form, not unlike that used to represent the bonding between atoms in complex chemical molecules.

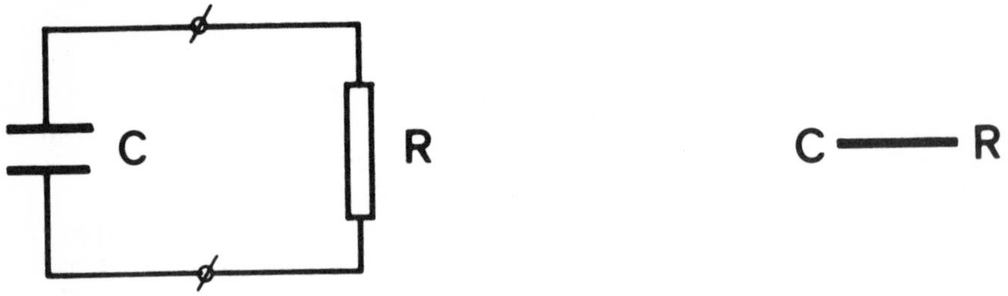

Fig. 2. Circuit and corresponding bond graph.

With each bond was associated an effort and a flow variable corresponding with the across and through terminology, except for mechanical systems where their meaning is interchanged.

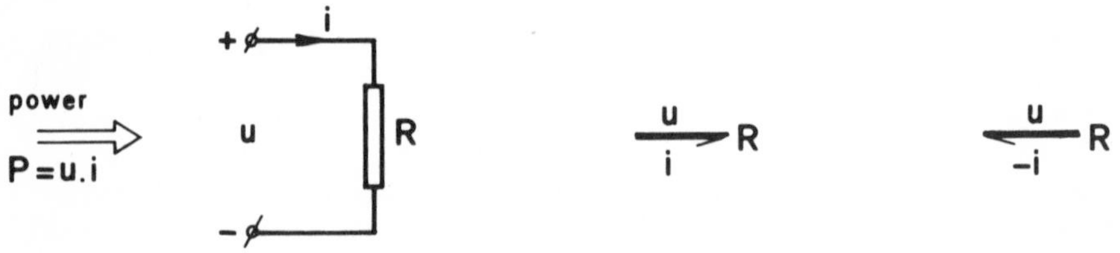

Fig. 3. (a) Sign conventions at a one-port; (b) load convention or (c) source convention.

By orienting each bond with a half arrow the assumed positive direction of power flow was introduced by Rosenberg, necessary for network-like calculations or for numerical computation. Both kinds of calculation are described by Karnopp and Rosenberg in their book devoted to the bond graph techniques (1968) [28].

Paynter's graphical symbolism is extremely economical and efficient. By a single line he represents the power bond and the energy ports of interacting elements of a system. Two-ports, such as transformers and gyrators, were simply represented by two bonds, and eventually, in 1959, the general multi-port junctions were proposed: the 0-junction to represent the parallel electrical connection in which the "effort" is shared, and the 1-junction representing the series electrical connection in which the "flow" is shared.

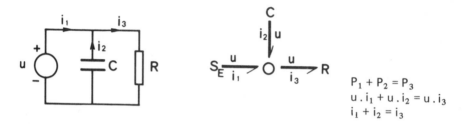

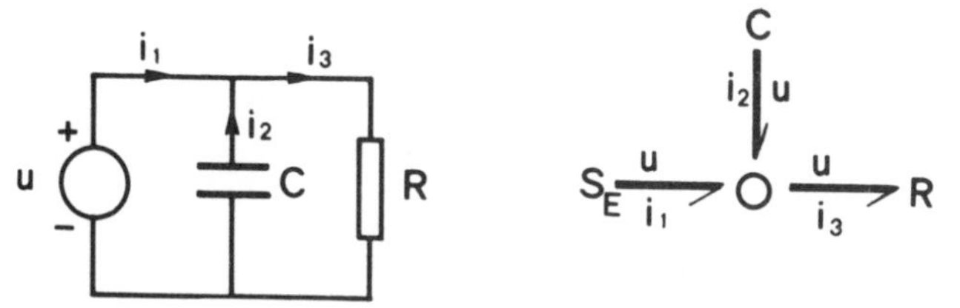

$$P_1 + P_2 = P_3$$
$$u \,.\, i_1 + u \,.\, i_2 = u \,.\, i_3$$
$$i_1 + i_2 = i_3$$

Fig. 4. Parallel or o-junction; use of orientation-(half) arrows.

It is interesting to compare these concepts with the nodes and converse nodes of Polder's very effective "Network theory for variable epicyclic gear trains" (1969), [30] discussed by the author in his contribution.

Multi-port components

Another breakthrough came with the introduction of multi-port components for the general non-electrical system: the modulated multi-port transformer, the multi-port storage components ("fields") and multi-port dissipators. In 1968 Karnopp formulated the concept of the multi-port power conserving transformation [31] which can be considered as a generalization of the electrical multi-winding transformer concept of Belevitch [4], or a general form of the non-energic constraint discussed by Adams. This also provides a positive link with the work of Kron particularly his device of the connection matrix (1939) [32]. By means of this new device it is now possible to represent non-linear kinematic constraints in two- or three-dimensional systems.

Three contributions in this volume are particularly concerned with the bond graph representation of physical systems. Thoma introduces the bond graph aspects of entropy and entropy flow, and draws comparisons with the more conventional notions of electrical charge. Polder gives a thorough account of the design of mechanical transmission systems by bond graph methods. Van Dixhoorn demonstrates the advantageous use of the multi-port transformer, and the general enhanced efficiency of the bond graph technique for the modelling and simulation of a three-dimensional system, by an application to vehicle suspension.

Bond graph methods are becoming applied to an increasingly diverse class of problem, and a useful collection of such applications can be found in the special issue of the *Journal of Dynamic Systems Measurement and Control* (1972) [29].

Gabriel Kron

No account of the development of network modelling could be complete without the inclusion of the name of Gabriel Kron, for the broad scope of his work alone can only suggest that the network as a useful working concept is far from exhausted. There is some evidence that the basic theory, and more sophisticated structure built around Kron's original machine and circuit models, will become of increasing importance, especially in relation to spatially distributed continuous systems. Part of that evidence is given in this book by Lynn and Russell, and the introduction to their contribution outlines the principles and historical evolution of the orthogonal network analysis first proposed by Kron in his "Tensor analysis of networks" in 1939 [32]. His original circuit model re-emphasized the inseparable natures of Kirchhoff's laws and Maxwell's field equations. He proposed a novel approach to circuit analysis by a piecewise method that he named "Diakoptics", and a rather unconventional use of the tensor calculus that enabled him to exploit the concepts of transformation, groups and invariance. Kron was amongst the first to demonstrate the engineering applications of combinatorial topology and differential geometry. The amazing extent of the applications of Kron's methods can only be fully appreciated by examination of his published work, much of which is collected in one volume "Diakoptics" [33]. It is a most remarkable fact that a single conceptual idea can provide a unifying framework for such an apparent diversity of problems that have been shown to include, apart from those mentioned by Lynn, scattering theory, least squares estimation, linear optimal control and filtering.

Kron's name was mentioned often during the course of the meeting, and it has been recorded that Paynter [34], the originator of the bond graph concept, has acknowledged his debt to "the penetrating but Olympian tensorial methods of Kron". Also Karnopp, as stated earlier, has shown [35] that the junction structure of the bond graph technique is directly related to the Kronian connection matrices. So for this, and other, reasons there is no doubt that Kron's work has influenced much that appears in this volume.

3. NETWORK MODELLING AND THERMODYNAMICS

The multi-port concept, mentioned earlier, and its associated power bonds form a direct connection to fairly longstanding, and classical, thermodynamic ideas. It is central to the study of thermodynamics (in essence thermostatics) to consider the nature of physical variables, and interesting to follow the development of dynamic systems analysis in relation to this.

Multi-port energy storage

The energy U, accumulated in a real physical component, in contact with its surroundings, has for long been described in classical equilibrium thermodynamics by the sum of the thermal, mechanical, expansion, electric, magnetic, chemical, and other energetic components. In differential form:

$$dU = T_{21}dS + Fdx_{21} + v_{21}db + p_{21}dV + u_{21}dq + id\phi_{21} + \mu_{21}dn + \ldots \quad (1)$$

or written as a power *P*:

$$P = T.\phi_s + F.v + v.F + p.\phi_v + u.i + i.u + \mu.\phi_n + \ldots \quad (2)$$

Equation (1) has the form: Σ intensity . *d* (extensity).

Paynter considers every term of eqn. (1) or (2) to describe the energy transfer at a separate port. So eqn. (2) defines the power to an ideal multi-port storage component. In bond graph terminology such components are called a "field". They are simply symbolized by a letter, indicating a storage component, to which several bonds are attached. Examples are given by Thoma and in references [28] and [29]. The total internal energy, as well as the intensities at every port, are functions of all stored extensities, so the components are essentially non-linear with respect to the extensities.

Multi-port dissipation and chemical thermodynamics

The concept of multi-port dissipation was the missing link in the network modelling of chemical thermodynamics.

In 1972 Auslander *et al.* [36] succeeded in designing a bond graph notation for chemical reactions and for coupled chemical reaction and diffusion as encountered in biochemistry. Following De Donder and Van Rysselberghe, [38], they used reaction rate J^R for the reaction through-variable; the difference between forward and reverse affinity A_{fr} for the reaction across-variable, their product being the rate of change of Gibbs' free energy *G* in a reaction: $\frac{dG}{dt} = A_{fr}J^R$. Near reaction equilibrium a linear one-port dissipator *R* can characterize the reaction: $J^R = \frac{1}{R} . A_{fr}$. Far from equilibrium J^R depends in a non-linear (exponential) way on both A_f and A_r separately. A two-port dissipator is the tool needed to model this relation. A thorough discussion of this very promising development is given in reference [37].

In search of network variables for thermal energy conversion and transport

The problem to find a network representation for thermal energy conversion processes, as encountered in steamboiler-turbine units, is not yet solved. Two interesting approaches are given in following chapters by Thoma and Rietman. Both authors reject the generally found classification of thermal variables, in which temperature difference T_{21} is the across- and heat flow ϕ_H the through-variable. These variables have an application only, where heat behaves in the old "phlogiston" way of a conserved quantity.

They are unsuitable to describe reversible thermal energy conversion because the product $T_{21} . \phi_H$ is not the power at the port. Thoma remedies this by taking entropy flow $\dot{S}$ for the through-variable.

An application of this choice is shown in Fig. 5, showing a transformer representing either a thermocouple or a Peltier-cooling element.

Thoma presents in his contribution the modelling of irreversibilities by a modulated transformer, while in an earlier paper [39] his approach to heat convection is discussed. There he retains entropy flow but uses enthalpy temperature for the across-variable.

Rietman has a different answer to the definition of variables, more in line with Oster and Perelson's chemical thermodynamic approach. He advocates for all kinds of systems the use of available power, also called exergy flow, instead of total power or energy flow. In mechanical or electrical systems the energy is theoretically always

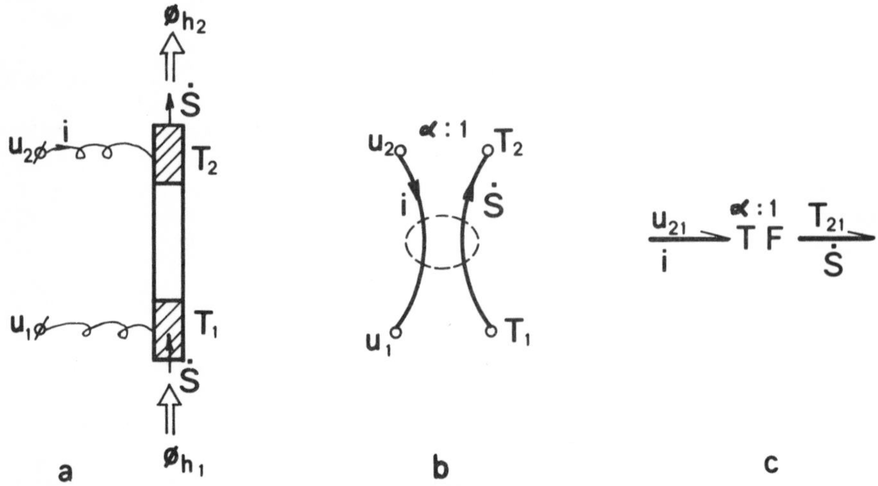

$u_{21} = \alpha . T_{21}$ (transformer)

$u_{21} . i = T_{21} . \dot{S}$ (power conservation)

$\dot{S} = \alpha . i \quad \alpha$ = Seebeck coefficient

$\phi_{H_2} = T_2 . \dot{S} = T_2 . \alpha . i = \pi_2 . i$

$\phi_{H_1} = T_1 . \dot{S} = T_1 . \alpha . i = \pi_1 . i$

π_j = Peltier coefficient at T_j

Fig. 5. Network graph (b) and bond graph (c) of a reversible thermoelectric energy conversion element (a).

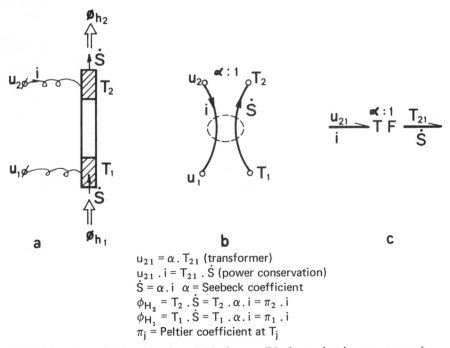

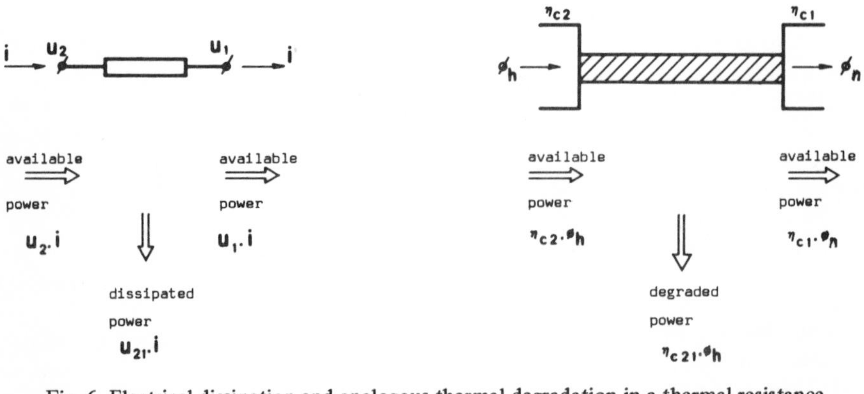

Fig. 6. Electrical dissipation and analogous thermal degradation in a thermal resistance.

completely available and exergy equals energy, but in thermal systems only that part of the energy which is above ambient temperature T_0 is available:

$$\text{available power} = \text{exergy flow} = \frac{T - T_0}{T} \cdot \phi_H = \eta_c \cdot \phi_H$$

So Rietman proposed [40,41] using in thermal conduction systems heat flow ϕ_H for the through-variable and the Carnot factor difference η_{c21} for the across-variable.

While an electrical resistance produces a voltage drop, a thermal resistance produces a temperature difference so accordingly a Carnot factor difference. Exergy is dissipated or degraded to unavailable energy as in the electrical case. Due to the non-linear relation between T and η_c the thermal resistance of a conductor expressed in η_c and ϕ_H is not a constant any more, but slightly dependent on temperature [42]. The same applies to a heat capacitance.

The main emphasis in Rietman's contribution, however, is on exergy-based variables for heat convection, like steam flow. In 1969 [40] he introduced mass-flow for the through and specific exergy difference for the across-variable. He succeeds in splitting up specific exergy difference in two independent potential differences. This clearly corresponds with the two-port nature of gases having two independent state variables; gas storage can indeed be described by a two-port capacitance.

The future application of these ideas in bond graphs with double bonds might disclose an important field.

4. THERMODYNAMICS AND CLASSICAL MECHANICS

Newtonian mechanics were of a causal form in the sense that movement resulted from the application of forces, and being concerned, as it was, with particle dynamics the notion of dissipative non-conservative systems did not assume importance. Resulting from D'Alembert's reinterpretation of Newton's second law, as mentioned earlier, arose the wholly new view of dynamic behaviour in conservative systems that we call "Classical Mechanics" today, and at first formulated by Euler and Lagrange. It is one of the most intuitively attractive ideas that natural systems behave in a simple "economic" manner to minimize some particular function. To formulate dynamic behaviour in this functional manner also requires some attention to be given to the nature of physical variables. It is therefore not surprising that some formal links can be made between the two disciplines of thermodynamics and classical dynamics. On the other hand the central problem of classical thermodynamics was to understand the equilibrium behaviour of dissipative non-conservative systems. This was achieved by the concept of entropy. Finally, in the last phase of this historical development modern irreversible thermodynamics, as developed by Onsager [43], Prigogine [44], De Groot [45], Truesdell [46] and others, is bringing together the necessary ingredients for a physical interpretation of the dynamic behaviour of the general class of non-linear, non-conservative dissipative system. Certain structural aspects of this recent work appear to be reflected in well-known features of network forms. Some difference of opinion still prevails concerning what might be the eventual significance of some of these corresponding forms, but this should not unduly constrain further speculative work in this area, but only serve to encourage it. In an earlier context, Kron's persistence in the face of continued scepticism, eventually brought the recognition of validity from mathematicians and physicists for his unconventional use of their tools.

Although not mentioned in the previous section of this introduction, the functional analysis of networks can be said to have been initiated by Cherry [47] and Millar [48] in 1952, by their definition of content and co-content. These being power functions particularly, but not exclusively, applicable to dissipative devices, it is not unexpected that these functions might perform an important role in the integration of thermodynamic (i.e. entropic) and variational ideas, within the framework of the more recent irreversible thermodynamics.

A large part of modern systems theory is devoted to the solution of problems in which some performance function has been explicitly stated and which must be maximized/minimized, either in the static or dynamic situation. Pontryagin's formulation of the optimal control problem has a direct correspondence with the classical variational dynamic problem. Methods of stability analysis for non-linear systems, developed in recent years from the original work of Liapunov at the turn of the century, also rest firmly on the functional approach.

Modern systems theory, from the point of view of formulation, is now inseparably linked to a description in terms of first order state space equations of a differential or difference character. The form of these equations arises inevitably from the basic nature of physical storage devices, and the expression of conservation principles. To be able to initiate the examination of a dynamic system from a network description is clearly of some advantage, but it is not one on which the analyst can depend. However, it is by bringing to bear on the problem all the resources of network theory, thermodynamics, and variational methods that the highest degree of insight and understanding is assured. This investigational procedure can be considered as a mapping of one problem onto another, and this is the nature of most of the contributions in the second part of this volume.

Before describing briefly how certain of these contributions do, in fact, relate to the interpretation just outlined, one further topic of some interest is worthy of note, which although not mentioned there, is also relevant to the previous discussion on network models. It is one of the main unifying conclusions of network theory, and provides a bridge from the combinational aspects of networks to the functional aspects. It is Tellegen's theorem and was first formulated by him in 1952 [49]. It is remarkable that virtually every energy distribution theorem for networks, particularly from a thermodynamic point of view, as well as several results from nonequilibrium thermodynamics, can be derived from this powerful theorem. It is purely a topological result which states, regardless of any constitutive relationship, and based only on Kirchhoff's laws, that over any network the summation of the products of the effort and the flow (i.e. voltage and current) for all branches is zero. It is therefore effectively a statement of the conservation of power within the system. Since 1883, when Heaviside himself first used a form of this theorem to derive Maxwell's minimum heat theorem (minimum entropy production), there have been several versions of it, but it remained to Tellegen to exploit its true generality.

In the field of network thermodynamics it was Meixner [50] who made some of the earlier contributions to the subject, and recognized in about 1960 that the so-called Onsager–Casimir conditions could be related to network forms. Willems, in his chapter discusses these conditions in more general terms.

Recently Oster, Perelson and their co-workers [36, 37] have reported as already mentioned their work which brings to bear on certain chemical and biological systems problems, in an exciting and hopeful way, the combined techniques of bond graphs, network thermodynamics and Tellegen's theorem.

Jones, Holding and Evans give an interpretation of Sewell's formulation for the generalized variational problem, which is itself given in Sewell's contribution. This approach brings the advantage of a systematic method of selecting sets of independent generalized coordinates, and this, in turn, allows the scalar functions to be derived in unambiguous form. This suggests that the attractive symmetry of Sewell's structure is a reflection of basic network properties. Jones also provides another link between lumped and distributed field problems by his generalized classification of variables.

It would appear that Korn, Holding and Evans' contribution is complementary to that of Willems. In the former certain more intuitive ideas are presented in an attempt to formulate an approach to a unified theory of dynamic systems fundamentally starting from the physical point of view. Willems, however, starting from a system theoretic framework develops a theory for a class of defined dissipative system, in a rigorous analytical manner, that emphasizes the significant relationship between the dissipative properties and the internal symmetry. It is to be hoped that this type of approach will eventually form a sound basis from which to proceed into fundamental generalizations concerning non-linear systems.

The general applicability of both the network and thermodynamic concepts is further suggested by Franksen, where fundamental analogies are established between both of these fields and that of economics. Both this paper and the final contribution by English, which also discusses the relevance of mechanical and thermodynamic concepts in economic theory, can be firmly placed within the same framework of the preceding contributions, which are concerned with more tangible engineering systems.

It is now a distinct possibility that a more integrated theory of dynamic systems behaviour, embracing the mechanism of linear, non-linear, lumped and distributed systems will slowly evolve from what may appear to be, at first sight, a rather uncoordinated attack on the problem. It is rare that any basically useful or essential concept in one particular field of physics (e.g. entropy) will never be exploitable in another. This fact alone is a reflection of the general "common-ness" of physical laws, which can be the only foundation for any truly generalized inter-discipline of systems analysis.

REFERENCES

1. Whitney, H. "Non-separable and planar graphs". *Trans. Am. Math. Soc. 34*, 1932, 339–362.
2. Whitney, H. "2'-Isomorphic graphs". *Am. J. Math. 55*, 1933, 245–254.
3. Veblen, O. *Analysis Situs.* Am. Math. Soc. Cambr. Colloquium Publications, 1931.
4. Belevitch, V. *Classical Network Theory.* Holden-Day, San Francisco, 1968.

5. Seshu, S. and Reed, M. B. *Linear Graphs and Electrical Networks.* Addison-Wesley, Reading, Mass., 1961.
6. Roe, P. H. O. N. *Networks and Systems.* Addison-Wesley, Reading, Mass., 1966.
7. Thorn, D. C. *An Introduction to Generalized Circuits.* Wadsworth Publ. Co., 1963.
8. Rohrer, R. A. *Circuit Theory – an introduction to the state space approach.* McGraw-Hill, New York, 1970.
9. "Models and analogies for demonstrating electrical principles, parts I–XIX". *Engineer, 142*, 1926.
10. Nickle, C. A. "Oscillographic Solution of Electro-mechanical Systems". *Trans. A.I.E.E. 44,* 1925, 844–856.
11. Olson, H. F. *Dynamical Analogies.* Van Nostrand, Princeton, 1943.
12. Murphy, G., Shippy, D. J. and Luo, H. L. *Engineering Analogies.* Iowa State Univ. Press, 1963.
13. Maxwell, J. C. *Treatise on electricity and magnetism.* 1873.
14. Darrieus, M. "Les modèles mécaniques en électrotechnique. Leur application aux problèmes de stabilité". *Bull. Soc. Franc. Electric. 36,* 1929, 794–809.
15. Hähnle, W. "Die Darstellung elektromechanischer Gebilde durch rein elektrische Schaltbilder". *Wissenschaftliche Veröffentl. Siemens Konzern, 11,* 1932, 1–23.
16. Firestone, F. A. "A new analogy between mechanical and electrical systems". *J. Acoust. Soc. A. 4*, 1933, 249–267.
17. Trent, H. M. "Isomorphisms between oriented linear graphs and lumped physical systems". *J. Acoust. Soc. A., 27*, 1955, 500–527.
18. Koenig, H. E. and Blackwell, W. A. *Electromechanical System Theory.* McGraw-Hill, New York, 1961.
19. Shearer, J. L., Murphy, A. T., Paynter, H. M. and Karnopp, D. C. *Report on the dynamic systems workshop.* M.I.T. Press, Cambridge, Mass., 1961.
20. Shearer, J. L., Murphy, A. T. and Richardson, H. H. *Introduction to System Dynamics.* Addison-Wesley, Reading, Mass., 1967.
21. Macfarlane, A. G. J. *Engineering Systems Analysis.* Harper & Co., 1964.
22. Harman, W. W. and Lytle, D. W. *Electrical and mechanical networks.* McGraw-Hill, New York, 1962.
23. Seely, S. *Dynamic Systems Analysis.* Reinhold, 1964.
24. Seely, S. *An Introduction to Engineering Systems.* Pergamon Press, 1972.
25. Martens, H. R. and Allen, D. R. *Introduction to Systems Theory.* Merrill Publ. Co., 1969.
26. Wheeler, H. A. and Dettinger, D. *Measuring the efficiency of a superheterodyne converter by the input impedance circle diagram.* Wheeler monograph no. 9, Wheeler Lab's, Great Neck, New York, 1949. *See also:* Wheeler, H. A. "Ten years of multiport terminology for networks". *Proc. IRE, 47,* 1959, 1786.
27. Paynter, H. M. *Analysis and design of engineering systems.* M.I.T. Press, Cambridge, Mass., 1960.
28. Karnopp, D. and Rosenberg, R. C. *Analysis and simulation of multiport systems.* M.I.T. Press, Cambridge, Mass., 1968.
29. "Special issue on bond graphs". *J. Dyn. Syst., Meas., Control, Trans ASME, series G, 94,* Sept. 1972.

30. Polder, J. W. *A network theory for variable epicyclic gear trains.* Ph.D. Thesis, Eindhoven, Netherlands, 1969.
31. Karnopp, D. "Power-conserving transformations: Physical interpretations and applications using bond graphs". *J. Franklin Inst. 288*, 1969, 175–201.
32. Kron, G. *Tensor analysis of networks.* Wiley, New York, 1939.
33. Kron, G. *Diakoptics.* Macdonald, London, 1963.
34. Paynter, H. M. "Bond graphs and diakoptics". *Matrix and Tensor Quarterly, 19*, 1969, 104–107.
35. Karnopp, D. C. "On bond graph theory and the penetrating but Olympian methods of Kron". *Matrix and Tensor Quarterly, 20,* 1970, 120–124.
36. Auslander, D. M., Oster, G. F., Perelson, A. S. and Clifford, G. "On systems with coupled chemical reaction and diffusion". *J. Dyn. Syst., Meas., Control, Trans ASME, Series G, 94,* 1972, 239–248.
37. Oster, G. F., Perelson, A. S. and Katchalsky, A. "Network thermodynamics: dynamic modelling of biochemical systems". *Quarterly Review Biophysics, 6,* 1973, 1–134.
38. De Donder, Th. and Van Rijsselberghe, P. *Affinity.* Stanford Univ. Press, Stanford, 1936.
39. Thoma, J. U. "Bond graphs for thermal energy transport and entropy flow". *J. Frankl. Inst. 292,* 1971, 109–120.
40. Rietman, J. and Vinke, W. "The boiler as a part of the power generating system". *Proc. Symp. Internal Corrosion in Boilers,* Royal Naval College, Den Helder, Netherlands, 1968, 9–16.
41. Rietman, J. "Thermodynamic principles of thermal energy converters in view of modern systems analysis" in *Design and Econ. Consids. Shipbuilding and Shipping.* Veenmans, Wageningen, Netherlands, 1969, 527–550.
42. Vinke, W. "Propulsion systems viewed as energy transforming and transporing systems" in *Design and Econ. Consids. Shipbuilding and Shipping.* Veenmans, Wageningen, Netherlands, 1969.
43. Onsager, L. "Reciprocal relations in irreversible processes". *Phys. Rev. 37, 38,* 1931.
44. Prigogine, I. *Introduction to the Thermodynamics of Irreversible Processes.* Charles C. Thomas, 1955.
45. Groot, S. R. de. *Thermodynamics of Irreversible Processes.* Interscience, New York, 1952.
46. Truesdell, C. *Rational Thermodynamics.* McGraw-Hill, New York, 1969.
47. Cherry, C. "Some general theorems for non-linear networks possessing reactance". *Phil. Mag., 42,* 1951, p. 1161.
48. Millar, W. "Some general theorems for non-linear networks possessing resistance". *Phil. Mag., 42,* 1951, p. 1150.
49. Tellegen, B. D. H. "A general network theorem with applications". *Philips Res. Rep. 7,* 1952, 259–269.
50. Meixner, J. "Thermodynamics of electric networks and the Onsager–Casimir reciprocal relations". *J. Math. Phys., 4,* 1963.

MODELS, BOND GRAPHS AND ENTROPY

J. U. Thoma
Consulting Studio
Zug
Switzerland

SUMMARY

Both theories of physics and computation schemes are models of experimental facts. Bond graphs as a versatile tool for modelling, alternative to equivalent circuits and network graphs, are a kind of graphical theory of physics. For thermal systems, bond graphs can employ entropy flow and absolute temperatures as effort (potential). Entropy as a kind of thermal charge gives the Carnot efficiency formula and underlines the similarity with electrical and mechanical engineering systems represented by bond graphs, to which the concept of disorder and information is compared. Finally some striking bond graph representations of electrical components are given.

1. RELATIONS BETWEEN THEORY AND MODELS

The object of this contribution is to discuss possible models solving a range of engineering and economic problems. This task has usually been assigned to theory, especially in physics and this is why we first treat the relationship between theory and models.

The path from an open problem to numerical results can be described by Figs 1 or 2.

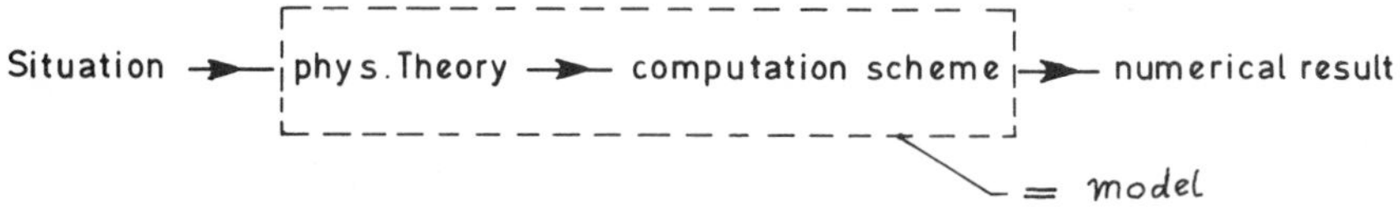

Fig. 1. Different concepts of system analysis.

(*a*) Figure 1 shows in general how one can come from a situation, that is from a number of given facts, first to a theory of physics, then to a computation scheme and finally to numerical results. A physical theory must include all accepted effects of physics, whilst a computation scheme can be in contradiction with some of them for the ease of handling. However, if the scheme is then outside its intended range, it will lead frequently to errors, especially with the qualitative considerations of practical engineering. Both physical theory and computation scheme are a model of the situation.

(*b*) Figure 2 shows more specifically for a real situation or machine how one can first set up a bond graph or an equivalent circuit, then go to the block diagram known from control engineering and from this to numerical results. As a short cut, the ENPORT program allows one to obtain simulation and numerical results directly from the bond graphs. An alternative and frequent method in electrical engineering is to set up a network graph from the equivalent circuit and obtain the simulation from it.

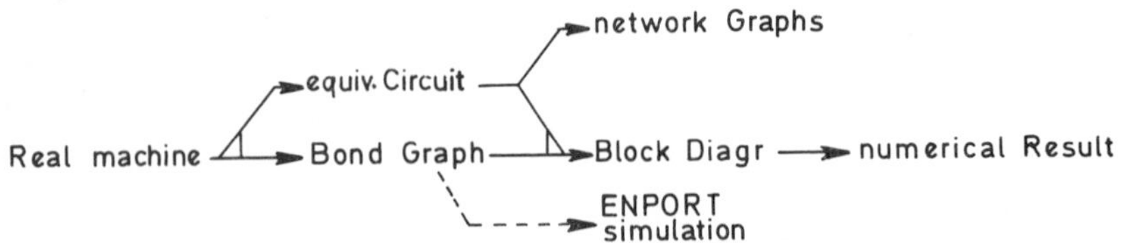

Fig. 2. Relation between bond graphs and other graphical aids.

All representations or graphs have the disadvantage that they lose progressively similarity with the real machine. This should be allowed only reluctantly and against tangible advantages for computation.

The following remarks can be made about the elements of Figs 1 and 2.

1. Bond graphs are a method of representing engineering systems on paper by symbols defined elsewhere [1, 2]. Originally invented by Paynter after intensive study and use of block diagrams they were brought to their present perfection by Karnopp and Rosenberg. Bond graphs are a systematic way to set up block diagrams in different causalities for engineering systems using power conservation as ordering principle. A bond graph begins as an acausal representation, and then the causalities are chosen systematically observing certain rules [1, 2], leading to a causal representation. The process of choosing causalities and positive power directions is called augmenting a bond graph.
2. With the ENPORT program, the elements of a bond graph and their numerical values can be directly typed in a computer. The computer assigns causalities and simulates the actual system behaviour for different inputs. Only the positive power directions must be selected by the user.
3. A block diagram is the basic tool of (classical) control engineering. It is a causal representation [3], giving a network of causes and effects represented by blocks with one or more inputs and one output and by connections. The signal flow graphs of control engineering are a related but not so useful representation in the case of non-linear systems.
4. Equivalent circuits come from electrical power engineering in electronics, where real components are represented by fictive elements with idealized behaviour in an otherwise normal circuit. As an example, a real voltage generator can be replaced by an ideal generator with voltage independent of current and by a series resistance. This is an equivalent circuit, since the series resistance has no hardware equivalent, but it models well the voltage/current characteristics of a real generator. The disadvantage with elaborate equivalent circuits is that they retain little resemblance with the actual circuit or lay-out.

5. Network graphs [4] are based on the classification of variables as through- and across-variables. They show essentially the points where and how the variables are measured and help to set up equations.

The author is especially fond of bond graphs since they are universally applicable in many different engineering disciplines. They unite the viewpoints of control engineering with its emphasis on causality, time dependence, frequency response and stability, and of power engineering, where power is approximately conserved, the power loss described by the notion of efficiency. Bond graphs retain some similarity with the real machine and can be made to account for all accepted effects of physics, so that a properly set up bond graph is a graphical theory of physics. Furthermore, they seem to be especially suitable for the simulation and design of energy or power conversion schemes with high efficiency and adaptability to variable load and other circumstances. This should become apparent from the contents of Section 3 below.

2. POTENTIAL AND FLOW IN BOND GRAPHS

Table 1 contains the main symbols for bond graphs as given in more detail in [2]. We shall describe here how the notion of potential and flow in the context of this book fits to bond graphs.

Generally, all power transporting variables are divided into potentials, called efforts in the bond graph literature, and into flows, such that their product is the power

$$E = ef \qquad e = \text{effort (potential)} \qquad (1)$$
$$f = \text{flow}$$

In addition, the generalized momentum as the time integral of effort, and the displacement as time integral of flow are important in bond graph theory. They can be shown in what is called by many "Paynters Tetragon", Fig. 3.

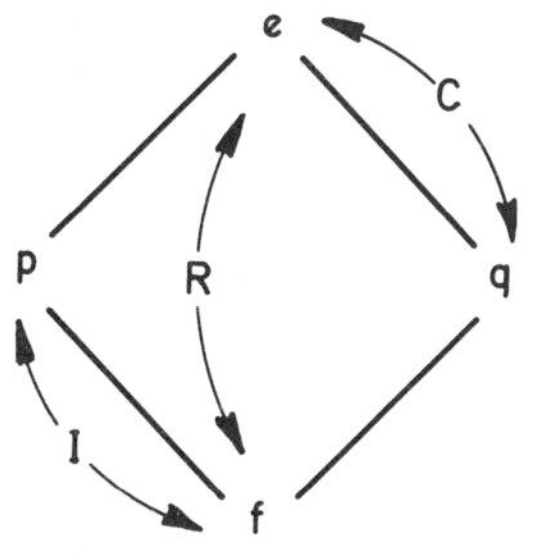

Fig. 3. Paynters Tetragon for effort, flow, displacement and momentum.

In this figure, a (possibly non-linear) R-element gives a static but otherwise the arbitrary relation between effort and flow, a C-element between displacement and effort, and an I-element between flow and momentum.

In electrotechnology the effort is identified with potential or voltage, the flow

TABLE 1
Table of Bond Graph Elements

$\frac{e}{f}$ ———	simple bond, effort on top, flow on bottom
———⇀	half arrow for positive power flow to the right
———|	cross-stroke for causality: effort to right, flow left
|↼———	bond with positive power flow and effort acting to left
———➤	activated bond, one variable neglected
	One power port elements
———R	resistance element, power dissipating
———I	inertia element, energy storing
———C	capacitance element, energy storing
Se———	effort source, power producing
Sf———	flow source, power producing
	Note: There exist also R-fields, I-fields and C-fields similar to the one-ports but with two or more ports
	Two power port elements
$\frac{e_1}{f_1}$ TF $\frac{e_2}{f_2}$	transformer, power conserving $e_2 = \lambda e_1 \quad f_1 = \lambda f_2$
$\frac{e_1}{f_1}$ GY $\frac{e_2}{f_2}$	gyrator (overcrossed transformer), power conserving $f_2 = me_1 \quad f_1 = me_2$
——MTF——	modulated transformer resp. gyrator
——MGY——	(λ or m adjustable)
	Three power port elements
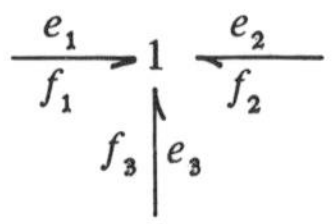	1-junction,* power conserving $e_1 + e_2 + e_3 = 0 \quad f_1 = f_2 = f_3$
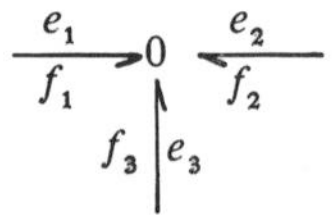	0-junction,* power conserving $e_1 = e_2 = e_3 \quad f_1 + f_2 + f_3 = 0$ *Note:* Junctions can have more than three ports, being then power conserving multiports

*Following a request of the editor, the author has used the American notation 0-junction and 1-junction for what he normally calls *p*-junction and *s*-junction. The latter are generally preferable since they are more easily memorized from the relation with parallel and series circuits and more easily programmed on a computer (Rosenberg, private communication).

with current, and in incompressible hydraulics with pressure and volume flow, respectively. In other fields, the identification depends on the analogy used:

1. In the direct analogy, preferred by bond graphers and the author, efforts in mechanics are force and torque, and flows are velocity and rotation frequency (rotative speed). The C-element is a spring and the I-element a mass or moment of inertia.

2. In the inverse analogy, velocity and rotation frequency are efforts, force and torque are flows. Consequently the I-elements are springs and the C-elements masses or moments of inertia.

The inverse analogy has the advantage of preserving the classification of across- and through-variables of network graphs, so that all flows, that is current, volume flow, torque and force are through-variables, and all efforts that is voltage, pressure, velocity and rotation frequency are across-variables. This classification refers to the measuring process, since an across-variable is measured from 2 points, whilst a through-variable measurement requires the cutting and inserting of an instrument or meter, as for instance a wire for electric current. It is therefore prefered for network graphs as for example in [4].

The direct analogy corresponds better to the Faraday/Maxwell interpretation of current as electric charge flow, and of voltage as tension (energy per displacement) without movement. Furthermore momentum is in the mind of many associated with movement and displacement without it, corresponding to the direct analogy and to the usual engineering terminology. In the inverse analogy momentum as time integral of effort is an angle or position and displacement the time integral of force or torque. One could therefore say that the inverse analogy is at variance with accepted effects of physics and reduces the model to a computation scheme in the sense of Fig. 1. This contribution will be entirely based on the direct analogy.

Bond graphs allow a classification of elements similar to, but more precise than, the energic and nonenergic elements of network theory. In fact, transformers, gyrators and junctions are power conserving, R-elements are power dissipating (excluding thermal energy from the universe of the discourse) and C- and I-elements are energy storing but not dissipating. In fact, each capacitor gives back its energy when it is discharged to its original state.

Thermodynamics is the extension of physics and engineering to include thermal effects, whence energy is always conserved as stated by the first law of thermodynamics. As described more in detail below, the second law of thermodynamics requires that there exists a variable called entropy, the flow of which being a bond graph flow variable such that the product with the absolute temperature as effort represents power transport in thermal conduits [5].

$$\dot{E}_{th} = T\dot{S} \qquad (2)$$

T = absolute temperature
$\dot{S}$ = entropy flow

With this extension, bond graphs allow to set up models as physical theories from a large range of linear and non-linear engineering systems, including electrical, fluid, mechanical and thermal effects. Also, since by inclusion of thermal effects no power is lost, all R-elements will be replaced by conversion devices like transformers, gyrators or by negative flow sources.

Bond graphs are an alternative to equivalent circuits, a method well established in electrotechnology, where bond graphs are consequently less needed. In composite or interdisciplinary systems, e.g. in electroacoustics, equivalent circuits have the disadvantage that a new set of symbols must be introduced for each discipline, whilst bond graphs need only the set of Table 1. Also equivalent circuits for non-ideal elements are must less well established outside of electronics, so that bond graphs with their systematic introduction of sign conventions and causality are a powerful tool.

3. BOND GRAPH TREATMENT OF ENTROPY

3.1. Entropy conservation and generation

The indentification of entropy flow as flow variable in bond graphs is comparatively recent, [5], and involves some special points. Thermal energy will be called heat, and the difference between heat conduction and convection is very important. The latter is associated with a macroscopic transport of matter, e.g. in hot water or steam pipes.

Entropy is a variable similar and analogous to electric charge. It is conserved in a reversible process, a somewhat misleading but usual terminology of thermodynamics. It is better to say that entropy is conserved except for the following effects:

1. Heat conduction under finite temperature difference.
2. Mechanical and electric friction (Joule losses).
3. Mixing of fluids or fluid flows.

An irreversible process in the sense of thermodynamics is then producing entropy by one of the above effects. Heat conduction under finite temperature differences is really the basic entropy generator, to which friction can be reduced [5]. The same applies to combustion and electro-chemical processes that comprise a conversion process followed by friction and/or heat conduction. Essential in this interpretation is that a finite heat flux can be the product of infinite temperature and zero entropy flow, so that it carries no entropy.

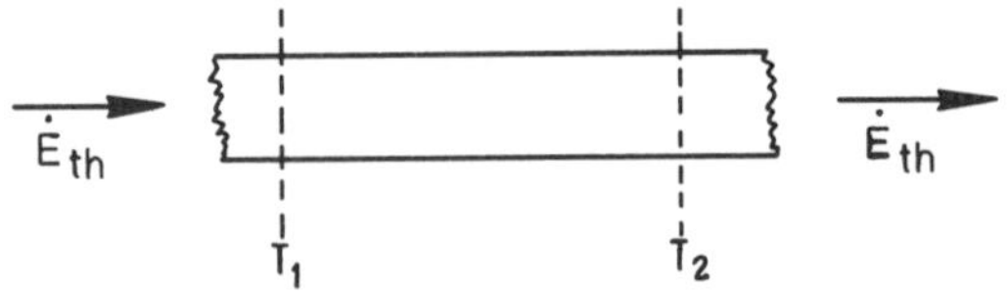

Fig. 4. Scheme of heat conduction.

Figure 4 shows heat conduction in a homogeneous bar between 2 control areas having the temperatures T_1 and T_2, where thermal energy is conserved. The efforts (temperatures) are reduced and entropy flow is increased as represented by the transformer of Fig. 5 and by the equation:

$$E_{th} = T_1\dot{S}_1 = T_2\dot{S}_2 \tag{3}$$

Here T_2 is always smaller than T_1, so that $\dot{S}_2$ is increased. Contrary to an ideal transformer in the sense of bond graphs, the reduction rate depends on the energy transport (power) since in normal conduction the temperature difference between two given points is proportional to it. Therefore heat conduction is more correctly modelled by a modulated transformer, i.e. by the temperature T_1 as modulating signal shown by the bond graph of Fig. 5 (right). This figure contains an activated bond with T_1 that carries no power.

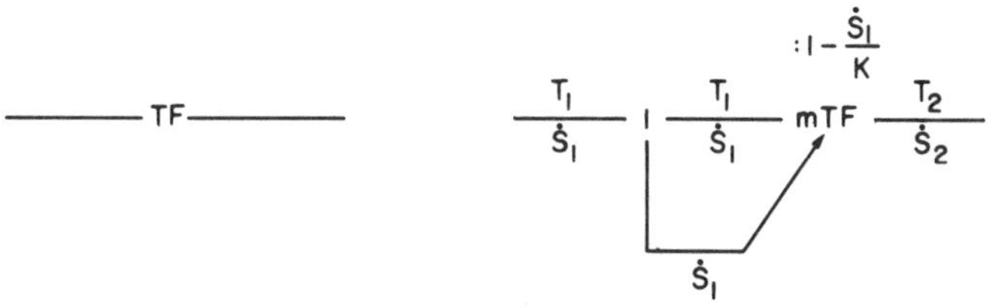

Fig. 5. Heat conduction as simple transformer (left) and as modulated transformer (right).

The second law of thermodynamics states that entropy is never destroyed but it can be created. Therefore all processes run in the direction of creating entropy, i.e. heat conduction always takes place from higher to lower temperature increasing thereby the entropy flow. It should be particularly noted, that equations 2 and 3 are valid also in irreversible, that is entropy generating processes. Thermodynamic equilibrium is reached, if the temperatures in all parts of the systems are equalized, so that no more heat conduction or entropy generation occurs. In this sense, a system with a constant heat source, e.g. a burner, is not in equilibrium, although the variables can be time independent.

The zeroth law of thermodynamics refers to the character of temperature and states that if a body *A* has the same temperature as the body *B*, and the body *B* has the same temperature as a body *C*, then *A* has also the same temperature as *C*. In other words, the relation of equal temperature is transitive.

The third law, for completion (Nernst theorem, 1905) states, that at the temperature of absolute zero all bodies have zero entropy. More precisely, all bodies have the same entropy irrespective of their chemical nature, and this is usually set equal to zero.

The introduction of a modulated transformer for heat conduction, where the transforming factor depends on one of the flows, is similar to such transformers in mechanics, especially with large rotations of mechanical members like connecting rods [2, chapter 12]. It appears well suited to computation with state variables.

On the other hand, the modulation limits the usefulness of the entropy transformer for practical computations, but represents properly the physics of the situation. Therefore a different representation involving available energy and the Carnot factor (see eqn. 7 below) is introduced by Rietman [11] in this volume.

The present interpretation with conserved entropy is especially useful since:

1. All machine designs aim to make entropy production as small as possible.
2. Near the state of thermodynamic equilibrium entropy is conserved, equilibrium being calculated by the condition that entropy is a maximum.

The usual definition of entropy can be obtained by integrating eqn. 1 over time

$$dE = T\dot{S}\,dt = T\,dS \tag{4}$$

where it refers to heating of a piece of matter. Reversing now and adding the subscript rev. to state that in the process entropy generation must be avoided, gives the usual definition of entropy:

$$S = \int_{T_1}^{T_2} \frac{dE_{rev}}{T(E)} \tag{5}$$

As a corollary, electric charge could be explained similarly with charging of a capacitor, or filling it up with electrical energy:

$$dE = ui\,dt = u\,dq$$

$$q = \int_{E_1}^{E_2} \frac{dE_{isol}}{u(E)} \tag{6}$$

Here q is the electric charge, u = voltage and E = electrical energy, whilst the subscript isol means that the capacitor must be perfectly insulated against charge loss.

In the author's opinion the explanation of power transport as product of temperature or voltage and of entropy or electrical charge flow is much clearer. Nevertheless one can in principle also speak of loading a capacitor with electric energy and consider the electric charge as a consequence.

3.2. Carnot energy conversion

The Carnot engine [6] converting heat into mechanical energy is idealized in the sense that no entropy is generated, so that entropy flows are conserved and obey the Kirchhoff law. As shown on the bond graph in Fig. 6, the thermal power is supplied by the entropy flow $\dot{S}_1$ at the temperature T_1 and must be disposed of at

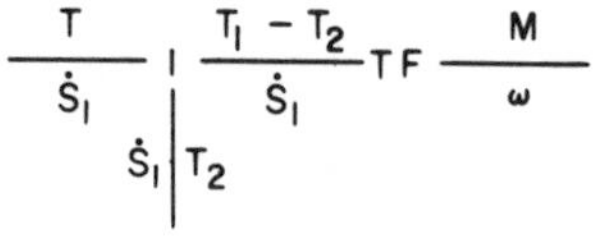

Fig. 6. Bond graph for Carnot conversion engine (entropy conserving).

the temperature T_2 of the entropy sink. The mechanical interface, represented by the transformer, carries away no entropy, only torque and rotation frequency. The power needed for disposal is $T_2\dot{S}_1$, so that only the difference $(T_1 - T_2)\,\dot{S}$ can be converted. This leads immediately to the efficiency η_{cT}

$$\eta_{cT} = \frac{\text{converted power}}{\text{input power}} = \frac{\dot{S}_1(T_1 - T_2)}{\dot{S}_1 T_1} = \frac{T_1 - T_2}{T_1} \tag{7}$$

Equation 7 is the celebrated Carnot efficiency formula, derived here solely from conservation of entropy flow without any details of the process. The fraction on the right is also called Carnot factor. In practice, all machines generate some entropy flow increasing the power needed to dispose it at the entropy sink and reducing the efficiency.

Similarly as an electric d.c.-motor, that requires a return wire to avoid to become flooded with electric charge, a Carnot conversion engine requires a reconduction of entropy flow, that is an entropy sink if it is to operate for any length of time. Depending on the temperature of the sink, a power is associated with it, that can sometimes be put to other uses, e.g. room heating (so-called total energy systems). The same applies to the electric motor if the return wire is not at ground potential, only that the return electrical power is generally more useful.

The identification of entropy as thermal charge has been developed by Job [10] from the point of view of physical chemistry. Consequently thermodynamics becomes much simpler and the associated energies (enthalpy, free energy and free enthalpy) are no longer essential but useful computational aids.

3.3 Conduction and convection

Thermal energy transport is complicated by the frequent use of convection. Here matter is charged with entropy according to its constitutive relation between temperature and entropy.

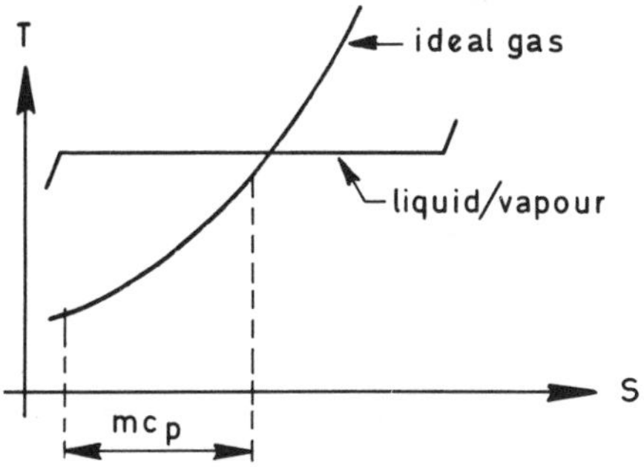

Fig. 7. Characteristics between entropy and temperature for ideal gas (a), for water/steam mixture at constant pressure (b).

Figure 7 shows this relation for an ideal gas, and for a liquid/vapour (water/steam) equilibrium, where temperature remains constant. The energy deposited in the matter is equal to the integral below the characteristics (the zero-point of entropy is arbitrary and coincides with the zero point of energy). Only in the case of constant temperatures over the charging range is the energy equal to the product of temperature and deposited entropy, otherwise it is smaller. Therefore for convection, a form factor α (smaller than one) must be included in the relation between energy transport and entropy flow

$$\dot{E} = \alpha T \dot{S} \tag{8}$$

Figure 8 contains as a corollary a conveyor band for electric energy convection. The band transports capacitors from a charging to a discharging station. With normal linear capacitors the voltage is proportional to charge and thus energy in each capacitor one half the product of final voltage and charge

$$U = \frac{1}{C} \cdot q$$

$$E = \int U(q)\, dq = \frac{Uq}{2} \tag{9}$$

If one defines current as the charge of each capacitor times the transport frequency, that is equal to the current needed by the charging station, then the energy transport is only one half the voltage times current. Thus also here a form factor α is needed

$$\dot{E} = \alpha U \dot{q} = \alpha U i \tag{10}$$

Only if voltage is independent of charge, as in lead accumulators, becomes the form factor α in eqn. 10 equal to one.

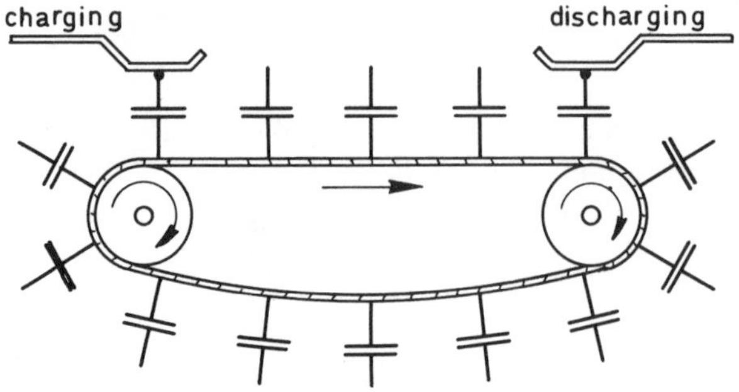

Fig. 8. Electricity convection with conveyor band for capacitors.

Such an electrical convection scheme illustrates well the similarity between electric and thermal energy, but it is much less practical, although it has a certain resemblance with the Van de Graaf generator.

If charging takes place from a constant voltage generator, one half of the energy is lost due to sparks, spurious resistances or oscillations. This can be avoided by using an adjustable voltage source (Variac with rectifier) to progressively increase the voltage of each capacitor. This remark illustrates only, that certain precautions are needed also in the electrical case to avoid loss, just as for thermal energy entropy generation must be avoided.

3.4 Energy, entropy and available energy

In thermal transport processes such as steam pipes or with heat conduction, only entropy flow and energy transport have physical reality. The energy transport is composed of the convection of internal energy with the mass flow, and of the displacement energy, that is the pressure and volume flow. Whilst the latter is definitely transport of mechanical energy, the internal energy of matter is, in the light of the first law not thermal nor mechanical, but just universal energy, that can be taken out at the thermal or mechanical interface. The sum of internal and displacement energy is the enthalpy of thermodynamics.

Since in all continuous processes, entropy has to be disposed of with consequent energy need, one defines as available energy E_{av} (or exergy in German literature) of a substance

$$E_{av} = U - T_{env} S$$

T_{env} = temperature of environment
U = internal energy

The available energy is the internal energy less the energy needed for disposal which depends on the temperature of the environment. Consequently it appears to the author as somewhat artificial, similar to an accountants reserve for the cost of disposal of undesirable goods, like garbage. It is hoped that this cross connection to economics will be made more explicit in the future.

The dependence of available energy of a substance, as in a steam pipe, on the temperature of the cooling water outside makes it unsatisfactory as a concept of physics, although the sink temperature varies less than 10% in ordinary engineering practice. The available energy transported in a fluid line decreases as the fluid undergoes entropy generating processes or when energy is taken out with entropy conservation, as in an adiabatic turbine.

3.5 Matter as C-field

Matter behaves essentially as a capacitor with a definite relation between effort and displacement at the mechanical interface and between temperature and stored entropy at the thermal interface with the environment. Since charging and discharging always involves heat conduction, a transformer similar to that shown in Fig. 5 can be added if the consequent entropy generation is significant, see Fig. 9. Heating without entropy generation, called reversible heating in thermodynamics, must proceed so slowly as to produce neglectable temperature differences at the interface and within the piece of matter.

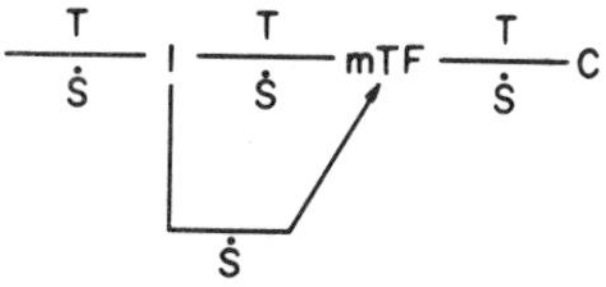

Fig. 9. C-element with modulated transformer representing irreversible heating of matter.

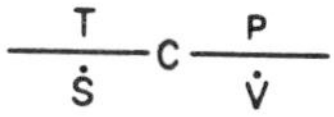

Fig. 10. Matter as C-field for energy convection.

More generally, a piece of matter is a C-field in the sense of bond graphs with 2 bonds as in Fig. 10 in the case of a fluid. It has a relation between entropy and temperature at one bond, between pressure and volume at the other bond, but there are cross-coupling effects such as the known thermal expansion and the temperature increase if the volume is reduced at constant entropy. Only this cross-coupling allows taking out at the mechanical interface some of the heat put in and this is the basis of all heat/mechanical energy conversion engines. Entropy is conserved at best and must always be taken out again at the thermal interface, just as volume at the mechanical interface.

Without the cross-coupling, the piece of matter would behave as 2 separate C-elements, one thermal and one mechanical, each energy conserving for itself. A piece of matter acting as a C-field is thus the heart of all heat conversion engines independently whether it is an ideal gas, water/steam, or a solid. In fact, Carnot sets out in his classic book [6] to examine the suitability of substances other than water for heat engines. He concludes that steel, for example, with its thermal expansion would be usable in principle but is impractical.

Other examples of C-fields are electric capacitors with movable plates, where force and displacement of the plates constitute the mechanical bond, and a water tank with a float for level control, if the forces and the cross-section of the float are not neglectable [7].

4. DIFFERENT ASPECTS OF ENTROPY

Entropy has really three different aspects in the literature:

1. As a macroscopic thermal charge or massless fluid as explained above.
2. As a measure of disorder derived in statistical thermodynamics.
3. As a concept of economics.

In the author's opinion entropy as thermal charge is the basic concept and the interpretation as disorder a kind of theory of structure. It follows from the Boltzmann formula between entropy and statistical probability W of a certain configuration or state of matter [8]:

$$S = k \ln W \tag{11}$$

where k is the Boltzmann constant, having the same dimensions E/T. Related to it is the use of entropy as a measure of information, where

$$I = -S = -K \ln W \tag{12}$$

In eqn. 12, W is the product of the single probabilities of the characters of a message, corresponding to the microstates of matter. In communication theory, the constant K is usually taken as dimensionless and the information taken in bits (where $K = 1.45$). In order to connect to thermodynamics, K must be identified with Boltzmann constant k. In this sense, k can be taken as an elementary quantity (quantum) of entropy. The thermodynamic entropy of mixing of 2 ideal gases, is then equal to the nondimensional information required (1 bit per molecule) to separate it again, multiplied by the (dimensional) Boltzmann constant.

If the concept of entropy in economics is to obtain a legitimate place, it should be shown how it relates to its other general properties. The author hopes that this is done elsewhere.

5. FURTHER BOND GRAPH EXAMPLES

This section contains a few examples from other fields, that show strikingly bond graphs as graphical theories of physics, or of machines applying physical principles.

In an electric d.c.-motor, torque is proportional to armature current and voltage proportional to rotation frequency (rotative speed), the proportionality depending on the exciting field and the field current. The motor is thus a gyrator modulated by the field current as shown on Fig. 11 with the field coil in series at the left and in parallel at the right. In the parallel case, the field ohmic resistance R_f is important and must be included, since it determines the field current. This bond graph represents with the proper numerical values the known torque/speed characteristics of series and parallel exited d.c.-motors.

On a more basic level Fig. 12 describes a simple magnetic circuit with a gyrator and a C-element, such as an inductor or the exiting field of the d.c.-motor [2, Chapter 14].

Fig. 11. d.c.-motor as modulated gyrator; series field (left) and parallel field (right).

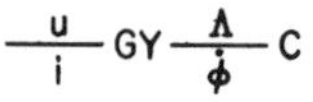

Fig. 12. Simple magnetic circuit with winding (inductor) as gyrator and C-element.

The gyrator connects the voltage u with the magnetic flux derivative $\dot{\phi}$ and the current i with the magnetic tension Λ (magnetic motoric force = m.m.f). The connection ratio is n/γ where n is the number of turns and γ Fleischmann constant, that has to be introduced to establish a correct dimensional system of electrical variables [9]. With normal (S.I.) electrical units, γ is dimensionless and equal to one, whence the magnetic tension is simply equal to the ampere-turns.

With negligible hysteresis, the magnetic circuit is a C-element having a generally non-linear relation between magnetic flux and tension. Any hysteresis destroys the character of an energy conserving C-field, but can be represented somewhat artificially by an R-element [2] modulated by a suitable variable, e.g. the magnetic tension.

An AC-transformer has two windings with mostly common flux, except for the stray fluxes. The magnetic tensions are additive, the main flux is common, as represented by the 1-junction of Fig. 13, leading to the C-element giving the relation between magnetic tension and common flux. The stray fluxes are taken away by

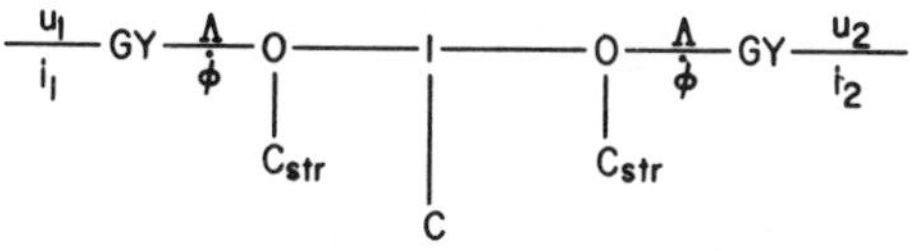

Fig. 13. Transformer with stray fluxes with gyrators for each winding.

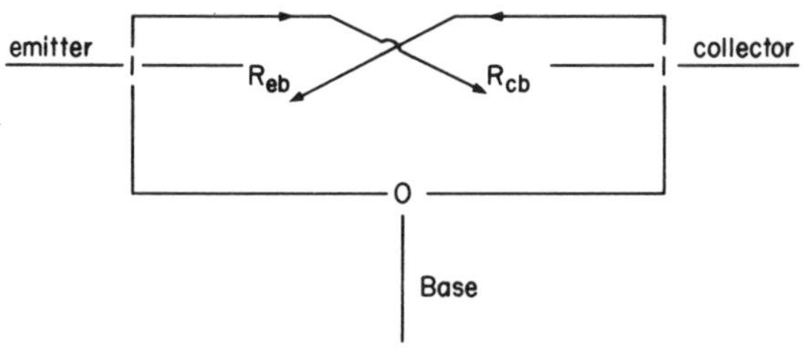

Fig. 14. Bond graph for bipolar transistor using controlled R-elements.

the small C-elements with 0-junction on each side. They have the same size (values) as a consequence of the reciprocal properties of the induction laws.

Figure 14 shows a basic bond graph for an electronic bipolar transistor, consisting essentially of two R-elements between emitter resp. collector and base. They are modulated by the current in the other bond as shown by the activated bonds, in reality by the minority carriers swept across the base region. This is still a very crude representation, but it gives some idea of the operation of a transistor. It can presumably be refined to display the movement of majority carriers more in detail.

6. ENTROPY GENERATION BY R-FIELDS IN IRREVERSIBLE THERMODYNAMICS

Electrical and mechanical losses take place in R-elements and are turned into heat, that is into entropy flow and temperature. Including thermal effects, a simple resistor becomes a RS-field (Fig. 15) where the thermal source supplies the same power as absorbed by the normal R-element. Consequently the RS-field is power conserving (excluding storage effects in unsteady conditions) whilst R-fields of bond graphs are power absorbing.

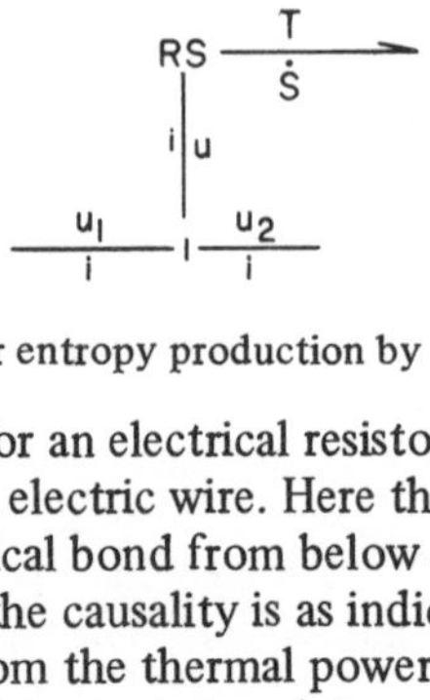

Fig. 15. RS-field for entropy production by an electric resistor.

Figure 15 shows a RS-field for an electrical resistor inserted in series (1-junction) in a lower bond representing an electric wire. Here the current and voltage drop enter the field through the vertical bond from below and entropy flow and temperature leave at the right. Usually the causality is as indicated where the temperature determines the entropy flow from the thermal power flow. The temperature also has often an influence on the characteristics of the non-thermal bond, e.g. where the electrical resistance depends on temperature.

Entropy generation by conduction can be represented after Job [10] by stating, that the power loss obtained from the entropy flow and the temperature difference appears as additional entropy flow injected at the low temperature side. The bond graph in Fig. 16 shows this with a thermal RS-field, where the loss power enters

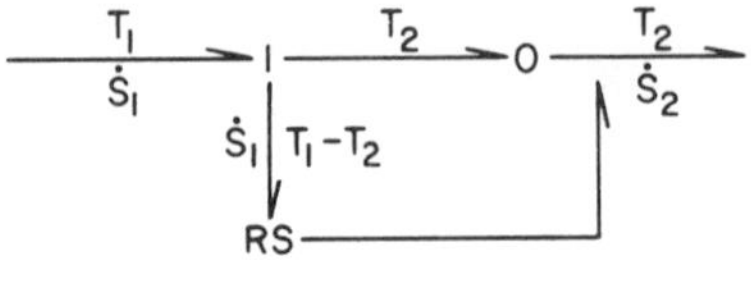

Fig. 16. RS-field for entropy production by heat conduction.

through the vertical bond. The corresponding entropy flow leaves the RS-field similarly as in Fig. 15, but it joins the outgoing flow through the 0-junction at right.

This description is alternative and sometimes preferable to the modulated transformer of Fig. 5 above.

The bond graphs in Figs. 15 and 16 show the similarity between electrical (or mechanical) and thermal losses. The difference is that with the former the new entropy flow remains separate whilst with the latter it joins the outgoing entropy flow.

RS-fields represent also several coupled flow processes with entropy generation, as for coupled heat and electrical conduction in a piece of matter (Peltier and Seebeck effects) in Fig. 17. This RS-field consists of a R-field with 2 coupled bonds,

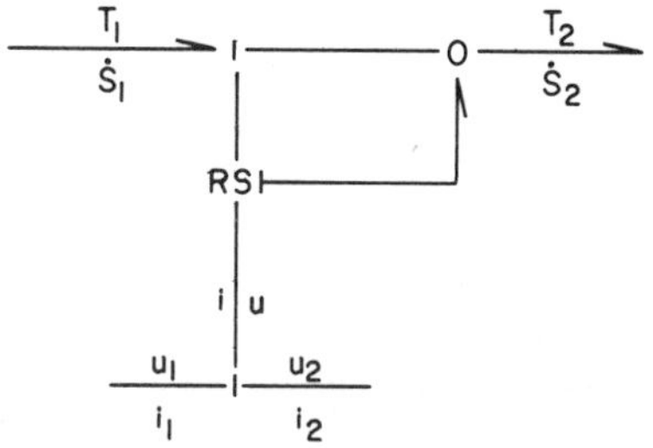

Fig. 17. RS-field for entropy production with coupled electric current.

which have the symmetry discovered by Onsager and named after him (in the appropriate causality, [2, chapter 9]), and of a source part delivering an entropy flow according to the losses similarly as in Figs. 15 and 16. Due to the coupling, an entropy flow in the upper bond can drive an electric current in the lower bonds and vice-versa.

The theory is developed for small signals, that is in the proportional range, where the 4 variables, voltage and temperature drop, entropy flow and current can be connected by a matrix with 4 elements. They are determined from the three properties:

1. Electrical resistance at constant temperature.
2. Entropy resistance at vanishing electric current.
3. Entropy flow entrained by electric current at constant temperature.

and from Onsager's symmetry condition.

It is to be noted, that the power flow in one vertical bond of Fig. 17 can reverse, as when the device can supply electrical power as long as more thermal power is absorbed by the other vertical bond. Then the produced entropy flow remains positive in accordance with the second law of thermodynamics.

Some electrical (and mechanical) components reduce their voltage drop with increasing current in part of their characteristics, as often described by a negative resistance which can excite oscillations. The total dissipation of such a device remains positive and it always produces entropy.

Figure 18 shows a bond graph, where the falling characteristic is represented by a constant voltage source at left with a negative R-element in parallel. Electrical power is supplied by the voltage source at right over the 0-junction, that carries further I-, C- and R-elements. The arrangement will oscillate if the value of the positive R-element at right is less than the negative resistance of the RS-field in the centre.

The RS-field supplies electrical power in the circuit taken from an entropy flow and consequently destroys entropy. However, the voltage source at left is driven in reverse and absorbs always more power and produces a larger entropy flow. It is represented therefore in Fig. 18 as a SS-field (power conserving source field). The new entropy flow enters partly into the RS-field after passing the 0-junction, partly it is disposed into the environment over the lower bond at right.

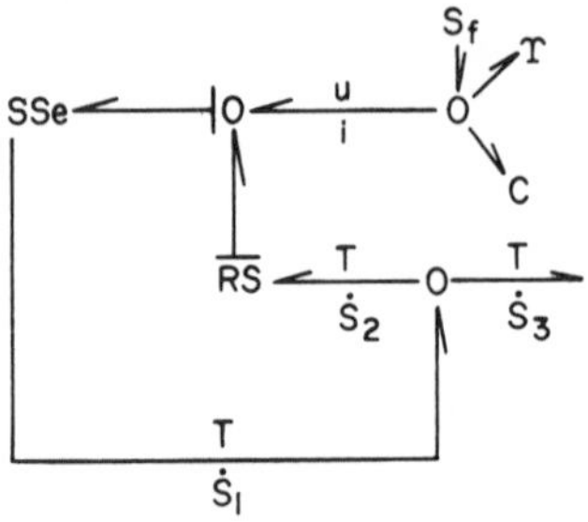

Fig. 18. Bond graph with RS- and SS-field for electrical components with negative resistance.

The apparent contradiction of the entropy destruction in the negative resistance is resolved by the source field, that is needed to describe the device properly. Consequently the net entropy production of the device is always positive in accordance with the second law of thermodynamics.

Concluding flow process thermodynamics, R-elements and R-fields become RS-fields, where the source part represents the entropy generation. They are then power conserving since by including thermal effects power is universally conserved. Entropy generation by heat conduction is also describable by a RS-field and a RS-field with several bonds represents the coupling between electrical and thermal effects in matter, including the consequent entropy production.

The above applications conclude the present contribution on models and bond graphs for entropy and other engineering economic problems. The author hopes that the discussions at the Twente Colloquium 1973 will open new ways of communication and assistance between different disciplines.

REFERENCES

1. Karnopp, D. and Rosenberg, R. C. *Analysis and Simulation of Multiport Systems.* M.I.T. Press, Cambridge, Mass., 1968.
2. Karnopp, D. and Rosenberg, R. C. *System Dynamics, a unified Approach*, 1972, available as lecture notes from M.S.U. Bookstore, International Center, East Lansing, Mich. 48 823, U.S.A.
3. Thoma, J. and Karnopp, D. "Simulation stetiger Systeme durch Diagramme", in A. Schöne (ed.), *Simulation Stetiger Systeme*. Carl Hanser Verlag, 1973.
4. Shearer, J., Murphy, A. and Richardson, H. *Introduction to System Dynamics*. Addison-Wesley, Reading, Mass., 1967.
5. Thoma, J. "Bond graphs for thermal energy transport and entropy flow". *J. Franklin Inst. 292*, 1971, 109–120.

6. Carnot, Sadi. *The motive power of fire.* Original French 1824, edited by E. Mendoza 1960, Dover Publications.
7. Karnopp, D., private communication, 1972.
8. Lay, J. E. *Thermodynamics*. Merrill Publ. Co., 1963.
9. Fleischmann, R. "Die Struktur des physikalischen Begriffsystems". *Zeitschrift für Physik, 129*, 1951, 377–400; *see also* ibid. *138*, 1954, 301–308.
10. Job, G. *Neudarstellung der Wärmelehre.* Akademische Verlagsges., 1972.
11. Rietman, J. "New system variables for the flow of thermal energy based on the concept of exergy", in Van Dixhoorn, J. J. and Evans, F. J. (eds), *Physical Structure in Systems Theory*. Academic Press, London, 1974 (this volume).

Although not specifically referred to, many bond graph articles are found in *Journal of Dynamic Systems, Measurement, and Control, Trans. ASME, 94 Series G,* No. 3, Sept. 1972.

NOMENCLATURE

c_p	specific heat at constant pressure
e	effort
f	flow
i	electric current
k	Boltzmann constant, (heat conductivity in Fig. 5)
m	mass
N	number of turns
p	momentum or pressure
q	displacement or electric charge
u	voltage
C	capacity
E	energy
$\dot{E}$	energy flow, power
K	constant of entropy formula
M	turning moment, torque
S	entropy
$\dot{S}$	entropy flow
T	temperature
T_{env}	temperature of environment
U	internal energy
$\dot{V}$	volume flow
W	probability
α	form factor in equation 8
γ	Fleischmann constant
ϕ	magnetic flux
Λ	magnetic tension

Note added in proof:
The Onsager Symmetry condition of the RS-field of Fig. 17 is necessary to avoid that either the efficiency of entropy pumping against temperature drop or of electric power generation can become larger than one, hence a consequence of the second law of thermodynamics. Ref. Thoma, J. U., Entropy and Mass Flow for Energy Conversion, Journal of the Franklin Institute, Philadelphia, early 1975.

For a short introductory book on bond graphs in German, see Thoma, J. U., Bonddiagramme, GT2o, Girardet Publisher, Essen, 1974.

NEW SYSTEM VARIABLES FOR THE FLOW OF THERMAL ENERGY BASED ON THE CONCEPT OF EXERGY

J. Rietman
Royal Naval College
Den Helder
The Netherlands

SUMMARY

The purpose of this paper is to investigate the possibility of defining variables for thermal systems, analogous to the way this already has been done in the case of mechanical, hydraulic and electrical systems. Application of the concept of exergy (available energy) and non-reversible, non-stationary thermodynamics offers possibilities in this respect. Via derivation of an exergy balance equation and its reticulation the system variables appear quite naturally. As the specific exergy itself is not sufficient to determine the state of a fluid, two partial potentials are introduced, namely the temperature potential, solely dependent on temperature and the pressure potential, solely dependent on pressure.

1. INTRODUCTION

The unifying concept in the analysis of the behaviour of complex engineering systems is the concept of energy. The energy of a state of a physical object is the measure of its capacity for producing changes in its own state or in the states of other physical objects. These changes can be considered to occur by flow of energy between different components in a system.

Energy flows always appear as the product of two quantities, called system variables. One of these variables has the character of a potential difference between two points in space and is called across-variable or trans-variable. In many cases one of the two points is a reference point. A second type of variable is the quantity describing the rate of flow caused by the trans-variable, called the through-variable or per-variable and which specification only requires one point in space.

The following table gives a survey of the system variables in some physical systems together with dimensions.

In various papers in the field of system theory much attention has been paid to mechanical, electrical and hydraulic systems. However, components of which the behaviour must be described by thermodynamic relations, have been rarely considered.

TABLE 1

System	Across-variable	Through-variable
Mechanical, translating	velocity, v (m/s)	force, F (N)
Mechanical, rotating	angular velocity, ω (s^{-1})	torque, M (Nm)
Hydraulic	pressure difference, Δp (N/m^2)	volume flow rate, ϕ_V (m^3/s)
Electrical	voltage, e (V)	current, i (A)

This may partly be caused by the complexity of thermal components and the restricted possibilities of describing thermal processes in terms of simple analytical relations.

Perhaps, even more important, is the difference in frame work of classical thermodynamics as a macroscopical physical theory as compared to the structure of the physical theories underlying the operation of mechanical, electrical and hydraulic systems such as classical mechanics, the theory of electromagnetic phenomena and fluid mechanics. Considering the latter three their variables are defined on a basis of a continuum theory. For a point of a system any arbitrary physical quantity x can be represented by the expression

$$x = x(\mathbf{r}, t) \tag{1.1}$$

In mechanical, electrical and hydraulic systems the physical quantities are functions of space and time coordinates in general and one speaks of system dynamics in the case the time dependence plays a role of interest.

In this sense thermodynamics has nothing to do with dynamics although its name may suggest this. The time dimension does not even show up in thermodynamics. Also the space dependence of thermodynamic functions is a point of no concern. Therefore thermodynamics as a basis of many heat engines has in virtue of its structure no room for space and time dependency of quantities of state like pressure p, density ρ, internal energy u, enthalpy h, entropy s, etc. Also this restriction is noticeable quite clearly in the description of flow engines, in which one would prefer, in analogy with other systems, already mentioned, to use the mass flow rate ϕ_m rather than the discrete quantity of mass, Δm, entering or leaving the system in a time interval, Δt.

It is this fundamental difference in frame work which makes it impossible to apply thermodynamics in its classical form to a general system theory. However, during the last decade interesting developments took place which imply great possibilities, namely the rise of a thermodynamics of irreversible, non-stationary processes. In a number of publications concerning this subject an extensive survey is given of the present theory [2, 3].

This modern branch of thermodynamics offers the fundamental possibility to be fitted into a physical system theory, because quantities of state are defined on a basis of a continuum theory as functions of space and time, and may be better characterized as state variables henceforward.

The equations which underly the system theory are all of the balance equation type. If we consider an arbitrary physical system with volume V, surrounded by a closed surface A, then generally the following integral balance-equation can be stated for an arbitrary intensive, specific state variable x of a fluid with density ρ inside V

$$\int_V \frac{\partial(\rho x)}{\partial t} dV = -\int_A \mathbf{\Phi}_x'' \cdot \mathbf{n}\, dA + \int_V \phi_x''' \, dV \tag{1.2}$$

This equation applies to a continuous system. $\mathbf{\Phi}_x''$ is the flux of x, ϕ_x''' is the production rate density of x.

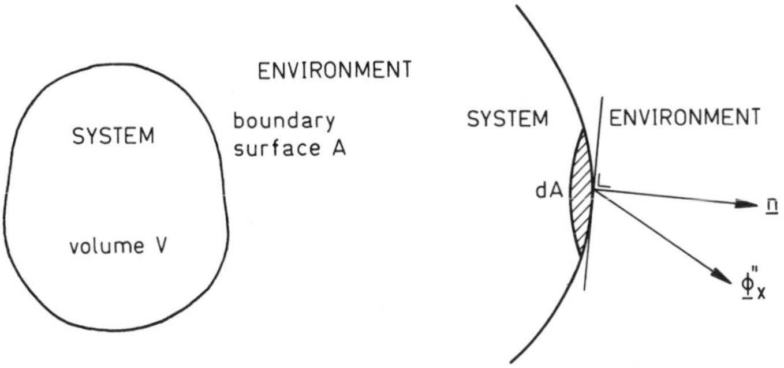

Fig. 1. Representation of a system and transport across the system boundary.

In practical applications we deal mostly with systems in which various physical processes are thought to be lumped in discrete regions. Although this is an idealization, it is a very practical and necessary one, once one tries to analyse the behaviour of actual systems or components, (for instance, an electrical network). In order to do so we have to reticulate the balance equations. For a system with l storage elements, m discrete transfer areas and n source elements, the reticulation of (1.2) leads to

$$\sum^{l} \frac{d}{dt}(m_i x_i) = -\sum^{m} \epsilon_j \phi_j + \sum^{n} \phi_k \tag{1.3}$$

If x_i is interpreted as a specific energy, then (1.3) is the reticulated energy balance equation. The first term of the right part of (1.3) especially describes the various kinds of discrete energy flow through the system boundary. In that case ϵ_j is the so-called across-variable and ϕ_j the so-called through-variable. These quantities show up automatically after derivation of equation (1.3) for any particular system, transferring energy to or from its environment.

Before we are able to fit thermal systems into the general system theory with the aid of an equation like (1.3), we first have to solve another problem, namely the fact that, different from, for instance, mechanical and electrical energy, heat, internal energy and enthalpy are kinds of energy which cannot be converted completely into work. This difference is of great significance as soon as we try to define an across-variable for a thermal system. An across-variable of a fluid, being a kind of potential, should express its capability to perform work. In addition the potential should decrease due to dissipation processes. The electrical potential, for instance, satisfies these requirements.

One could consider as an across-variable the specific enthalpy of a fluid. In fact, from the description of various thermal systems in the technical literature, one observes the application of enthalpy as such quite frequently. However, there is a direct objection against this use of enthalpy considering the flow of a gas through a throttling valve. In this irreversible process the enthalpy remains practically constant, whereas the capability to perform work decreases noticeably.

Using a suggestion of Rant [14] that part of the thermal energy which can be converted into work completely, taking into account the prevailing environmental conditions, given by the temperature T_0 and the pressure p_0, is called *exergy* (available energy). Specific exergy of a fluid agrees very well with the notion of an across-variable. When the fluid flows through a throttling valve the exergy decreases corresponding to the analogous phenomenon of an electrical current passing through a resistor in which the electrical potential also decreases.

From the above it may appear that the combination of the quantity specific exergy with the thermodynamics of non-reversible, non-stationary processes offers possibilities to develop a basis for a system theory of thermal systems. As the behaviour of systems, mentioned in Table 1, is based on energy balance equations, it is obvious to base the behaviour of thermal systems on an exergy balance equation.

The present paper is partly a reexposition of the authors former work [7, 8, 9]. A new element has been added by the introduction of an exergy-based pressure and temperature potential, together forming the components of the previously introduced across-variable:specific energy

2. THE CONCEPT OF EXERGY IN THERMODYNAMICS

Thermodynamics tells us that different kinds of energy are not equal. On one side we have valuable kinds of energy like mechanical and electrical energy, which are easy to handle and, apart from dissipation effects as friction and electrical resistance, can be converted into each other unlimited.

On the other hand heat and internal energy are less valuable kinds of energy which are of restricted use only and cannot be converted into other types of energy completely. Using a suggestion of Rant, as already mentioned above, we now introduce the concept of *exergy* (available energy).

By definition, the exergy of a system is the maximum available work to be obtained from the system if it is brought into equilibrium with its environment via a reversible process.

In order to derive an expression for the exergy of the internal energy, we consider a closed system (a system is said to be *closed* when it can exchange heat and work with its environment without exchanging any quantity of matter with its environment) containing a homogeneous fluid of one component in one single phase. The system can exchange heat with the environment and can do work on the external world.

If we suppose the system to undergo an infinitesimal transition, we obtain by applying the first law of thermodynamics,

$$dW = -dU + dQ \tag{2.1}$$

dW being the work done on the environment and dQ being the heat received from the environment. Applying to this equation the always valid relation

$$TdS = dU + pdV \tag{2.2}$$

and substitution of this equation into (2.1) we get

$$dW = dQ - TdS + pdV \tag{2.3}$$

Suppose the state of the fluid inside the system is depicted as position 1 in the *T–S*-diagram in Fig. 2.

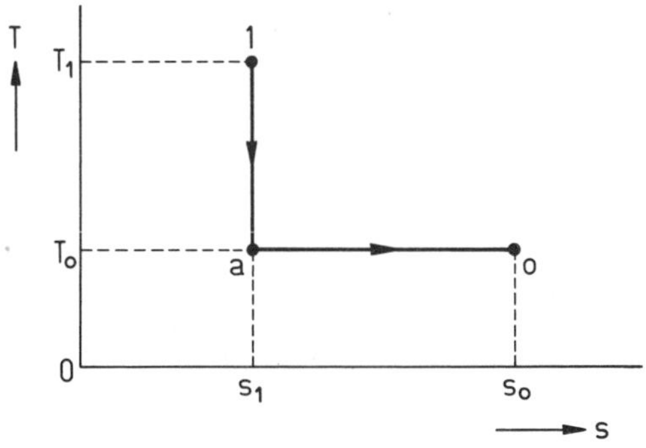

Fig. 2. Process to obtain the maximum available work or exergy.

Now the exergy of the system, by definition, equals the work done on the external world by the system if we bring it into equilibrium with the environment at pressure p_0 and temperature T_0 via a reversible process.

In order to prevent irreversibilities during the mentioned process, we must avoid heat exchange with the environment at any temperature $T \neq T_0$. As a consequence the process should proceed as indicated in the *T–S*-diagram by the line connecting positions 1 and 0.

Now consider the process 1-a in Fig. 2. No heat is exchanged with the surroundings, no entropy is being produced inside the system, thus

$$\begin{aligned} dQ &= 0 \\ dS &= 0 \end{aligned} \tag{2.4}$$

Integration along path 1-a now gives

$$W_{1a} = \int_1^a pdV \tag{2.5}$$

However, because of the pressure p_0 of the environment, the maximum available work of this process is

$$\hat{W}_{1a} = \int_1^a (p - p_0)\, dV \tag{2.6}$$

or

$$\hat{W}_{1a} = \int_1^a pdV + p_0(V_1 - V_a) \tag{2.7}$$

Now, from (2.2)

$$\int_1^a p dV = U_1 - U_a \tag{2.8}$$

so

$$\hat{W}_{1a} = U_1 - U_a + p_0(V_1 - V_a) \tag{2.9}$$

In the next process $a - 0$, heat is exchanged in thermal equilibrium with the environment at a temperature T_0, no entropy is being produced inside the system, thus

$$dQ = T_0 dS \tag{2.10}$$

and

$$W_{a0} = \int_a^0 p dV \tag{2.11}$$

and again

$$\hat{W}_{a0} = \int_a^0 p dV + p_0(V_a - V_0) \tag{2.12}$$

From (2.2) follows

$$\int_a^0 p dV = U_a - U_0 - T_0(S_a - S_0) \tag{2.13}$$

and we obtain, observing that $S_a = S_1$

$$\hat{W}_{a0} = U_a - U_0 - T_0(S_1 - S_0) + p_0(V_a - V_0) \tag{2.14}$$

Considering now the process $1-0$ as a whole, we arrive at the following expression for the exergy E_u of a closed system with internal energy U

$$\hat{W}_{10} = E_u = U_1 - U_0 - T_0(S_1 - S_0) + p_0(V_1 - V_0) \tag{2.15}$$

As the fluid inside the system is homogeneous we calculate the specific exergy of it by dividing (2.15) by the total mass (a constant) m.

$$m = \rho_1 V_1 = \rho_0 V_0 \tag{2.16}$$

and we obtain for the exergy of the specific internal energy u,

$$e_{u_1} = u_1 - u_0 - T_0(s_1 - s_0) + p_0\left(\frac{1}{\rho_1} - \frac{1}{\rho_0}\right) \tag{2.17}$$

Dropping the indices in (2.17), we arrive at a general expression for the specific exergy e_u of the specific internal energy u of an arbitrary fluid

$$\boxed{e_u = u - u_0 - T_0(s - s_0) + p_0\left(\frac{1}{\rho} - \frac{1}{\rho_0}\right)} \tag{2.18}$$

This equation in its specific form (per unit of mass) also applies to an open system. (A system is said to be *open* when it can exchange quantities of matter with its environment, in addition to energy exchanges.) For an open system it is of interest to know the specific exergy e_h of the specific enthalpy h, in which

$$h = u + \frac{p}{\rho} \tag{2.19}$$

Now the exergy of $\frac{p}{\rho}$ is equal to $\frac{1}{\rho}(p - p_0)$, thus in connection with (2.18) we get

$$\boxed{e_h = h - h_0 - T_0(s - s_0)} \tag{2.20}$$

The exergy E_Q of a quantity of heat can be found considering a general heat engine, operating between two reservoirs, one with constant temperature T and one with a constant environmental temperature T_0.

If we take $T > T_0$ the maximum efficiency, or so-called Carnot factor, of this engine is

$$\eta_c = \frac{T - T_0}{T} = 1 - \frac{T_0}{T} \tag{2.21}$$

This Carnot factor determines the value of the heat Q entering the reversibly operating engine at a temperature T. Of this amount of heat only the fraction $\eta_c Q$ can be obtained as work. The remainder, the fraction $(1 - \eta_c)\,Q$, has to be transferred to the reservoir with temperature T_0 and cannot be converted into work. On account of the foregoing, we may conclude

$$E_Q = \left(1 - \frac{T_0}{T}\right) Q = \eta_c Q \tag{2.22}$$

or the specific exergy e_q of a heat quantity q

$$\boxed{e_q = \left(1 - \frac{T_0}{T}\right) q = \eta_c q} \tag{2.23}$$

3. THE INTRODUCTION OF AN EXERGY-BASED TEMPERATURE AND PRESSURE POTENTIAL

Considering the specific exergy of the enthalpy according to (2.20), it is clear that this quantity can be used as a potential of a flowing fluid.

In order to be a useful quantity, the potential of a fluid should also give information about its state. However, the exergy of a fluid does not give sufficient information about the state of the fluid. This information is anyway necessary to determine the way the work is to be obtained from the fluid. The exergy of the specific enthalpy is represented in a $h-s$-diagram by the line $11'$ for a fluid of which the state is depicted by point 1 in the diagram. The state of the environment is depicted by point 0 and the line, on which point $1'$ is located, is the so-called environment line. This line is tangent in 0 to the environment isobar.

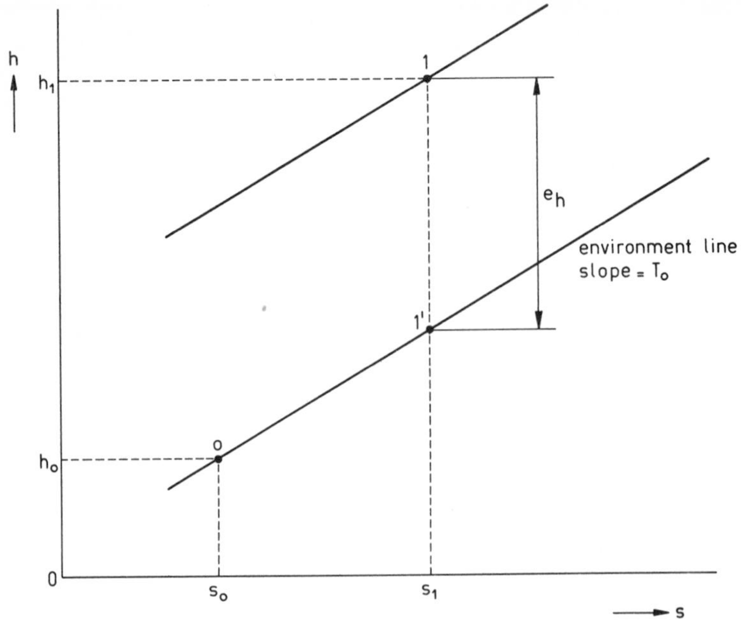

Fig. 3. Specific exergy in a h–s-diagram.

If we now draw a line through 1, parallel to the environment line, then it is clear that this line is the set of points with the same exergy. However, the states of the fluid along this line are different. Hence it is any way necessary to know a second state variable to fix the state of the fluid. It is also possible, of course, to give two state variables from which the exergy can be calculated.

If we consider the specific exergy according to (2.20), then a first possibility is to determine e_h, as a function of the two state variables h and s. If h and s are known at any instant, then the state of the fluid is also known at the same instant. The individual changes of h and s indicate the course of the process and according to (2.20) also the change of e_h. However, from practical considerations, it is desirable that the system variables can be measured directly by a measuring instrument. Would it be possible to measure the exergy as a system variable, then the use of it would become considerably more important.

From this point of view the use of h and s as variables seems less satisfactory. Thermodynamics, however, also has state variables which can be measured directly, namely temperature and pressure. So it makes sense to express e_h as a function of two independent variables in T and p. This also agrees very well with the custom to fix the state of the environment in terms of T_0 and p_0.

As point of departure relation (2.20) will be best. In order to express the right-hand-side of this equation in T and p, we look for thermodynamic functions which are functions of T and p by nature.

Such a function is for instance the specific free enthalpy g, defined by

$$g = h - Ts \tag{3.2}$$

With the generally valid equation

$$dh = Tds + \frac{1}{\rho}\, dp \tag{3.3}$$

we can conclude from (3.2) and (3.3)

$$dg = -sdT + \frac{1}{\rho}\, dp \tag{3.4}$$

From (3.4) follows

$$s = -\left(\frac{\partial g}{\partial T}\right)_p \tag{3.5}$$

$$\frac{1}{\rho} = \left(\frac{\partial g}{\partial p}\right)_T \tag{3.6}$$

Combining (3.2) and (3.5)

$$h = g - T\left(\frac{\partial g}{\partial T}\right)_p \tag{3.7}$$

As we can write for (2.20)

$$e_h = h - T_0 s - g_0 \tag{3.8}$$

substitution of (3.7) and (3.5) into (3.8) gives

$$e_h = g - g_0 - (T - T_0)\left(\frac{\partial g}{\partial T}\right)_p \tag{3.9}$$

Herewith the most general thermodynamic relation has been found for e_h as a function of T and p. We obtain for the total differential

$$de_h = \left(\frac{\partial e_h}{\partial T}\right)_p dT + \left(\frac{\partial e_h}{\partial p}\right)_T dp \tag{3.10}$$

It is convenient to write this in the form

$$de_h = \epsilon_T dT + \epsilon_p dp \tag{3.11}$$

defining

$$\epsilon_T = \left(\frac{\partial e_h}{\partial T}\right)_p \tag{3.12}$$

$$\epsilon_p = \left(\frac{\partial e_h}{\partial p}\right)_T \tag{3.13}$$

Then from (3.9) follows

$$\epsilon_T = -(T - T_0)\left(\frac{\partial^2 g}{\partial T^2}\right)_p \tag{3.14}$$

and

$$\epsilon_p = \left(\frac{\partial g}{\partial p}\right)_T - (T - T_0)\frac{\partial^2 g}{\partial T \partial p} \tag{3.15}$$

For an arbitrary fluid we can write

$$-T\left(\frac{\partial^2 g}{\partial T^2}\right)_p = c_p(T, p) \tag{3.16}$$

introducing the specific heat capacity c_p, which generally is a function of T and p. From (3.14) and (3.16) follows

$$\epsilon_T = \left(1 - \frac{T_0}{T}\right) c_p(T, p) \tag{3.17}$$

Therefore ϵ_T has the character of a specific heat capacity reduced by the Carnot factor, defined in (2.21).

From (3.6) and (3.15) follows

$$\epsilon_p = \frac{1}{\rho} - (T - T_0)\left(\frac{\partial \frac{1}{\rho}}{\partial T}\right)_p \tag{3.18}$$

so ϵ_p has the character of a specific volume.

In dealing with actual fluids, particularly gases, it is advantageous to express the equation of state of a real gas by the compressibility factor Z, which is defined by the relation

$$\frac{p}{\rho} = RTZ\,(T, p) \tag{3.19}$$

For ideal gases $Z = 1$. Deviations from $Z = 1$ are mainly caused by two effects. At *low* temperatures and *high* pressures the dimensions of the molecules and their mutual attractive forces become important. At *high* temperatures and *low* pressures dissociation and ionization processes occur.

From (3.18) and (3.19) follows for ϵ_p

$$\epsilon_p = T_0 R \frac{Z(T, p)}{p} - R(T - T_0)\frac{T}{p}\frac{\partial Z(T, p)}{\partial T} \tag{3.20}$$

Suppose we have de_h in the form of an equation like (3.11) and we want to obtain e_h by integration of (3.11) between the state (T, p) and the state of the environment (T_0, p_0).

Within the limits of many practical applications, for instance gas turbine systems, we may assume a gas to be ideal in so far, that c_p is a function of T only and Z practically equal to one.

Hence, taking $c_p = c_p(T)$ and $Z = 1$,

$$\epsilon_T = \left(1 - \frac{T_0}{T}\right) c_p(T) \tag{3.21}$$

$$\epsilon_p = T_0 \frac{R}{p} \tag{3.22}$$

and it appears that now ϵ_T is a function of T only and ϵ_p a function of p only, so that the desired integration can be executed quite easily

$$e_h = \int_{T_0}^{T} \left(1 - \frac{T_0}{T}\right) c_p(T)\, dT + T_0 R \ln \frac{p}{p_0} \tag{3.23}$$

As a consequence of the foregoing result we may define a function $e_T(T)$, solely dependent on temperature, by

$$\boxed{e_T(T) = \int_{T_0}^{T} \left(1 - \frac{T_0}{T}\right) c_p(T)\, dT} \tag{3.24}$$

In the same way we define a function $e_p(p)$, solely dependent on pressure, by

$$\boxed{e_p(p) = RT_0 \ln \frac{p}{p_0}} \tag{3.25}$$

Considering now the potential character of e_h, we shall call that part of e_h which is exclusively dependent on pressure, namely e_p, the *pressure potential* of the concerning fluid.

We shall call e_T, being exclusively dependent on temperature, the *temperature potential* of the fluid.

The preceding part of this paper has tried to find a quantity for a fluid, expressing the potential to perform work on the external world. This implies that a part of the energy indicated by this potential, of the fluid can be converted into useful work completely, provided, of course, the process of this conversion proceeds reversibly. Possible irreversibilities in a process will result in a decrease of this potential.

As a first result the *specific exergy* e_h has been found as an appropriate potential. However, the specific exergy had the drawback of giving no information about the state of a fluid. The state of a fluid is determined by two independent state variables as, for instance, temperature and pressure.

A second important result, which has been obtained now for a fluid, satisfying the conditions:

c_p is a function of T only,
$Z = 1$,

is the possibility to divide the specific exergy e_h into two independent partial potentials, namely a *temperature potential*, e_T, only dependent on temperature and a *pressure potential*, e_p, only dependent on pressure. Hereby we have met the requirements that the state of a fluid is being fixed by two independent state variables $e_T(T)$ and $e_p(p)$ and the potential e_h by

$$\boxed{e_h(T, p) = e_T(T) + e_p(p)} \tag{3.26}$$

4. THE TEMPERATURE AND PRESSURE POTENTIALS IN THE CASE OF IDEAL GASES

On account of the equation of state

$$\frac{p}{\rho} = RT \tag{4.1}$$

and therefore Z being equal to one and the fact that c_p = constant, we arrive at a very simple relation for e_T and e_p.

If in (3.24) c_p = constant, we obtain

$$e_T = c_p T_0 \left(\frac{T}{T_0} - 1 - \ln \frac{T}{T_0}\right) \tag{4.2}$$

and if $Z = 1$, we obtain, like (3.25),

$$e_p = RT_0 \ln \frac{p}{p_0} \tag{4.3}$$

Defining temperature and pressure ratios

$$T_r = \frac{T}{T_0} \tag{4.4}$$

$$P_r = \frac{p}{p_0} \tag{4.5}$$

we obtain

$$e_T = c_p T_0 (T_r - 1 - \ln T_r) \tag{4.6}$$

$$e_p = RT_0 \ln P_r \tag{4.7}$$

Both potentials can be depicted in a h–s-diagram as is shown in Fig. 4.

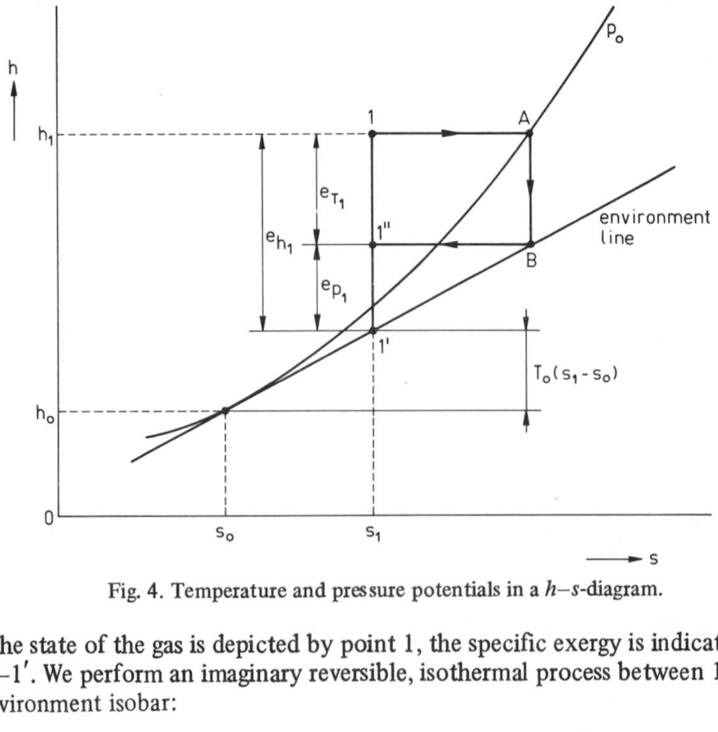

Fig. 4. Temperature and pressure potentials in a h–s-diagram.

If the state of the gas is depicted by point 1, the specific exergy is indicated by line 1–1′. We perform an imaginary reversible, isothermal process between 1 and the environment isobar:

Arriving at point A, e_T has not been changed, but e_p has become zero, consequently e_T is indicated by the line AB. Going back from B to the line $1-1'$, parallel to the isotherm $1A$, we divide $1-1'$ into two parts corresponding with e_T and e_p.

It is also possible to depict the potentials in a T–s-diagram, as shown in Fig. 5.

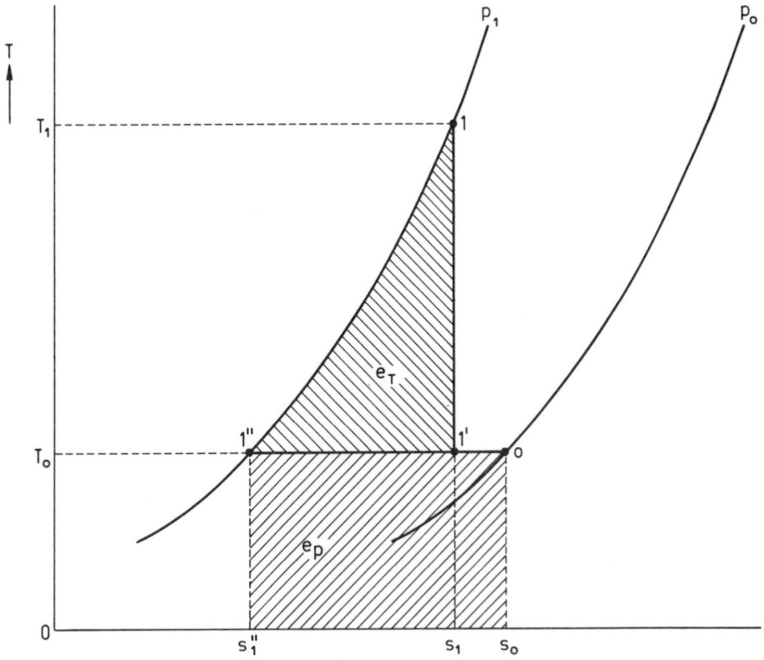

Fig. 5. Temperature and pressure potentials in a T–s-diagram.

From (3.3) and the relation $dh = c_p dT$ follows

$$c_p dT = T ds + \frac{1}{\rho} dp \tag{4.8}$$

If we consider this change along the isobar $p = p_1$, then we get

$$c_p dT = T ds \ (p = \text{constant}) \tag{4.9}$$

so that

$$de_T = \left(1 - \frac{T_0}{T}\right) c_p dT = (T - T_0)\, ds \tag{4.10}$$

and

$$e_{T_1} = \int_{T_0}^{T_1} (T - T_0)\, ds \tag{4.11}$$

so e_T corresponds with the surface $1-1'-1''$ in Fig. 5.

If we now consider a change according to (4.8) along the isotherm $T = T_0$, we get

$$-\frac{1}{\rho}dp = T_0\,ds \tag{4.12}$$

In this case

$$\frac{1}{\rho}dp = T_0 R\frac{dp}{p} = de_p \tag{4.13}$$

so

$$e_{p_1} = -\int_{p_0}^{p_1} T_0\,ds \tag{4.14}$$

and e_p proves to correspond with the surface $S_1''-S_0-0-1''$ in Fig. 5. The specific exergy of the gas in state 1 is equal to the sum of both mentioned surfaces.

5. DIFFERENTIAL BALANCE EQUATIONS

5.1 The general form

As has been pointed out in section 1 the basic equations, underlying system theory, are all of the balance equation type. The general balance equation in integral form is given by (1.2). This equation applies to any arbitrary physical system with volume V, bounded by a closed surface A, containing a fluid with density ρ. Applying Gauss' theorem from vector analysis to (1.2), we may transform it into a differential or local balance equation for a volume element dV of a continuous system

$$\frac{\partial(\rho x)}{\partial t} = -\operatorname{div}\boldsymbol{\phi}_x'' + \phi_x''' \tag{5.1}$$

The fundamental equations of mechanics, electricity and fluid-mechanics are in agreement with models (1.2) or (5.1). Sometimes, however, when the considered equation is a conservation law, the source term ϕ_x''' is absent. Irreversible thermodynamics also uses equations of this kind to describe irreversible, non-stationary processes. Some of these equations, like the balance equations for mass, momentum and total energy, are similar to those used in fluid mechanics. In addition to these we will find important expressions like the entropy balance equation and the exergy balance equation. It is the aim of this section to obtain this last one especially.

Considering the flux $\boldsymbol{\phi}_x''$ more closely, it appears that it can often be split up into a *convective* part, describing the transport of x through the surface dA by the fluid itself, and a *contact* part:

$$\boldsymbol{\Phi}_x'' = \rho \mathbf{v} x + \boldsymbol{\phi}_{x_c}'' \tag{5.2}$$

$\mathbf{v}$ is the average velocity of the volume element dV passing dA. Without further information nothing can be said about the contact flux $\boldsymbol{\phi}_{x_c}''$.

Equation (5.1) can be rewritten now

$$\frac{\partial(\rho x)}{\partial t} = -\operatorname{div}(\rho \mathbf{v} x + \boldsymbol{\phi}_{x_c}'') + \phi_x''' \tag{5.3}$$

The mass balance equation follows from (5.3) by putting $x = 1$ and considering that only convection of mass, $\rho\mathbf{v}$, and no contact transport takes place, while conservation of mass requires a zero source term. So we arrive, at the *differential mass balance equation*

$$\boxed{\frac{\partial \rho}{\partial t} = -\operatorname{div} \rho\mathbf{v}} \tag{5.4}$$

From (5.3) follows

$$\rho \frac{\partial x}{\partial t} + x \frac{\partial \rho}{\partial t} = -x \operatorname{div} \rho\mathbf{v} - \rho\mathbf{v} \,.\, \operatorname{grad} x + - \operatorname{div} \boldsymbol{\phi}''_{x_c} + \phi'''_x \tag{5.5}$$

Combining (5.5) with (5.4) gives

$$\rho \frac{\partial x}{\partial t} = -\rho\mathbf{v} \,.\, \operatorname{grad} x - \operatorname{div} \boldsymbol{\phi}''_{x_c} + \phi'''_x \tag{5.6}$$

For a fluid in motion an arbitrary property $x = x(\mathbf{r}, t)$ changes by two effects. First, the rate of change of x with time t *at a fixed point* is the *local time* derivative $\frac{\partial x}{\partial t}$. Secondly, if we follow the motion of a *particular fluid element*, we will have to know how x changes as the element moves. The rate of change due to this effect is called *convective time derivative.* The total time derivative, following the motion of the fluid, is called $\frac{Dx}{Dt}$, i.e.

$$\frac{Dx}{Dt} = \frac{\partial x}{\partial t} + \text{convective time derivative}$$

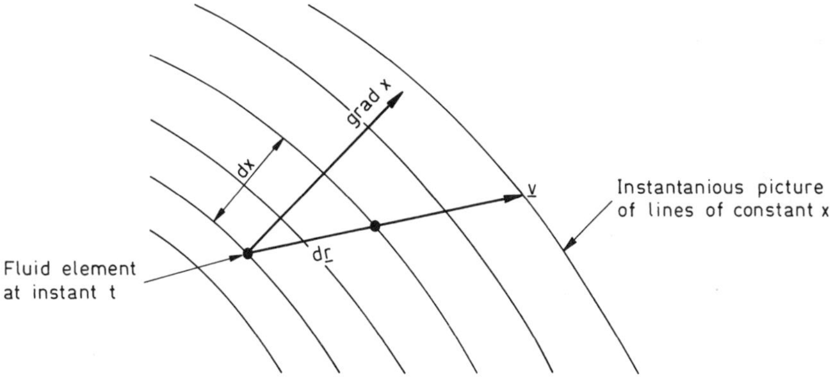

Fig. 6.

If at a time t the element is in the position $\mathbf{r}$, at time $t + dt$ it will be at $\mathbf{r} + d\mathbf{r}$ or $\mathbf{r} + \mathbf{v}dt$. The change in x due to this displacement is given by the scalar product of $d\mathbf{r}$ with grad x (see Fig. 6), i.e.

$$dx = d\mathbf{r} \, . \, \text{grad} \, x = \mathbf{v} \, . \, \text{grad} \, x \, dt$$

and the convective time derivative = $\mathbf{v}$. grad x, therefore

$$\frac{Dx}{Dt} = \frac{\partial x}{\partial t} + \mathbf{v} \, . \, \text{grad} \, x \tag{5.7}$$

Hence we can write instead of (5.6)

$$\rho \frac{Dx}{Dt} = -\text{div} \, \boldsymbol{\phi}''_{xc} + \phi'''_x \tag{5.8}$$

5.2 Differential mechanical, internal and total energy balance equations

The *differential mechanical energy balance equation* is obtained by putting x equal to a specific mechanical energy

$$\rho \frac{D}{Dt} (\tfrac{1}{2} v^2 + gz) = -\text{div} \, (p\mathbf{v} - \boldsymbol{\sigma} \, . \, \mathbf{v}) + \phi'''_{me} \tag{5.9}$$

with the source term

$$\phi'''_{me} = \rho p \frac{D}{Dt} \left(\frac{1}{\rho} \right) - \boldsymbol{\sigma} : \text{grad} \, \mathbf{v} \tag{5.10}$$

$\rho \dfrac{D(\frac{1}{2} v^2)}{Dt}$ is the total rate of increase of kinetic energy per unit volume, $\rho \dfrac{D(gz)}{Dt}$ is the total rate of increase of potential energy per unit volume, $-\text{div}(p\mathbf{v} - \boldsymbol{\sigma} \, . \, \mathbf{v})$ is the power delivered by the pressure and the viscous stresses (represented by the stress tensor $\boldsymbol{\sigma}$ on the volume element), $\rho p \dfrac{D}{Dt}\left(\dfrac{1}{\rho}\right)$ is the rate of reversible expansion work due to conversion of internal energy per unit volume; it can be either positive or negative, depending on an expansion or a compression of the fluid. $\boldsymbol{\sigma}$: grad $\mathbf{v}$ is the rate of irreversible conversion into internal energy due to viscous dissipation inside dV; it is always positive.

If u represents a specific internal energy, we obtain the *differential internal energy balance equation*

$$\rho \frac{Du}{Dt} = -\text{div} \, \boldsymbol{\phi}''_H + \phi'''_u \tag{5.11}$$

In this equation $\boldsymbol{\phi}''_H$ represents the heat flux to the environment by contact transport.

The source term ϕ'''_u describes the conversion of mechanical energy into internal energy.

Hence

$$\phi'''_u = -\phi'''_{me} = -\rho p \frac{D}{Dt} \left(\frac{1}{\rho} \right) + \boldsymbol{\sigma} : \text{grad} \, \mathbf{v} \tag{5.12}$$

Adding together (5.9) and (5.11) results in the *total differential energy balance equation*

$$\boxed{\rho \frac{D}{Dt} (\tfrac{1}{2} v^2 + gz + u) = -\text{div} \, (p\mathbf{v} - \boldsymbol{\sigma} \cdot \mathbf{v} + \phi''_H)} \tag{5.13}$$

Because of the conservation of total energy this equation has no source terms. We also may introduce the property specific enthalpy, defined by

$$h = u + \frac{p}{\rho} \tag{5.14}$$

into (5.13). An important point should be born in mind, however. The term $\frac{p}{\rho}$ only represents energy in case of a fluid entering or leaving an open system. The fluid inside a system with volume V has a certain pressure and a certain density and a certain internal energy u. The expression $\frac{p}{\rho}$ can be calculated and added to the internal energy u to determine the enthalpy of the fluid. However $\frac{p}{\rho}$ is not an energy quantity (although it has the units of energy) and therefore the enthalpy of the fluid inside V is not an energy quantity. The only energy of the fluid besides $\frac{1}{2}v^2$ and gz is its internal energy u.

When the fluid crosses the boundary of an open system the expression $\frac{p}{\rho}$ represents flow energy and the enthalpy of this fluid represents the sum of the internal energy and the flow energy. Therefore, *the enthalpy of a fluid represents an energy quantity only when this fluid crosses the boundary of an open system.*

We write (5.13) in the form of (5.3)

$$\frac{\partial}{\partial t}\,[\rho(\tfrac{1}{2}v^2 + gz + u)] = -\,\mathrm{div}\,[\rho\mathbf{v}(\tfrac{1}{2}v^2 + gz + u) + p\mathbf{v} - \boldsymbol{\sigma}\,.\,\mathbf{v} + \boldsymbol{\Phi}''_H] \tag{5.15}$$

Considering that $p\mathbf{v} = \rho\mathbf{v}\frac{p}{\rho}$, we get, after introduction of the enthalpy

$$\frac{\partial}{\partial t}\,[\rho(\tfrac{1}{2}v^2 + gz + u)] = -\,\mathrm{div}\,[\rho\mathbf{v}(\tfrac{1}{2}v^2 + gz + h) - \boldsymbol{\sigma}\,.\,\mathbf{v} + \boldsymbol{\Phi}''_H] \tag{5.16}$$

5.3 Differential entropy balance equation

We now consider the thermodynamic equation

$$Tds = du + pd\left(\frac{1}{\rho}\right) \tag{5.17}$$

In the thermodynamics of irreversible, non-stationary processes this equation is assumed to be still valid.

In which case we may write this equation as follows

$$T\frac{Ds}{Dt} = \frac{Du}{Dt} + p\,\frac{D}{Dt}\left(\frac{1}{\rho}\right) \tag{5.18}$$

From (5.18) we obtain by substitution of (5.11) and (5.12)

$$\rho T\frac{Ds}{Dt} = -\,\mathrm{div}\,\boldsymbol{\Phi}''_H + \boldsymbol{\sigma} : \mathrm{grad}\,\mathbf{v} \tag{5.19}$$

Now

$$\mathrm{div}\,\boldsymbol{\Phi}''_H = T\,\mathrm{div}\,\frac{\boldsymbol{\Phi}''_H}{T} + \boldsymbol{\Phi}''_H\,.\,\frac{\mathrm{grad}\,T}{T} \tag{5.20}$$

so from (5.19) and (5.20):

$$\rho \frac{Ds}{Dt} = -\mathrm{div}\, \frac{\boldsymbol{\Phi}_H''}{T} + \frac{1}{T}\left(\frac{\boldsymbol{\Phi}_H''}{T} \,.\, (-\mathrm{grad}\, T) + \boldsymbol{\sigma} : \mathrm{grad}\, \mathbf{v}\right) \tag{5.21}$$

This is the *differential entropy balance equation;* more concisely written:

$$\boxed{\rho \frac{Ds}{Dt} = -\mathrm{div}\, \boldsymbol{\Phi}_{s_c}'' + \phi_s'''} \tag{5.22}$$

defining the *entropy contact flux*

$$\boldsymbol{\Phi}_{s_c}'' = \frac{\boldsymbol{\Phi}_H''}{T} \tag{5.23}$$

and the *entropy production rate density*

$$\phi_s''' = \frac{1}{T}\left[\frac{\boldsymbol{\Phi}_H''}{T} \,.\, (-\mathrm{grad}\, T) + \boldsymbol{\sigma} : \mathrm{grad}\, \mathbf{v}\right] \tag{5.24}$$

The entropy production density term clearly shows the kind of irreversible processes which occur in the considered volume element dV, such as heat conduction and viscous dissipation. According to the second law of thermodynamics the source term satisfies the relation

$$\phi_s''' \geqslant 0 \tag{5.25}$$

The equality sign applies to the case of reversible transitions.

5.4 Differential exergy balance equation

In section 2 we introduced the concept of exergy. In a fluid in which non-stationary, irreversible processes occur, the specific exergy may differ from one point to another and is a function of time. With the aid of the above balance equations we now can derive the differential exergy balance equation. Consider for this purpose the entropy balance equation (5.21), written in the form of (5.3)

$$\frac{\partial(\rho s)}{\partial t} = -\mathrm{div}\left(\rho \mathbf{v} s + \frac{\boldsymbol{\Phi}_H''}{T}\right) + \phi_s''' \tag{5.26}$$

Multiplying each term of this equation with T_0 and subtracting the result from the total energy balance equation (5.16) we get

$$\frac{\partial}{\partial t}\left[\rho(\tfrac{1}{2}v^2 + gz + u - T_0 s)\right] = -\,\mathrm{div}\left[\rho \mathbf{v}(\tfrac{1}{2}v^2 + gz + h - T_0 s) + \right.$$
$$\left. -\boldsymbol{\sigma}\,.\,\mathbf{v} + \left(1 - \frac{T_0}{T}\right)\boldsymbol{\Phi}_H''\right] - T_0\,\phi_s''' \tag{5.27}$$

As p_0 is a constant, we may write for the left-hand side of (5.27)

$$\frac{\partial}{\partial t}\left[\rho\left(\frac{1}{2}v^2 + gz + u_0 - T_0 s + \frac{p_0}{\rho}\right)\right]$$

Next we consider the mass balance equation (5.4) and multiply it with the constant expression $u_0 - T_0 s_0 + \frac{p_0}{\rho_0} = h_0 - T_0 s_0$

$$\frac{\partial}{\partial t}\left[\rho\left(u_0 - T_0 s_0 + \frac{p_0}{\rho_0}\right)\right] = -\operatorname{div}\left[\rho \mathbf{v}(h_0 - T_0 s_0)\right] \tag{5.28}$$

Now extract (5.28) from (5.27) and we obtain

$$\frac{\partial}{\partial t}\left[\rho\left(\tfrac{1}{2}v^2 + gz + u - u_0 - T_0(s - s_0) + p_0\left(\frac{1}{\rho} - \frac{1}{\rho_0}\right)\right)\right] =$$

$$= -\operatorname{div}\left[\rho\mathbf{v}(\tfrac{1}{2}v^2 + gz + h - h_0 - T_0(s - s_0)) - \boldsymbol{\sigma}\cdot\mathbf{v} + \left(1 - \frac{T_0}{T}\right)\boldsymbol{\Phi}''_H\right] +$$

$$- T_0\phi'''_s \tag{5.29}$$

Substitution of (2.18) and (2.20) gives

$$\frac{\partial}{\partial t}\left[\rho(\tfrac{1}{2}v^2 + gz + e_u)\right] = -\operatorname{div}\left[\rho\mathbf{v}(\tfrac{1}{2}v^2 + gz + e_h) + \right.$$

$$\left. - \boldsymbol{\sigma}\cdot\mathbf{v} + \left(1 - \frac{T_0}{T}\right)\boldsymbol{\Phi}''_H\right] - T_0\phi'''_s \tag{5.30}$$

This is the *differential exergy balance equation*, more concisely written as

$$\boxed{\frac{\partial(\rho e_s)}{\partial t} = -\operatorname{div}\boldsymbol{\Phi}''_e + \phi'''_e} \tag{5.31}$$

In this equation e_s is the total specific exergy of a stagnant fluid

$$e_s = \tfrac{1}{2}v^2 + gz + e_u \tag{5.32}$$

and $\boldsymbol{\Phi}''_e$ is the total exergy flux

$$\boldsymbol{\Phi}''_e = \rho\mathbf{v}(\tfrac{1}{2}v^2 + gz + e_h) - \boldsymbol{\sigma}\cdot\mathbf{v} + \left(1 - \frac{T_0}{T}\right)\boldsymbol{\Phi}''_H \tag{5.33}$$

ϕ'''_e is the exergy production rate density,

$$\phi'''_e = -T_0\phi'''_s \tag{5.34}$$

Because of (5.25)

$$\phi'''_e \leqslant 0 \tag{5.35}$$

Upon introducing the Carnot factor $\eta_c = 1 - \frac{T_0}{T}$, the expression $\frac{T_0}{T}\frac{\boldsymbol{\Phi}''_H}{T}\cdot(-\operatorname{grad} T)$ can be transformed into $\boldsymbol{\Phi}''_H\cdot(-\operatorname{grad}\eta_c)$. Therefore we can write, using (5.24),

$$\phi'''_e = -\boldsymbol{\Phi}''_H\cdot(-\operatorname{grad}\eta_c) - \frac{T_0}{T}\,\boldsymbol{\sigma} : \operatorname{grad}\mathbf{v} \tag{5.36}$$

expressing the exergy loss due to heat conduction and viscous dissipation.

That which has been said about the entalpy, h, also applies to the exergy of the enthalpy, e_h, i.e. the exergy of a stagnant fluid is e_u and the exergy of a flowing fluid, crossing a boundary, is e_h, therefore we shall call e_u *specific stagnant exergy* and e_h *specific flow exergy*.

Between e_u and e_h exists the following relation

$$e_h = e_u + \frac{1}{\rho}(p - p_0) \tag{5.37}$$

6. DIFFERENTIAL TEMPERATURE AND PRESSURE POTENTIAL BALANCE EQUATION

In section 3 we introduced the temperature and pressure potentials e_T and e_p. We shall investigate now whether it is possible to derive differential balance equations for both potentials. To do this we consider the internal energy balance equation

$$\rho \frac{Du}{Dt} = -\operatorname{div} \boldsymbol{\Phi}''_H - \rho p \frac{D}{Dt}\left(\frac{1}{\rho}\right) + \boldsymbol{\sigma} : \operatorname{grad} \mathbf{v} \tag{6.1}$$

This equation can be transformed into an enthalpy balance equation with the relation

$$\rho \frac{D}{Dt}\left(\frac{p}{\rho}\right) = \rho p \frac{D}{Dt}\left(\frac{1}{\rho}\right) + \frac{Dp}{Dt} \tag{6.2}$$

resulting in

$$\rho \frac{Dh}{Dt} = -\operatorname{div} \boldsymbol{\Phi}''_H + \frac{Dp}{Dt} + \boldsymbol{\sigma} : \operatorname{grad} \mathbf{v} \tag{6.3}$$

Now

$$\frac{Dh}{Dt} = \left(\frac{\partial h}{\partial T}\right)_p \frac{DT}{Dt} + \left(\frac{\partial h}{\partial p}\right)_T \frac{Dp}{Dt} \tag{6.4}$$

From thermodynamics is known

$$\left(\frac{\partial h}{\partial T}\right)_p = c_p(T, p) \tag{6.5}$$

$$\left(\frac{\partial h}{\partial p}\right)_T = \frac{1}{\rho} - T\left(\frac{\partial \frac{1}{\rho}}{\partial T}\right)_p \tag{6.6}$$

Thus

$$\frac{Dh}{Dt} = c_p(T, p)\frac{DT}{Dt} + \left[\frac{1}{\rho} - T\left(\frac{\partial \frac{1}{\rho}}{\partial T}\right)_p\right]\frac{Dp}{Dt} \tag{6.7}$$

If we consider a gas with the properties that $c_p = c_p(T)$ and $Z \approx 1$, the expression between square brackets vanishes so

$$\frac{Dh}{Dt} = c_p(T)\frac{DT}{Dt} \tag{6.8}$$

Multiplying both sides by the Carnot factor $\eta_c = 1 - \frac{T_0}{T}$ results in

$$\eta_c \frac{Dh}{Dt} = \eta_c c_p(T) \frac{DT}{Dt} = \frac{De_T}{Dt} \tag{6.9}$$

Next we multiply (6.3) by η_c and obtain

$$\eta_c \, \rho \frac{Dh}{Dt} = -\eta_c \operatorname{div} \boldsymbol{\phi}_H'' + \eta_c \left(\frac{Dp}{Dt} + \boldsymbol{\sigma} : \operatorname{grad} \mathrm{v} \right) \tag{6.10}$$

Now

$$\eta_c \operatorname{div} \boldsymbol{\phi}_H'' = \operatorname{div} \eta_c \boldsymbol{\phi}_H'' + \boldsymbol{\phi}_H'' \cdot (-\operatorname{grad} \eta_c) \tag{6.11}$$

so

$$\eta_c \, \rho \frac{Dh}{Dt} = -\operatorname{div} \eta_c \boldsymbol{\phi}_H'' + \eta_c \left(\frac{Dp}{Dt} + \boldsymbol{\sigma} : \operatorname{grad} \mathrm{v} \right) - \boldsymbol{\phi}_H'' \cdot (-\operatorname{grad} \eta_c) \tag{6.12}$$

Combination of (6.9) and (6.12) yields

$$\rho \frac{De_T}{Dt} = -\operatorname{div} \eta_c \boldsymbol{\phi}_H'' - \boldsymbol{\phi}_H'' \cdot (-\operatorname{grad} \eta_c) + \eta_c \left(\frac{Dp}{Dt} + \boldsymbol{\sigma} : \operatorname{grad} \mathrm{v} \right) \tag{6.13}$$

From (6.2), (5.7) and (5.4) we can derive

$$\frac{Dp}{Dt} = \frac{\partial p}{\partial t} + \operatorname{div} p\mathrm{v} - \rho p \frac{D}{Dt} \left(\frac{1}{\rho} \right) \tag{6.14}$$

Combining this equation with (5.9) and (5.10) yields

$$\frac{Dp}{Dt} = \frac{\partial p}{\partial t} + \operatorname{div} \boldsymbol{\sigma} \cdot \mathrm{v} - \rho \frac{D}{Dt} \left(\tfrac{1}{2} v^2 + gz \right) - \boldsymbol{\sigma} : \operatorname{grad} \mathrm{v} \tag{6.15}$$

Now define

$$\phi_{me_r}''' = \frac{\partial p}{\partial t} + \operatorname{div} \boldsymbol{\sigma} \cdot \mathrm{v} - \rho \frac{D}{Dt} \left(\tfrac{1}{2} v^2 + gz \right) \tag{6.16}$$

then (6.15) becomes

$$\frac{Dp}{Dt} = \phi_{me_r}''' - \boldsymbol{\sigma} : \operatorname{grad} \mathrm{v} \tag{6.17}$$

Substitution of (6.17) into (6.13) gives the *differential temperature potential balance equation*

$$\boxed{\rho \frac{De_T}{Dt} = -\operatorname{div} \eta_c \boldsymbol{\phi}_H'' - \boldsymbol{\phi}_H'' \cdot (-\operatorname{grad} \eta_c) + \eta_c \phi_{me_r}'''} \tag{6.18}$$

From (3.25) we get

$$\rho \frac{De_p}{Dt} = \rho \frac{RT_0}{p} \frac{Dp}{Dt} = \frac{T_0}{T} \frac{Dp}{Dt} = (1 - \eta_c) \frac{Dp}{Dt} \tag{6.19}$$

Together with (6.17) we obtain the *differential pressure potential balance equation*

$$\rho \frac{De_p}{Dt} = (1 - \eta_c)(\phi'''_{me_r} - \boldsymbol{\sigma} : \text{grad } \mathbf{v}) \tag{6.20}$$

Comparing (6.18) and (6.20) we notice that ϕ'''_{me_r} expresses the rate of conversion per unit volume of various kinds of mechanical energy into exergy. Of this conversion only the fraction η_c can contribute to e_T, which also appears from (6.18) where ϕ'''_{me_r} shows up, multiplied by η_c. As mechanical energy, being a kind of valuable energy, is completely available, the fraction $(1 - \eta_c)$ cannot be lost. Therefore we find this fraction of ϕ'''_{me_r} in (6.20). Furthermore (6.18) contains a heat exergy transport term, $-\text{div } \eta_c \boldsymbol{\Phi}''_H$, and a heat exergy loss term, $-\boldsymbol{\Phi}''_H \cdot (-\text{grad } \eta_c)$.

The latter term is always > 0, because heat only flows in the direction of decreasing temperatures and consequently decreasing Carnot factor η_c.

Equation (6.20) contains a loss term, indicating the loss of mechanical energy due to viscous friction processes, $(1 - \eta_c)\, \boldsymbol{\sigma} : \text{grad } \mathbf{v}$. Only the fraction $(1 - \eta_c)$ is an exergy loss, because the fraction η_c is kept inside the system as dissipation heat.

For a gas which satisfies the following properties, i.e. $c_p = c_p(T)$ and $Z = 1$, we may draw the following *conclusions* from the foregoing analysis:

1. In the conversion of valuable energy (like mechanical energy) into exergy a fraction η_c of this energy attributes to the temperature potential and a fraction $(1 - \eta_c)$ attributes to the pressure potential.
2. Heat transport and heat loss manifest themselves as a change of the temperature potential.
3. Friction losses or generally dissipation of valuable energy result in a loss of pressure potential.

It will also be clear from this analysis that the division of e_h into e_T and e_p makes sense and that we have obtained a set of system variables, which can describe heat and mechanical effects of fluid flow separately.

The coupling of the two effects comes about by the quantity ϕ'''_{me_r}.

Adding (6.18) and (6.20) together gives a differential balance equation for the specific flow exergy, e_h

$$\rho \frac{De_h}{Dt} = -\text{div } \eta_c \boldsymbol{\Phi}''_H + \phi'''_{me_r} + \phi'''_e \tag{6.21}$$

This equation can be obtained directly from (5.30) if we replace e_u in this equation by e_h with (5.37), introduce ϕ'''_{me_r} via (6.16) and ϕ'''_e via (5.36).

7. RETICULATION OF THE GENERAL BALANCE EQUATION

The above equations have been derived for continuous systems. In practical applications the situation is often quite different. Rather than being continuously distributed throughout the whole region of the system volume V, various physical processes may be lumped in discrete regions. Although this is an idealization, it is a very practical and necessary one, if one tries to analyse the behaviour of actual systems or components.

In order to do so we will reticulate the balance equations. This amounts to the following assumptions:

1. That the system boundary is permeable to the passage of any physical quantity only on relatively few discrete areas, called ports;
2. That the accumulation of any physical quantity inside the system is not continuously distributed throughout its volume, but is rather lumped in discrete localities or regions;
3. That the sources or sinks of any physical quantity are also confined to discrete regions.

Considering an arbitrary specific physical quantity x, the foregoing assumptions may be depicted as indicated in Fig.7.

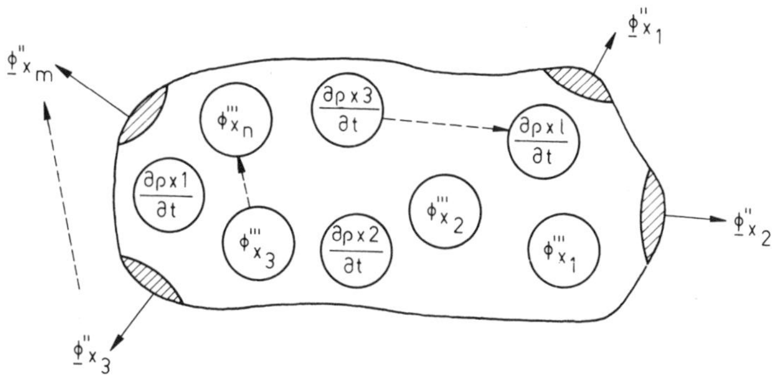

Fig. 7. Reticulated system.

As a starting point for reticulation we may use the balance equation in its integral form like the one in (1.2).

We first consider the reticulation of the storage of x. When the integration of the storage term is performed over the entire volume V there will be only a discrete number, say l, of regions V_i in which $\frac{\partial(\rho x)}{\partial t} \neq 0$. Thus:

$$\int_V \frac{\partial(\rho x)}{\partial t}\, dV = \sum^{l} \int_{V_i} \frac{\partial(\rho x)}{\partial t}\, dV_i \tag{7.1}$$

Now, a reticulation of the storage of x in V will have meaning only if the volume V_i in which the storage is supposed to occur is fixed in size and position. Assuming this to be the case, we may carry the time differentiation outside the integral and change it from partial to total giving:

$$\int_V \frac{\partial(\rho x)}{\partial t}\, dV = \sum^{l} \frac{d}{dt} \int_{V_i} \rho_i x_i \, dV_i \tag{7.2}$$

It is further assumed that V_i is small enough that x may be put equal to an average value. Then as V_i is fixed in size and position

$$\int_{V_i} \rho_i x_i \, dV_i = x_i \int_{V_i} \rho_i \, dV_i = m_i x_i \tag{7.3}$$

m_i being the total instantaneous mass inside V_i.

Hence we obtain

$$\int_V \frac{\partial(\rho x)}{\partial t} \, dV = \sum^{L} \frac{d}{dt} (m_i x_i) \tag{7.4}$$

Now we consider the transfer of x across the system boundary. As we can be seen from (5.2) there are two types of transfer to be considered. First the convective transfer of x, $\rho \mathbf{v} x$, and secondly the contact transfer, $\boldsymbol{\phi}''_{x_c}$. The latter one may be composed of several kinds of transfer. Very often each term may consist of some kind of flow quantity, usually a vector, generally written $\boldsymbol{\phi}''$ and a scalar quantity, generally written as ϵ, having the character of a potential. Thus quite generally

$$\boldsymbol{\phi}''_{x_c} = \boldsymbol{\phi}'' \epsilon \tag{7.5}$$

It is easy to see that $\rho \mathbf{v} x$ is of the same type as $\boldsymbol{\phi}'' \epsilon$, thus we may write, assuming that the number of discrete transfer areas equals m, any area being indicated by the index j,

$$\boldsymbol{\phi}''_x = \sum^{m} \boldsymbol{\phi}''_j \epsilon_j \tag{7.6}$$

It is further assumed that A_j is of sufficiently limited extent that the ϵ_j may be put equal to an average value. Then

$$\int_{A_j} \epsilon_j \boldsymbol{\phi}''_j \cdot \mathbf{n} \, dA_j = \epsilon_j \int_{A_j} \boldsymbol{\phi}''_j \cdot \mathbf{n} \, dA_j = \epsilon_j \phi_j \tag{7.7}$$

Hence the total transfer of x through the boundary of V is equal to

$$\int_A \boldsymbol{\phi}''_x \cdot \mathbf{n} \, dA = \sum^{m} \epsilon_j \phi_j \tag{7.8}$$

In physical system theory the quantities ϵ are generally called *across-variables*, the quantities ϕ *through-variables.*

Finally we will examine the reticulation of the source term ϕ'''_x. We assume that ϕ'''_x will only occur in a discrete number, say n, of restricted regions V_k. It is therefore evident that

$$\int_V \phi'''_x \, dV = \sum^{n} \int_{V_k} \phi'''_{x_k} \, dV_k \tag{7.9}$$

With the definition

$$\int_{V_k} \phi'''_{x_k} \, dV_k = \phi_k \tag{7.10}$$

we finally obtain

$$\int_V \phi_x''' \, dV = \sum^n \Phi_k \tag{7.11}$$

Combining the foregoing results, we may write the *reticulated balance equation* for an arbitrary specific physical quantity as follows

$$\boxed{\sum^l \frac{d}{dt}(m_i x_i) = -\sum^m \epsilon_j \phi_j + \sum^n \Phi_k} \tag{7.12}$$

This equation will be the basis of the practical analysis of thermal systems and components.

If the dimensions of the various terms in (7.12) conform to an energy flow rate, then the flow of energy across the system boundary is represented by $\sum^m \epsilon_j \phi_j$. As already noticed in section 1, flow of energy is characterized by the product of the two system variables ϵ_j and ϕ_j. The meaning of ϵ_j and ϕ_j in some systems is shown in Table 1.

8. NEW THERMAL SYSTEM VARIABLES

As already pointed out in section 1, we rather like to describe thermal systems in terms of exergy instead of energy. Therefore we shall now apply (7.12) to the exergy balance equation (5.31). This equation can be written as

$$\frac{\partial(\rho e_s)}{\partial t} = -\operatorname{div}(\rho \mathbf{v} e_f + \boldsymbol{\phi}_{ec}'') + \phi_e''' \tag{8.1}$$

e_s is the total specific stagnant exergy, according to (5.32), e_f is the total specific flow exergy, given by

$$e_f = \tfrac{1}{2}v^2 + gz + e_h \tag{8.2}$$

and $\boldsymbol{\phi}_{ec}''$ is the exergy contact flux,

$$\boldsymbol{\phi}_{ec}'' = -\boldsymbol{\sigma} \cdot \mathbf{v} + \eta_c \phi_H'' \tag{8.3}$$

In order to reticulate this equation we write it in integral form:

$$\int_V \frac{\partial(\rho e_s)}{\partial t} dV = -\int_A (\rho \mathbf{v} e_f + \boldsymbol{\phi}_{ec}'') \cdot \mathbf{n} \, dA + \int_V \phi_e''' \, dV \tag{8.4}$$

Reticulation leads to

$$\sum^l \frac{d}{dt}(m_i e_{s_i}) = -\sum^m \epsilon_j \phi_j + \sum^n \boldsymbol{\Phi}_{ek} \tag{8.5}$$

Considering in particular $\sum^{m} \epsilon_j \phi_j$, it can be said that this term describes the various types of exergy flow across the system boundary.

Some of these are of convective nature and are connected with $\rho \mathbf{v} e_f$ in (8.4), others are of contact nature and are connected with $\mathbf{\Phi}''_{ec}$ in (8.4).

The total exergy flux can be written as

$$\mathbf{\Phi}''_e = \rho \mathbf{v} e_f + \mathbf{\Phi}''_{ec} \tag{8.6}$$

or

$$\mathbf{\Phi}''_e = \rho \mathbf{v}(\tfrac{1}{2}v^2 + gz + e_h) + \eta_c \mathbf{\Phi}''_H - \boldsymbol{\sigma} \cdot \mathbf{v} \tag{8.7}$$

If we neglect in this expression the mechanical terms, being of no direct importance for the rest of this discussion, we obtain

$$\mathbf{\Phi}''_e = \rho \mathbf{v} e_h + \eta_c \mathbf{\Phi}''_H \tag{8.8}$$

For a system with m^+ discrete ports, through which convective exergy transport takes place and m° discrete ports, through which conductive exergy transport takes place, the reticulated term $\sum^{m} \epsilon_j \phi_j$ may be written as

$$\sum^{m} \epsilon_j \phi_j = \sum^{m^+} e^+_{h_j} \phi^+_{m_j} + \sum^{m^\circ} \left(1 - \frac{T_0}{T^\circ}\right) \phi^\circ_{H_j} \tag{8.9}$$

From this relation it is clear that transport of exergy manifests itself in two ways, namely

(*a*) convective exergy transport = $e_h \phi_m$
(*b*) conductive exergy transport = $\eta_c \phi_H$

Thus the thermal system variables could be defined as follows:

TABLE 2

Thermal system	Across-variable	Through-variable
convective	specific flow exergy, e_h (J/kg)	mass flow rate, ϕ_m (kg/s)
conductive	Carnot factor $\eta_c = 1 - \frac{T_0}{T}$ (dimensionless)	heat flow rate, ϕ_H (W)

In order to explain the kind of system variables which could be expected in thermal systems, we neglected the mechanical terms. As for gases the influence of viscous stresses is small, we may neglect the term $\boldsymbol{\sigma} \cdot \mathbf{v}$.

In real systems, however, like for example in turbo engines, the terms $\frac{1}{2}v^2$ and gz can play a very important role.

Moreover e_h consists of two partial potentials. So, a convective thermal exergy stream will generally have four partial across-variables summarized in one across-variable, the total specific flow exergy, e_f:

$$e_f = \tfrac{1}{2}v^2 + gz + e_T + e_p \tag{8.10}$$

and one through-variable, the *mass flow rate*

$$\phi_m = \int_A \rho \mathbf{v} \, . \, \mathbf{n} \, dA \tag{8.11}$$

As to (8.10) it is well to remember that e_f is an average value over the surface A.

From the analysis in the preceding sections it will be clear that the application of e_T and e_p makes sense only if they are independent of each other. That restricts their use to ideal gases or to real gases that can be considered ideal under the prevailing circumstances.

In the case of heat conduction, the exergy balance equation (5.30) reduces to the following equation

$$\frac{\partial(\rho e_u)}{\partial t} = -\operatorname{div} \eta_c \mathbf{\Phi}''_H - \mathbf{\Phi}''_H \, . \, (-\operatorname{grad} \eta_c) \tag{8.12}$$

In practical applications we usually consider *heat conduction in solids* and to a lesser extend in liquids. In these cases the density ρ may be considered a constant, so

$$e_u = u - u_0 - T_0(s - s_0) \tag{8.13}$$

Introducing the specific heat capacity c, we can write for e_u

$$e_u = cT_0 \left(\frac{T}{T_0} - 1 - \ln \frac{T}{T_0} \right) \tag{8.14}$$

Hence

$$\frac{\partial(\rho e_u)}{\partial t} = \rho c T_0 \frac{\partial}{\partial t} \left(\frac{T}{T_0} - 1 - \ln \frac{T}{T_0} \right) = \rho c \eta_c \frac{\partial T}{\partial t} \tag{8.15}$$

Now

$$\frac{\partial T}{\partial t} = \frac{T^2}{T_0} \frac{\partial}{\partial t} \left(1 - \frac{T_0}{T} \right) = \frac{T^2}{T_0} \frac{\partial \eta_c}{\partial t} \tag{8.16}$$

thus

$$\frac{\partial(\rho e_u)}{\partial t} = \rho c \frac{T^2}{T_0} \eta_c \frac{\partial \eta_c}{\partial t} \tag{8.17}$$

and the differential exergy balance equation in the case of pure heat conduction

$$\rho c \frac{T^2}{T_0} \eta_c \frac{\partial \eta_c}{\partial t} = -\operatorname{div} \eta_c \mathbf{\Phi}''_H - \mathbf{\Phi}''_H \, . \, (-\operatorname{grad} \eta_c) \tag{8.18}$$

For an infinitesimal small length dx of a solid wall with area A and heat conduction only in one direction, (8.18) can be written

$$A \rho c \frac{T^2}{T_0} \eta_c \frac{\partial \eta_c}{\partial t} dx = d\eta_c \phi_H + \phi_H \frac{\partial \eta_c}{\partial x} dx \tag{8.19}$$

Apparently the *specific heat exergy capacity per unit length*

$$c_e = A\rho c \frac{T^2}{T_0} \tag{8.20}$$

Equation (8.19) is analogous with an equation for an electric current i and voltage e in a conductor with homogeneously distributed resistance and capacity,

$$Ce \frac{\partial e}{\partial t} dx = d\, ie + i \frac{\partial e}{\partial x} dx \tag{8.21}$$

The only difference is that the electric capacity C is a constant and c_e is still dependent on temperature.

9. CONCLUSIONS

1. Application of system theory to various kinds of energy transport shows that energy flows manifest themselves as a product of two system variables, namely a so-called across-variable and a so-called through-variable. Reticulation of an integral energy balance equation discloses the kind of variables in a certain case of itself (see Table 1).
2. It is possible to fit thermal systems into a physical system theory. For that purpose thermodynamics of irreversible, non-stationary processes have to be used. Only from this new branch of thermodynamics it is possible to draw space- and time-dependent balance equations.
3. As regards thermal energy it has to be born in mind that it is principally impossible to convert this form of energy into useful work completely. This is an inherent property due to the second law of thermodynamics. In order to indicate which part of thermal energy can be converted into work the concept of exergy (available energy) has been introduced. The starting equation for reticulation therefore is an integral exergy balance equation instead of an integral energy balance equation.
4. Thermal energy transport appears generally in three different ways, namely:
   convection (fluid flow),
   conduction (heat conduction),
   radiation (electro-magnetic radiation).
   In this paper convection and conduction have been considered only.
5. In the case of thermal energy transport by means of convection reticulation of the integral exergy balance equation results in the specific exergy of the specific enthalpy

   $$e_h = h - h_0 - T_0(s - s_0) \tag{9.1}$$

   as across-variable and the mass flow rate

   $$\phi_m = \rho A v \tag{9.2}$$

   as through-variable.
6. In the case of a constant state of the environment (T_0, p_0) e_h is a function of two independent thermodynamic state variables. The knowledge of e_h only

therefore does not fix the state of a fluid. Considering a fluid with the properties:

c_p is a function of T only,
$Z = 1$,

e_h can be split up into a so-called *temperature potential* e_T, a function of T only,

$$e_T(T) = \int_{T_0}^{T} \left(1 - \frac{T_0}{T}\right) c_p(T)\, dT \tag{9.3}$$

and a so-called *pressure potential* e_p, a function of p only,

$$e_p(p) = T_0 R \ln \frac{p}{p_0} \tag{9.4}$$

while
$$e_h(T, p) = e_T(T) + e_p(p) \tag{9.5}$$

So e_h can be described as a function of two independent state variables.

7. An additional justification for the introduction of e_T and e_p has been found in the fact that heat losses result in a decrease of e_T, while losses of a mechanical nature result in a decrease of e_p. A separation of thermal and mechanical effects has been obtained by deriving differential balance equations for e_T and e_p.
8. In order to obtain information for control and protection of thermal systems, the appropriate signals are the state variables T and p. However, T and p as such do not supply information about the available energy in any location inside a system. In this respect the quantities e_T and e_p can supply us with much better information. Fortunately, e_T and e_p are related to T and p in a rather simple manner. Hence both requirements can be satisfied by measuring T and p and calculating e_T and e_p subsequently.
9. Measurement of e_T and e_p and of the mass flow rate ϕ_m (considering transport of a hot gas through a duct, for instance) makes it feasible to equip such a stream with a Watt-meter, indicating the available energy, E_H, passing a certain location inside the stream.

$$E_H = \phi_m(e_T + e_p) \tag{9.6}$$

10. In the case of thermal energy transport by means of conduction reticulation of the integral exergy balance equation results in the Carnot factor

$$\eta_c = 1 - \frac{T_0}{T} \tag{9.7}$$

to be the across-variable and the heat flow rate ϕ_H to be the through-variable.

11. The heat flow rate ϕ_H is physically analogous to the electric current i. Both quantities satisfy Kirchhoff's first law, i.e.

$$\left.\begin{aligned} \Sigma i &= 0 \\ \Sigma \phi_H &= 0 \end{aligned}\right\} \tag{9.8}$$

The temperature difference $\Delta T = T - T_0$ is generally considered to be the potential difference of a heat current, analogous to the electrical potential difference e. This is not correct. The product of the across- and through-variables should be an energy flow rate. The product $\phi_H \Delta T$ does not satisfy this criterion. However, the product $\phi_H \eta_c$ does, hence η_c is to be considered as analogous to e.

12. In the case of heat conduction Thoma suggests as system variables the temperature T as across-variable and the entropy flow rate

$$\phi_s = \frac{\phi_H}{T} \tag{9.9}$$

as through-variable [15, 16]. The product of both quantities is an energy flow rate, moreover a temperature as across-variable is an attractive proposition indeed. However, treating ϕ_s as a through-variable is very unattractive, because in the case of irreversible heat conduction ϕ_s increases continuously. This property does not fit into the concept of resistance in network theory. If a current passes through a resistor the potential difference will decrease but the current remains constant. With the temperature difference ΔT as across-variable Thoma's variable pair leads to the same result as the variable pair defined by the author, because

$$\phi_H \eta_c = \phi_s \Delta T \tag{9.10}$$

but the objections against ϕ_s remain.

It is the opinion of the author that the heat flow rate ϕ_H and the Carnot factor η_c as a variable pair fits into already existing conventions, derived from electrical network theory, much better.

13. The *final conclusion* of this paper is, that the introduced thermal system variables open the possibility to make network models of more complicated thermal systems. It may be advantageous to apply methods of analysis and theorems from network theory. Consequently the next step in the development of a thermal system theory will be the application of this theory to engineering systems for transportation and conversion of energy. In this respect a ship's propulsion system would be a very interesting object [10].

REFERENCES

1. Baehr, H. D. *Thermodynamik.* Springer Verlag, Heidelberg, 1962.
2. Groot, S. R. de. *Thermodynamik irreversibler Prozesse.* Bibliographisches Institut, Mannheim, 1960.
3. Haase, R. *Thermodynamik der irreversiblen Prozesse.* Steinkopff Verlag, Darmstadt, 1963.
4. Paynter, H. M. *Analysis and Design of Engineering Systems.* M.I.T. Press, Cambridge, Mass., 1961.
5. MacFarlane, A. G. J. *Engineering Systems Analysis.* Harrap, London, 1964.
6. Baehr, H. D. "Definition und Berechnung von Exergie und Anergie". *B.W.K. 17* (1965), 1–6.
7. Rietman, J. *Exergie en niet-omkeerbaarheidsthermodynamica met toepassingen in de systeemanalyse.* Publication Nr. 4, Oct. 1968, Royal Naval College, Den Helder, Netherlands.

8. Rietman, J. and en Vinke, W. "The Boiler as a Part of the Power generating System". *Proc. Symp. Internal Corrosion in Boilers.* Royal Naval College, Den Helder, Netherlands, 1968, 9–16.
9. Rietman, J. "Thermodynamic Principles of thermal Energy Convertors in View of Modern Systems Analysis", in *Design and Econ. Consids. Shipbuilding and Shipping.* Veenmans, Wageningen, Netherlands, 1969, 527–550.
10. Vinke, W. "Propulsion Systems Viewed as Energy Transforming and Transporting Systems", in *Design and Econ. Consids. Shipbuilding and Shipping* Veenmans, Wageningen, Netherlands, 1969.
11. Wylen, G. J. van. *Thermodynamics.* Wiley, New York, 1960.
12. Byron Bird, R. *et al. Transport Phenomena.* Wiley, New York, 1960.
13. Kay, J. M. *An Introduction to Fluid Mechanics and Heat Transfer.* University Press, Cambridge, 1963.
14. Rant, Z. "Exergie: ein neues Wort für technische Arbeitsfähigkeit". *Forsch. Ing. -Wes. 22* (1956), 36/37.
15. Thoma, J. U. "Bond Graphs for Thermal Energy Transport and Entropy Flow". *J. Franklin Inst. 292*, (1971), 109–120.
16. Thoma, J. U. "Models, bond graphs and entropy" in Dixhoorn, J. J. van and Evans, F. J. (eds), *Physical Structure in Systems Theory*. Academic Press, London, 1974 (this volume).

NOMENCLATURE

		Dimensions
A	surface enclosing a system, area of a cross-section	m^2
C	electric capacity per unit length	F/m
E_Q	exergy of a heat quantity	J
E_u	exergy of internal energy	J
F	force	N
M	torque	Nm
P_r	pressure ratio	–
Q	heat quantity transferred to a system	J
R	gas constant	J/kg K
S	entropy	J/K
T	absolute temperature	K
T_0	absolute temperature of the environment	K
T_r	temperature ratio	–
U	internal energy	J
V	volume of a system	m^3
W	work done by a system	J
Z	compressibility factor	–
c	specific heat capacity of a solid	J/kg K
c_e	specific heat exergy capacity per unit length	J/kg Km
c_p	specific heat capacity at constant pressure	J/kg K
e	electric voltage	V

e_f	total specific flow exergy	J/kg
e_h	specific exergy of the specific enthalpy h	J/kg
e_p	pressure potential	J/kg
e_q	specific exergy of a heat quantity q	J/kg
e_s	total specific stagnant exergy	J/kg
e_T	temperature potential	J/kg
e_u	specific exergy of the specific internal energy u	J/kg
g	gravitational acceleration,	m/s^2
	specific free enthalpy	J/kg
g_0	specific free enthalpy at temperature T_0 and pressure p_0	J/kg
h	specific enthalpy	J/kg
h_0	specific enthalpy at temperature T_0 and pressure p_0	J/kg
i	electric current	A
m	mass	kg
Δm	discrete amount of mass	kg
$\mathbf{n}$	unit vector normal to a surface element dA	–
p	absolute pressure	N/m^2
p_0	absolute pressure of the environment	N/m^2
q	specific heat quantity transferred to a system	J/kg
$\mathbf{r}$	radius vector	m
s	specific entropy	J/kg K
s_0	specific entropy at temperature T_0 and pressure p_0	J/kg K
t	time	sec
u	specific internal energy	J/kg
u_0	specific internal energy at temperature T_0 and pressure p_0	J/kg
$\mathbf{v}$	velocity vector	m/s
x	arbitrary physical quantity	–
z	height	m
Φ	production rate of an arbitrary physical quantity	–/s
Φ_e	production rate of exergy	W
ϵ	arbitrary across-variable	–
ϵ_p	reduced specific heat capacity	J/kg K
ϵ_T	reduced specific volume	m^3/kg
η_c	Carnot factor = $1 - T_0/T$	–
ρ	density	kg/m^3
ρ_0	density at temperature T_0 and pressure p_0	kg/m^3
$\boldsymbol{\sigma}$	viscous stress tensor	N/m^2
ϕ	arbitrary through-variable	–
ϕ_H	heat flow rate	W
ϕ_m	mass flow rate	kg/s
ϕ_v	volume flow rate	m^3/s
$\boldsymbol{\phi}''$	arbitrary flux	$-/m^2$ s

$\mathbf{\Phi}''_x$	flux of an arbitrary physical quantity x	$-/m^2$ s
$\mathbf{\Phi}''_{x_c}$	contact flux of an arbitrary physical quantity x	$-/m^2$ s
$\mathbf{\Phi}''_e$	total exergy flux	W/m^2
$\mathbf{\Phi}''_{e_c}$	exergy contact flux	W/m^2
$\mathbf{\Phi}''_H$	heat flux	W/m^2
$\mathbf{\Phi}''_{s_c}$	entropy contact flux	W/m^2 K
ϕ'''_x	production rate density of an arbitrary physical quantity x	$-/m^3$ s
ϕ'''_e	exergy production rate density	W/m^3
ϕ'''_{me}	mechanical energy production rate density	W/m^3
ϕ'''_{me_r}	conversions rate density of mechanical energy into exergy	W/m^3
ϕ'''_s	entropy production rate density	W/m^3 K
ϕ'''_u	internal energy production rate density	W/m^3
ω	angular velocity	rad/s

MODEL OF A MECHANICAL TRANSMISSION

J. W. Polder
Department of Mechanical Engineering
Eindhoven University of Technology
Eindhoven, The Netherlands

SUMMARY

The transformation of power in mechanical and other transmissions is performed in a complicated way (Olderaan's law). A network theory has been formulated by the author [10], greatly simplifying the representation and calculation of planetary gear trains and variable gear trains. Some essential concepts of this theory are presented, together with a bond graph model of a planetary gear train subject to dry friction.

1. HISTORICAL NOTES

Archimedes (287–212 B.C.) treated parallel forces exerted on a lever. Starting from two axioms: (*a*) equal weights at equal distances are in balance, (*b*) equal weights at unequal distances are not in balance, the farthest being descending, he deduced the equilibrium law for unequal weights on a lever [1]. In his proof, however, he

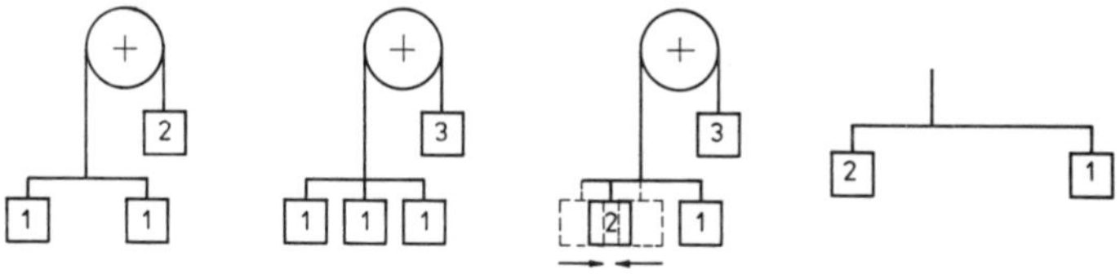

Fig. 1.

used the theorem to be proved [2]. The physical reasoning (Fig. 1) was insufficiently lucid to discover the missing axiom and a mathematical method failed to make the step from

$$\text{weight } A : \text{weight } B = \text{length } b : \text{length } a$$

via

$$A : B = b : a$$

to

$$Aa = Bb.$$

The equilibrium rules, given as proportions between forces and specified geometrical distances, were extended by our second Archimedes, Stevin (1548–1620), to non-parallel forces acting in one plane [3] (Fig. 2).

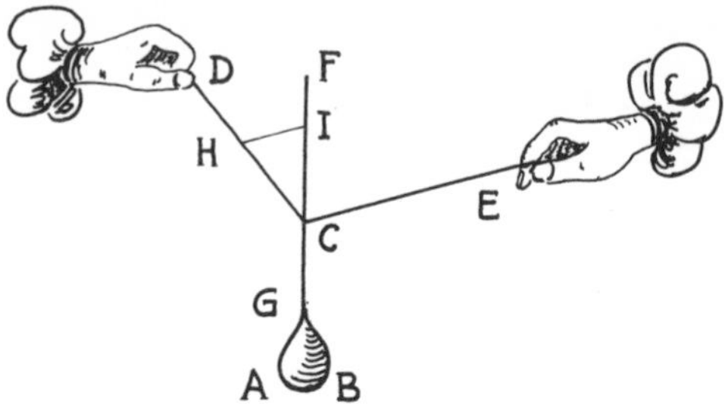

Fig. 2.

In Galilei's (1564–1642) publications the first trace is found of the concept of work or energy. Where other scientists observed forces and the geometry of their application points, Galilei paid attention to the possible displacements of the application points. He introduced the product of force and length of displacement. Huygens (1629–1695) often used that concept, yet without a strict definition. His contemporaries applied mechanical energy in simple calculation methods, but they seemed uninterested in a theoretical basis. Newton, who published his famous "Principia" in 1687, even avoided the treatment of energy. In the year 1725, Varignon [4] introduced an entirely new concept: the moment of a force about an axis. It is essential to the description of power flow in rotary mechanisms.

In subsequent centuries several branches of science and technology developed more or less separately, until in recent decades analogies were investigated. Power, defined as the product of an across-variable and a through-variable, is now a key in modern theories. In 1960, Olderaan [5] stated that any transmission of power to other levels of the across-variables or through-variables is achieved by *two* transformations of power. In the essential intermediate stage the power is described by an across-variable and through-variable pair of another nature.

Paynter [7] stressed the importance of the power concept in reticular systems by introducing the bond graph theory [8]. Belevitch [9] for electrical transformers and Polder [10] for mechanical transmissions stated the close relationship between transmissions and nodes (fixed connections) in the mathematical model. As Thoma explains in this volume, different systems may be related by a direct analogy or by an inverse analogy. The inverse analogy, often shortly called analogy [13], is prefered here in order to elucidate the properties of nodes (fixed connections) and transmissions in an easy way.

2. PHYSICAL MODEL OF A TRANSMISSION

The purpose of a transmission is the transformation of an input power with a certain value of its across-variable, into an output power with a given value of the through-variable. A belt drive, for instance, is used between a motor-driven shaft with a high angular velocity and a shaft which has to surmount a high torque. The high torque shaft may have a low angular velocity, determining an output power. In order to minimize power losses, the transmission has to reduce angular velocities and to increase torques with as well as possibly the same ratios. The main characteristic of a transmission is the same nature of input power and output power, both rotary, or both rectilinear, etc.

In 1960, Olderaan stated an important law [5]:

> Any transmission of power to other levels of the across-variables and through-variables is achieved by *two* transformations of power. In the essential *intermediate* stage the power is described by an across- and through-variable pair of *another nature.*

A lever, for instance, acts as a transmission of rectilinear power. A force exerted on a lever in an application point may move that point with a certain velocity, thus supplying an input power. This input power is transformed into a moment about the axis of the lever and an angular velocity of the lever; the intermediate stage shows rotational power. A second transformation yields the output power as a product of force and velocity of another application point.

In its simplest design (Fig. 3) a transmission for rotary power consists of four parts [6, 10]: input part (port); output part (port); conductive part and frame (port). Kinematically, a transmission is related to a four-bar mechanism. Dynamically, the conductive part not being connected to external elements, the transmission is considered a three-port. A port is a power transmitting connection. The frame is a port transmitting zero power since its angular velocity is zero. If a frame is not prevented from rotating, the transmission works either disastrously or like a planetary gear train. A rotatable frame is called a planet carrier.

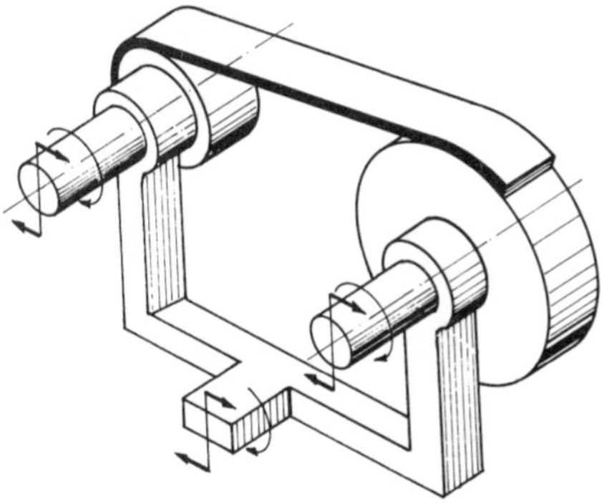

Fig. 3.

Both transformations of power in a transmission are subject to dissipative power, commonly caused by dry friction. Although these power losses seem rather small as compared with the transmitted power in the conductive part, it is unjustifiable to neglect it. Internally, in a transmission a complicated power flow may be acting, causing a high conductive power and high power losses. If the sum of dissipative powers tends to exceed the input power, the design is self-locking. Furthermore, the power losses depend on the orientation of the power flow. The efficiency of a given power flow cannot be used for direct calculation of the efficiency of the reversed power flow.

Instead of a chain drive or belt drive, a transmission for rotary power may consist of a toothed gear pair, a friction gear pair, an electric transmission (generator and motor) or a hydraulic transmission (pump and motor). The conductive part of a toothed gear pair or a friction gear pair exists only as a force acting on moving application points. For all transmissions Olderaan's law applies.

3. NODES AND TRANSMISSIONS

Each physical model may be abstracted to a mathematical model. In reverse, mathematical models can be interpreted in physical terms, commonly in several ways and sometimes in clumsy designs. In general, a mathematical model consists of a set of equations in several across-variables and through-variables. Any set of equations may be split up into:

1. A subset of equations, each equation with one across-variable and one through-variable, or possibly only one variable. These equations stand for passive components like moments of inertia, springs and dampers, or, in the case of one variable, for a source or a sink.
2. A subset of equations, each equation with either across-variables or through-variables. These equations, responsible for the configuration of the system, are readily recognized as Kirchhoff's laws or similar equations.

We will adopt one Kirchhoff law in an unchanged form: for a node the sum of through-variables is zero. The other Kirchhoff law: for a mesh the sum of across-variable differences is zero, will be presented in another form. Instead of across-variable differences along branches we refer to the across-variables in an absolute sense (potentials), and we observe that, starting from one connection of a node, going round a mesh and arriving at another connection of the same node, the across-variables in these connections are equal.

Furthermore, we restrict ourselves to three connections on every node. If more than three branches have to be connected, more nodes are used. The symbol for a node is given in Fig. 4. In the case of rotary mechanisms the across-variables are

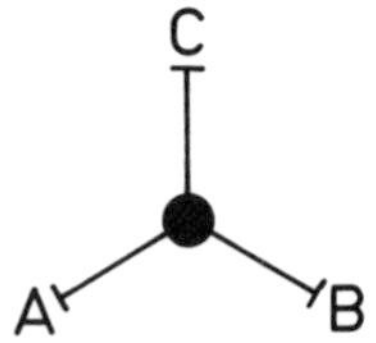

Fig. 4. Node.

angular velocities ω_A, ω_B, ω_C, and the through-variables are torques T_A, T_B, T_C. The equations for a *node* are

$$\begin{cases} \omega_A = \omega_B = \omega_C \\ T_A + T_B + T_C = 0 \end{cases} \tag{1}$$

The unequal number of equations for across- and through-variables raises the question: what is the meaning of a set consisting of one equation for across-variables and two equations for through-variables? The interchange of across- and through-variables in (1) yields a *converse node*

$$\begin{cases} \omega_A + \omega_B + \omega_C = 0 \\ T_A = T_B = T_C \end{cases} \tag{2}$$

represented in Fig. 5, which has no physical meaning. A slight extension, however, suffices to produce the equations of a transmission. This extension consists of the introduction into (2) of coefficients, subjected to a restriction.

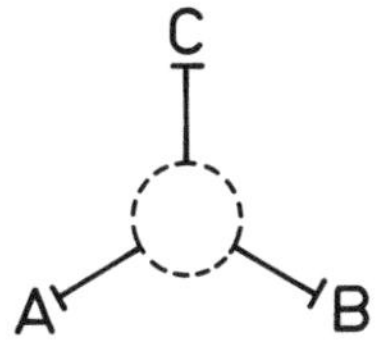

Fig. 5. Converse node.

Between the performance of an actual transmission, its physical model, and its mathematical model full harmony exists. To develop a theory of general validity it is recommendable to define a transmission by the properties of its mathematical model.

4. MATHEMATICAL MODEL OF A TRANSMISSION

A transmission comprising coaxial input shaft, output shaft and rotatable frame, may be presented here as the most general case of a transmission (Fig. 6). Transmissions with rotatable frames are known as planetary gear trains, and sometimes

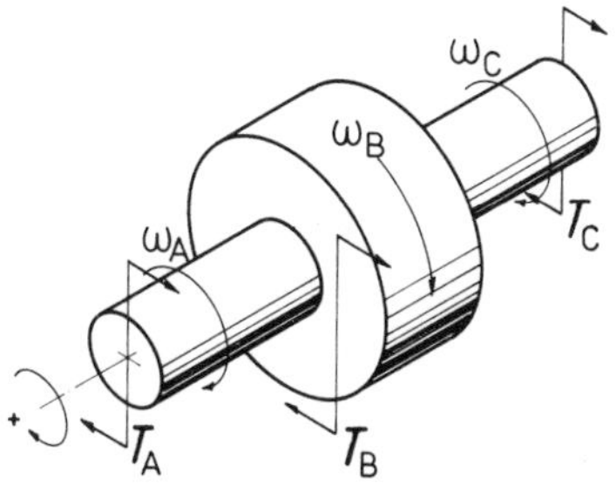

Fig. 6.

as special hydraulic drives. Commonly the frame of a transmission is prevented from rotating, but that case is accounted for by fixing the relative angular velocity to zero. The question whether a transmission may be treated as a black box is answered affirmatively since, paradoxically, engineers are very familiar with the internal design of all kinds of transmissions. This knowledge guarantees the validity of the model derived below.

A transmission is defined [12] by its number of equations and its relation to a node, thus:

a transmission (planetary gear train) is a three-port yielding one-linear equation for angular velocities and two for torques; under particular circumstances a transmission (planetary gear train) may degenerate into a node.

From this definition it is easy to derive *the equations* (3) *of a transmission* [10, 11, 12]. There are two parameters in the equations, one of them to be interpreted as a gear ratio $i_{A/B} = \omega_A/\omega_B$ for $\omega_C = 0$. The other one, a torque ratio, is commonly not given a symbol of its own, but is written as a product: $i_{A/B}\eta_{B/A}$. Thus, instead of a torque ratio, the efficiency $\eta_{B/A}$ is introduced as the second parameter.

$$\begin{cases} \omega_A - i_{A/B}\omega_B + (i_{A/B} - 1)\,\omega_C = 0 \\ i_{A/B}\eta_{B/A}\,T_A + T_B = 0 \\ (1 - i_{A/B}\eta_{B/A})\,T_A + T_C = 0 \end{cases} \tag{3}$$

The efficiency is written explicitly as

$$\eta_{B/A} = \frac{1}{i_{A/B}}\, i_{A/B}\eta_{B/A} = \frac{(\omega_B - \omega_C)}{(\omega_A - \omega_C)}\left(-\frac{T_B}{T_A}\right) = -\frac{T_B(\omega_B - \omega_C)}{T_A(\omega_A - \omega_C)} \tag{4}$$

Hence, the efficiency $\eta_{B/A}$ is not a relation between shaft powers P_A and P_B, but between what are known as planet powers, $P_a = T_A(\omega_A - \omega_C)$ and $P_b = T_B(\omega_B - \omega_C)$. A shaft power $P_A = T_A\omega_A$ is split up into a planet power $P_a = T_A(\omega_A - \omega_C)$ and a carrier power $T_A\omega_C$. Likewise, $P_B = T_B\omega_B$ is split up into planet power $P_b = T_B(\omega_B - \omega_C)$ and carrier power $T_B\omega_C$. The two carrier powers are in balance with the shaft power $P_C = T_C\omega_C$. The planet power P_a is transformed into a power in the conductive part, which is transformed again into planet power P_b (Fig. 7).

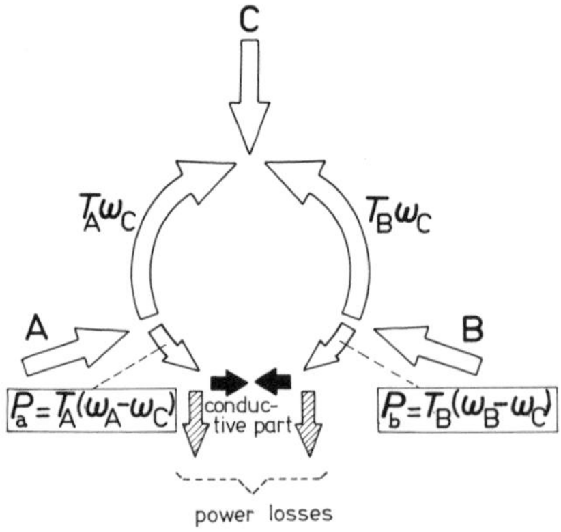

Fig. 7.

In an actual design the carrier powers are transmitted by coupling effects and the planet powers are transmitted by real transformations. The co-axiality of the three shafts causes a reiteration in the conductive part. However, for the sake of simplicity the various power losses of transformations, bearings, oil flow, etc., are condensed to one sum P_V (Fig. 8).

$$P_V = -P_a - P_b = -P_A - P_B - P_C \tag{5}$$

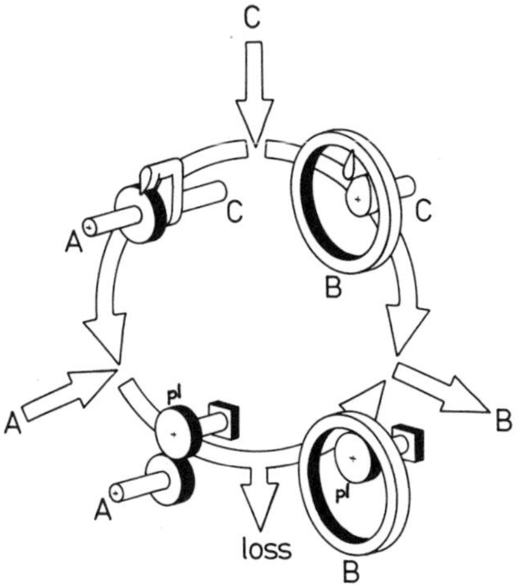

Fig. 8.

For most designs the efficiency $\eta_{B/A}$ may be treated as a constant, which establishes a convenient rating practice [10, 11, 12].

5. CONVERSE NODE

For the study of complicated designs as well as for theoretical considerations a simplification of a planetary gear train into the previously defined converse node and into some auxiliary transmissions may be helpful. For that purpose the equations (3) are rewritten as

$$\begin{cases} \omega_A^* + \omega_B^* + \omega_C^* = 0 \\ T_A^* = T_B^* \\ T_A^* = T_C^* \end{cases} \tag{6}$$

where

$$\begin{cases} \omega_A^* = \omega_A \\ T_A = T_A^* \end{cases} \quad \begin{cases} \omega_B^* = -i_{A/B}\omega_B \\ T_B = -i_{A/B}\eta_{B/A}T_B^* \end{cases} \quad \begin{cases} \omega_C^* = (i_{A/B} - 1)\,\omega_C \\ T_C = (i_{A/B}\eta_{B/A} - 1)\,T_C^* \end{cases} \qquad \text{(7a, b, c)}$$

After the transformation of a planetary gear train into a converse node it will have lost nearly all its properties, owing to transposition of moments of inertia and stiffnesses to the shafts [10] and of its gear ratio and efficiency to auxiliary transmissions. The auxiliary transmissions are two-ports, still subject to dissipative power, due to dry friction (Fig. 9). In the diagram a special symbol, having two parameters, is used. In the physical model converse nodes, supplementable with auxiliary transmissions, are interpreted as planetary gear trains. So they are fully acceptable. In theoretical considerations nodes and converse nodes play an equivalent part.

The theoretical equivalence of nodes and converse nodes leads towards a new viewpoint on dual networks, not suffering from the illogical restriction to planar networks: to construct a dual network of a given network, one has to maintain rigorously the same structure and to replace each component by its dual one and

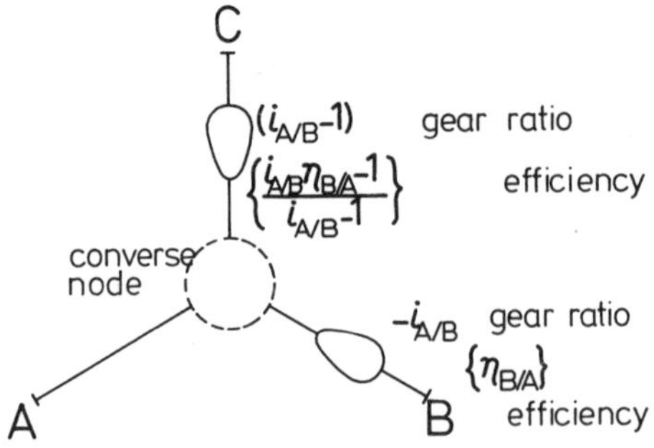

Fig. 9.

to change all nodes into converse nodes. Although the result may be very unpractical, it is still an authentic solution. A dual network has to be defined by an interchange of across-variables and through-variables.

6. DEGREES OF FREEDOM

To do justice to the equivalence of across- and through-variables, two degree-of-freedom numbers instead of one should be mentioned. In a survey of primitive three-ports (Fig. 10) and four-ports (Fig. 11), c_ω is the degree of across freedom and c_T the degree of through freedom.

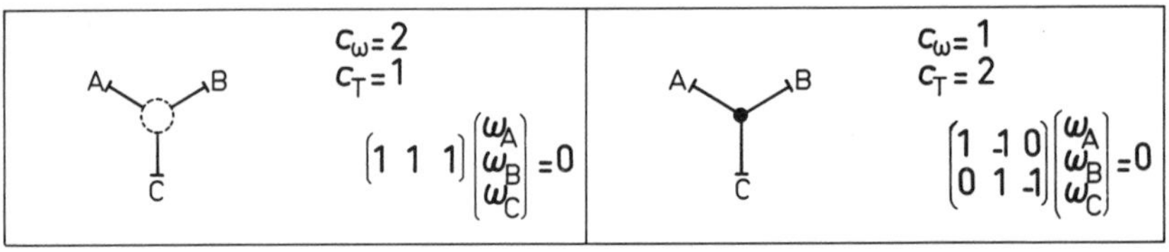

Fig. 10. Primitive three-ports.

Commonly, actual transmissions have one input shaft which is given an angular velocity by its coupling to a motor, $c_\omega = 1$, and one output shaft which has to surmount a certain torque, $c_T = 1$. To achieve such a transmission from a converse node, corresponding to $c_\omega = 2$, $c_T = 1$, one shaft can be blocked, yielding a *gear drive* in the most common sense. From a node, corresponding to $c_\omega = 1$, $c_T = 2$, one shaft has to be given a defined torque, say zero, yielding a *coupling* instead of a transmission. The primitive four-ports with $c_\omega = 3$, $c_T = 1$ and $c_\omega = 1$, $c_T = 3$ do not produce new transmissions, but those with $c_\omega = 2$, $c_T = 2$ are very useful. The latter are creating practical transmissions when two shafts are interconnected by a

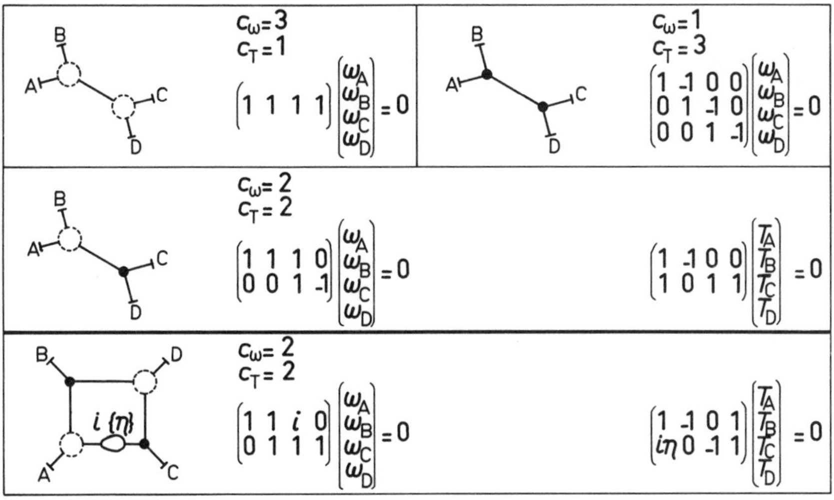

Fig. 11. Primitive four-ports.

variator. They may be called *variable shunts* (one planetary gear train, one node) or *variable bridges* (two planetary gear trains, two nodes).

To avoid a cumbersome treatment of input and output quantities, a *reticulator* [10] is introduced as a convenient mathematical artifice to replace the input and output shafts. A reticulator is a fictitious variator for which the interrelations between angular velocities and between torques are identical with those of the input and output shafts.

A *variable shunt* (Fig. 12) is a transmission comprising one variator (CD; gear ratio $x = \omega_C/\omega_D$), one node (BDE), and one planetary gear train (ACE; parameter

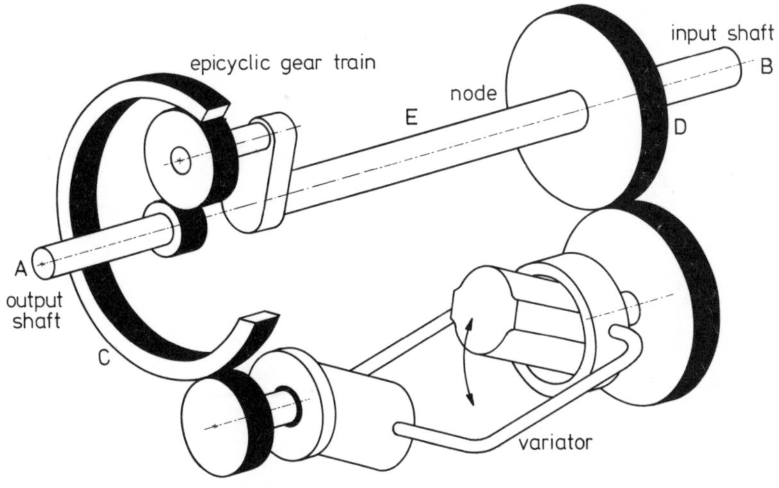

Fig. 12. Variable shunt.

i). An adapted oval between two shaft ends in the symbol of the planetary gear train indicates the direction for which the parameter is defined, $i = (\omega_A - \omega_E)/(\omega_C - \omega_E)$ (Fig. 13). In the network the input and output shafts are interconnected fictitiously by the reticulator (AB: gear ratio $y = \omega_A/\omega_B$). A mathematical treatment can be carried out in several ways. A representation comprising a converse node is useful for deriving formulae and for further theoretical analysis and synthesis (Fig. 14). For numerical calculations, however, the direction of the power flow in

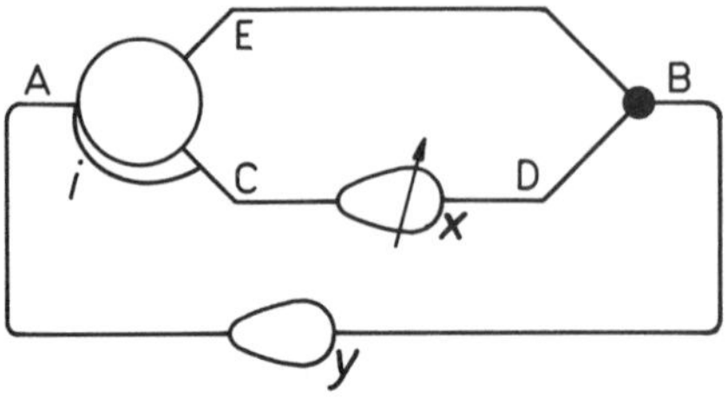

Fig. 13.

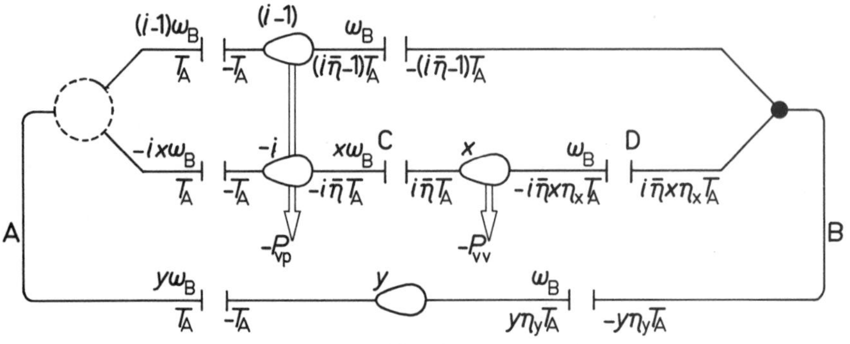

Fig. 14.

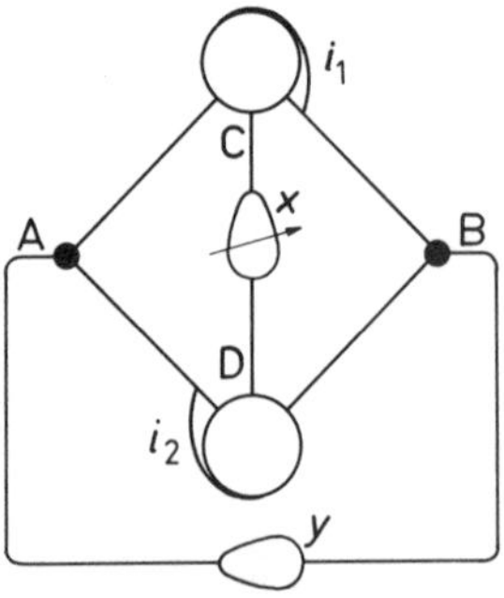

Fig. 15. Variable bridge.

the auxiliary transmissions may cause severe errors due to their interdependence. As soon as a decision has to be made about the value of the efficiency, the direction of the actual power flow must be determined. This is most easy if a model of a planetary gear train with carrier powers and planet powers is used.

A *variable bridge* may have a variator between two planetary gear trains and a reticulator between the nodes, or the reverse (Fig. 15). Both designs are described by one and the same mathematical model, which confirms again the equivalence of nodes and converse nodes in a general theory.

7. BOND GRAPH OF A PLANETARY GEAR TRAIN

In the previous sections elements of the author's network theory and network diagrams have been used to describe the relations in planetary gear trains. The bond graph symbolism [8] lends itself also very well to represent the structure of this transmission system. In both representations the dynamic aspects of a planetary gear train can be incorporated. The moments of inertia and stiffnesses, especially those of the planet gears, cause it to be a difficult dynamic system. Although it is possible to indicate moments of inertia within the bond graph of a planetary gear train [14], it is much easier to transpose all moments of inertia as well as stiffnesses to the connecting shafts [10]. The connecting shafts, including the frame, comprise series of moments of inertia and stiffnesses, which are left out of consideration here. So the bond graph of the essential part of a planetary gear train will have only two kinds of parameters: gear ratios i and efficiencies η. Gear ratios are accounted for by transformers, which transform the one planet power into the other. Efficiencies are accounted for by resistances and o-junctions combined with the above-mentioned transformers.* To obtain coaxiality of shafts, planet gears are designed as an intermediate gear with two gear engagements. Every engagement consists of two trans-

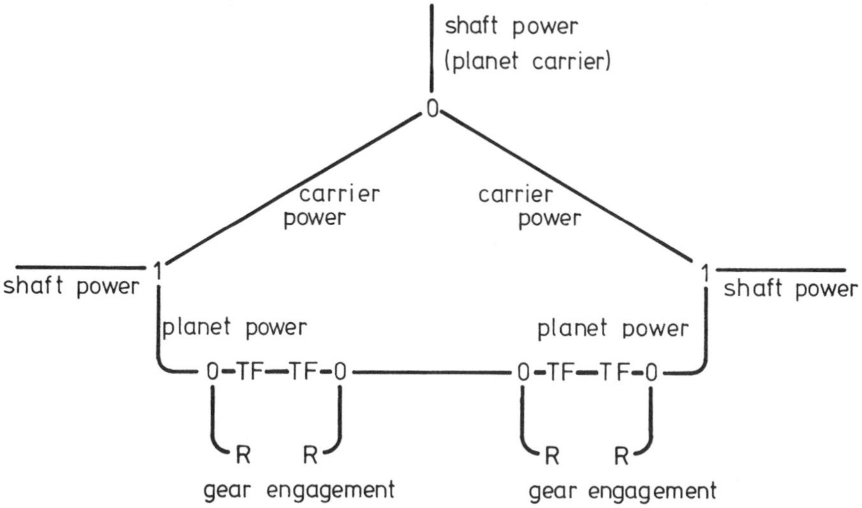

Fig. 16. Bond graph of a planetary gear train, corresponding to Olderaan's law.

*In accordance with other contributions in this volume across- and through-variables are used in preference to effort and flow variables, generally used in bond graph literature. This means that our mechanical bond graphs are dual to the conventional one's [8], which implies mainly an interchange of 0- and 1-junctions. Our 0-junction has a common ω, our 1-junction has a common T.

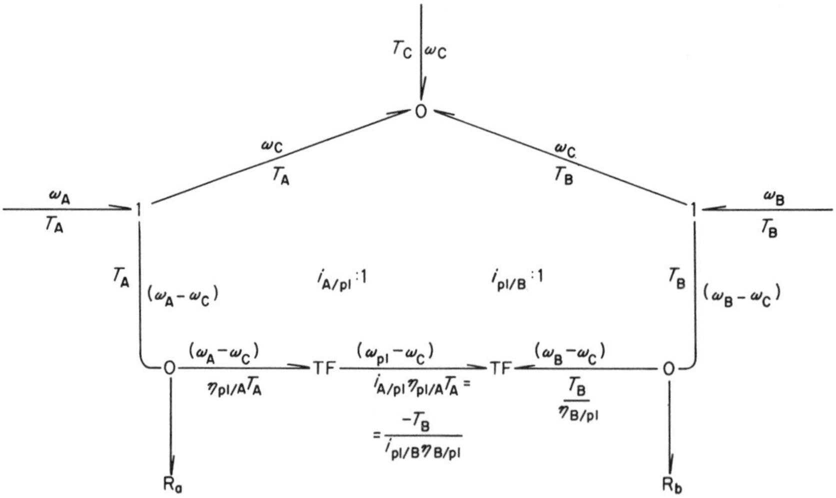

Fig. 17. Bond graph of a planetary gear train, free from moments of inertia and stiffnesses, with simplified calculation of power losses.

formations, according to Olderaan's law, each showing power dissipation, see Fig. 16. Since the conductive part of a toothed gear pair is immaterial, in practical calculations these two dissipative powers are not separated, but combined to one power loss. Hence, each gear engagement may be represented by one transformer and one power loss, as is shown in Fig. 17.

REFERENCES

1. Archimedes, *De Aequiponderantibus.*
2. Mach, E. *Die Mechanik in ihrer Entwicklung.* Brockhaus, Leipzig, 1912.
3. Dijksterhuis, E. J. *The principal works of Simon Stevin.* Swets and Zeitlinger, Amsterdam, 1955.
4. Varignon, P. *Projet d'une nouvelle mécanique.* 1725.
5. The excellent statement of Olderaan is applied in [6], unfortunately in a careless way. The author would be grateful to be informed about possible earlier references to this law.
6. Schlösser, W. M. J. and Olderaan, W. F. T. C. "An analogue theory for rotating drives". Dutch: *De Ingenieur, 73*, 1961, W99–W105. English: *Hydraulic Power Transmission, 8*, 1962, 89, 312–317, 347. German: *Oelhydraulik und Pneumatik, 5,* 1961, 12, 413–418.
7. Paynter, H. M. *Analysis and design of engineering systems.* M.I.T. Press, Cambridge, Mass., 1961.
8. Karnopp, D. and Rosenberg, R. C. *Analysis and simulation of multiport systems.* M.I.T. Press, Cambridge, Mass., 1968.
9. Belevitch, V. *Classical network theory.* Holden-Day, San Francisco, 1968.

10. Polder, J. W. *A network theory for variable epicyclic gear trains.* Thesis, Eindhoven, 1969.
11. Polder, J. W. "Nieuwe grondslagen voor de theorie van roterende overbrengingen". *De Ingenieur, 82,* 1970, W91–W103.
12. Polder, J. W. *A universal mathematical model for epicyclic gear trains.* ASME publication 72–PTG–50 (Symposium San Francisco, 1972).
13. Cannon, R. H. *Dynamics of physical systems.* McGraw-Hill, New York, 1967.
14. Brown, F. T. "Direct application of the loop rule to bond graphs". *J. Dyn. Syst., Meas., Control Trans. ASME*, September 1972, 253–261.
15. "Planetengetriebe. Begriffe, Symbole, Berechnungsgrundlagen". *VDI-Richtlinie 2157.* Düsseldorf, (at press).

NOMENCLATURE

c_T	degree of through freedom
c_ω	degree of across freedom
i, i_1, i_2	gear ratio
$i_{A/B} = \omega_A/\omega_B$	gear ratio*
P_A, P_B, P_C	shaft power
P_a, P_b	planet power
P_V	power loss
T_A, T_B, T_C	shaft torque (through-variable)
T_{pl}	planet torque
x	gear ratio
y	gear ratio
η, η_1, η_2	efficiency
$\eta_{B/A} = -P_B/P_A$	efficiency*
$\omega_A, \omega_B, \omega_C$	shaft angular velocity (across-variable)
ω_{pl}	planet angular velocity (absolute)
$(\omega_{pl} - \omega_C)$	planet angular velocity (relative to planet carrier C)

*Recently, it was proposed [15] to change the symbol $\eta_{B/A}$ (which had a theoretical origin) into η_{AB} (which is preferred for practical calculations). Similarly, $i_{A/B}$ changes into i_{AB}.

THE USE OF NETWORK GRAPHS AND BOND GRAPHS IN 3-D MECHANICAL MODELS OF MOTORCARS AND UNBALANCE

J. J. van Dixhoorn
Department of Electrical Engineering
Twente University of Technology
Enschede, The Netherlands

SUMMARY

Network graphs and bond graphs can be a logical first step in the, generally ill-structured, process of modelling. It is shown that in the domain of mechanical dynamics network methods can be extended beyond the usual simple one-dimensional cases.

Modulated transformers and other bond graph techniques are simultaneously used with their network counterparts in a number of examples. Frequency- and time-responses are shown, easily obtained from computer programs for network analysis and for digital simulation.

1. A TOPOLOGICAL BASIS FOR THE MODELLING PROCESS

In higher education and in interdisciplinary work a still not completely fulfilled need exists for a physical systems theory, embracing though not replacing the theory of electricity, theoretical mechanics, thermodynamics and reaction kinetics. Concurrently in both domains of higher education and of interdisciplinary work, another, more application-oriented, need is manifest: to have a systematic and logical basis for the process of modelling.

The Twente colloquium 1973, reported in this volume, brought together a number of people, who consider such a physical systems theory or such a form of physical systems engineering from a network viewpoint.

This paper is written from the systems engineering side and deals with the modelling process, restricted to mechanical systems. These systems are treated in an unconventional topological way, using network graphs and bond graphs. The purpose of the paper is two-fold: to be another eyeopener to mechanical engineers for these interdisciplinary methods and to demonstrate to systems analysts and electric network analysts the parallelism between network graphs and the generally unknown, though much more compact, bond graph symbolism.

The general aim of the paper is to show that these topological diagrams might provide a logical basis for the modelling process. This process of abstracting a physical reality to a mathematical model is difficult to teach because it lacks structure. It proceeds either intuitively or in an *ad hoc* way, using special intermediate steps stemming from special theories. The concept of power port and the related concepts of storing, transforming, transporting, dissipating and degrading elements can provide a logical first step to the modelling process.

Once this step has been taken and the network graph or bond graph of the physical problem has been set up, all kinds of well established models can be reached by algorithmic methods: differential equations, matrix differential equations, block diagrams, signal flow graphs, transfer functions, etc.

In the following a linear graph, if applied to model the relations in a network [1], will be called *network graph.* It is hoped that this adjective will prevent confusion with a much better known but differently applied linear graph, called the *signal flow graph.*

According to our goal of treating bond graphs and network graphs together, we have to classify physical variables in across- and through-variables. For mechanical systems velocity difference is the across-variable, being measured across two discrete points of an element; force is the through-variable, having the same value (continuity) throughout the element and at its connection points. In the original bond graph literature [7] variables are classified in effort- and flow-variables. This classification corresponds with the across- and through-variable classification for all physical systems except for mechanical ones, where the variables are interchanged. For people like the author, who want to use both the network and the bond graph symbolism, this is greatly to be deplored. Only one variable classification can be used in both graphs and this necessarily should be the across- and through-pair. The result is, that these *mechanical bond graphs are dual to the conventional mechanical bond graphs.* For this paper (as for Polder's paper [17] in this volume) this implies an interchange only of o- and 1-junctions: our o-junction is a spatial *parallel* connection of elements, having a common across-variable; our 1-junction is a spatial *series* connection of elements, having a common through-variable.

2. TWO-PORT TRANSFORMERS FOR MODELLING 2-DIMENSIONAL SYSTEMS

In many texts on system dynamics network graphs are used as the first step in modelling simple mechanical systems [1–6].

These mechanical systems however are generally restricted to one of the following classes:

1. Systems exhibiting translation in one direction only.
2. Systems in which all components rotate about parallel shafts; these shafts may either be fixed or have to pass through the mass centres of the components. Accordingly, movements, due to unbalance forces, do not occur.

There are of course examples of electrical analogues of more complicated mechanical systems [10, 11]. In those cases, however, the analysis often started by writing down the equations of the mechanical system. After their linearization and simplification an electrical network was drawn, satisfying these equations. Finally a circuit could be built and used for experiments. That kind of experimental approach has generally become obsolete now, because analogue and digital computers, programmed directly from the equations, offer excellent experimental possibilities. But apart from that, the procedure is the exact opposite from the above mentioned, which we are going to investigate and extend here. Equations were derived first, after that linearized networks were fitted to them. Our object is to draw the non-linear networks more or less by inspection, considering that the abstraction to the topological structure should be the *first* modelling step. After that an analysis of the network can be performed, using a variety of means, often without recourse to the equations. The originators of the bond graph methods had the same object and have solved most of the problems. Their approach however is generally not known to network people, though it is not restricted to bond graphs. So there still seem to be some good reasons to apply network graphs and bond graphs simultaneously to simple and more elaborate two- and three-dimensional problems.

The usual restriction of network methods to simple mechanical systems stems from the difficulty to model the numerous couplings between the translational and rotational movements due to kinematic constraints in constructions and in rigid bodies themselves. Consider for example the plane motion of a rigid body, like a physical pendulum, swinging about a fixed shaft through point 1. The translational displacements x_{21} and z_{21} of the masscentre 2 referred to the fixed point 1 are related to the rotational angle θ according to

$$x_{21} = -L \cos \theta \tag{1}$$

$$z_{21} = -L \sin \theta \tag{2}$$

Taking velocities as principal across-variables their relations are

$$\dot{x}_{21} = (L \sin \theta) \,.\, \dot{\theta} \tag{3}$$

$$\dot{z}_{21} = -(L \cos \theta) \,.\, \dot{\theta} \tag{4}$$

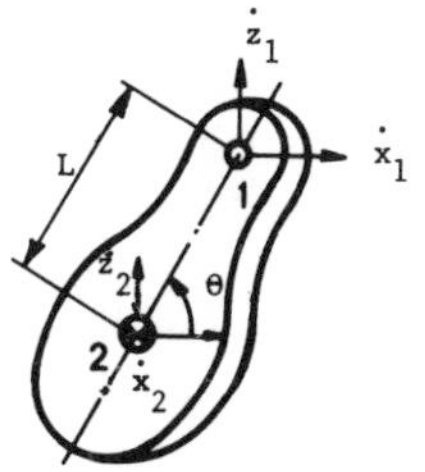

Fig. 1.

The three independent velocity coordinates ($\dot{x}_{21}$, $\dot{z}_{21}$, $\dot{\theta}$) of a free body in plane motion now show to be *coupled* or constrained by relations (3) and (4). Moreover the coupling coefficients $L \sin \theta$ and $-L \cos \theta$ are not constant but displacement (state) dependent.

2.1. Two port modulated transformers

It is still simple to model such constraints in a network sense by using ideal transformers with a variable transformation ratio. In this case two of these so-called *modulated transformers* (MTF's) are needed. The oriented network graph and bond graph diagrams of a MTF for eqn. (3) are shown in Fig. 2; diagrams for eqn. (3) and (4) together are shown in Fig. 4.

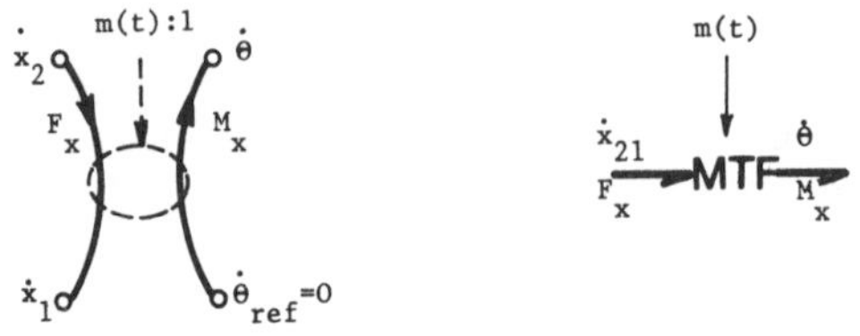

Fig. 2. Network graph and bond graph symbols of a 2-port modulated transformer.

It should be remarked, that this transformer has only two power ports. The MTF shows the ideal conversion of translational power (in the x-direction) to rotational power and vice versa. The non-constant, thus time dependent, transformation ratio $m(t)$ modulates this power conversion in a powerless way. So $m(t)$ figures as an "information input" to the transformer. In the network graph this information input is shown by a dotted, arrowed line. In the bond graph the unidirectional information input is depicted by a full arrow. To the power bonds, in contrast, one may add a half arrow, showing the direction of powerflow which in calculations is taken as the positive one.

So the above diagrams represent

1. a coupling between across-variables:

$$\dot{x}_{21} = m(t) \, . \, \dot{\theta} \tag{5}$$

2. ideal power conversion from translation to rotation. We have to introduce the through-variables F_x and M_x to state this power conversion:

$$\dot{x}_{21} \, . \, F_x = \dot{\theta} \, . \, M_x \tag{6}$$

The above equations imply that the same time dependent transformation ratio $m(t)$ acts between opposite through variables:

$$M_x = m(t) \, . \, F_x \tag{7}$$

In the example $m(t) = L \sin \theta$, θ being a state variable (extensity, displacement) of the system. The transformation ratio is state dependent. Introducing $m(t)$ in eqn. (7):

$$M_x = L \sin \theta \, . \, F_x \tag{8}$$

relating in an obvious way F_x and the torque M_x due to it.

2.2. Some 2-dimensional pendulum models using 2-port MTF's

1(a). Body having free motion in a vertical plane

The motion is characterized by 3 across-variables ($\dot{x}_{2\,0}$, $\dot{z}_{2\,0}$, $\dot{\theta}$) being the velocities of the mass centre with respect to constant reference velocities. The body

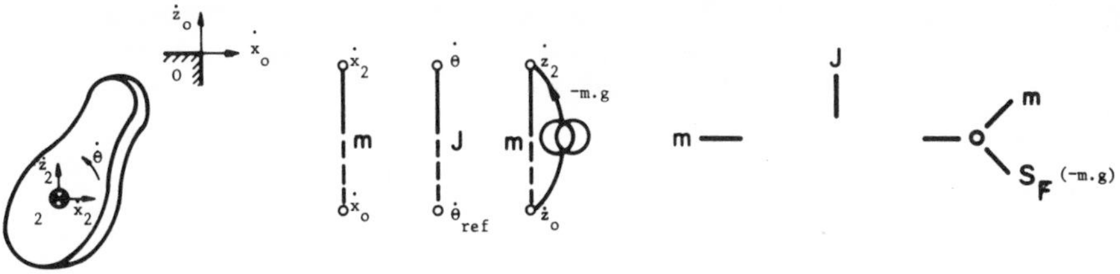

Fig. 3. Network graph and bond graph of free body only.

is characterized by its (translational) mass m and its principal moment of inertia J for rotation about a shaft through 2, corresponding with a principal axis of the body. The models show the mass-type storage elements m and J. In the $\dot{x}$ and $\dot{\theta}$ direction they are not connected to any internal or external influence. In the vertical ($\dot{z}$) direction a through-type (force) source $S_F = -m \,.\, g$ represents the external gravity force, working in a sense opposite to the indicated reference direction.

1(b). Kinematic constraints of the pendulum movement

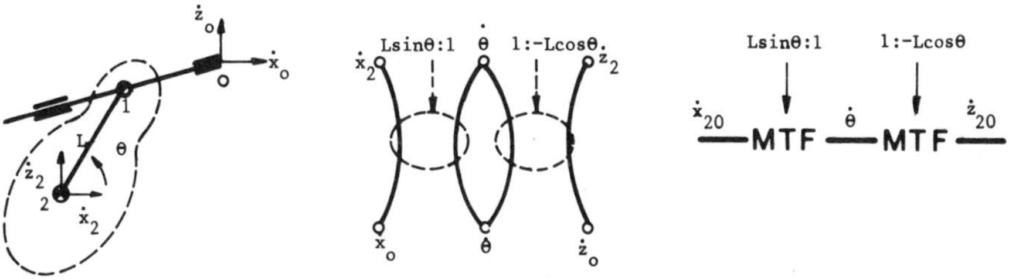

Fig. 4. Network graph and bond graph of constraints only.

Equations (3) and (4) describe the kinematic constraints of the "linkage" L, modelled by two MTF's.

1(c). Physical pendulum

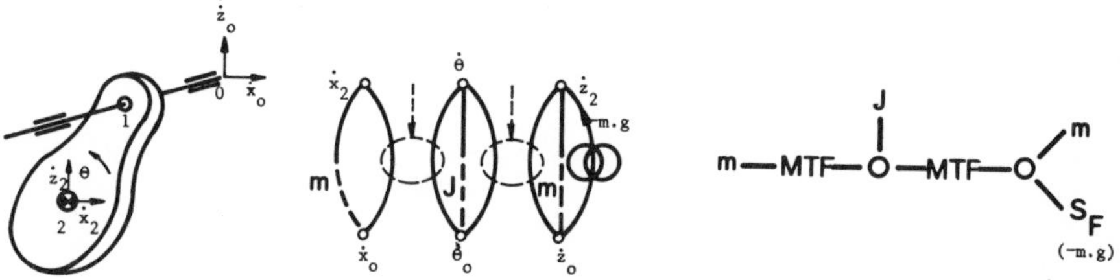

Fig. 5. Models of 2-dimensional pendulum.

The combined model shows the three mass-type buffers now being connected in parallel. So they are not independent. For every θ the stored energy or the initial velocity of one of the buffers determines those of both the others. Nonwithstanding both non-linear MTF's it is, in this case, possible to reduce both m's to the rotational side of the transformers. The resulting reduced moment of inertia $J_r = m \, . \, L^2$, so $J_{total} = J + J_r = J + mL^2$, being in accordance with Steiner's parallel axis theorem. One may also bring the source $-m \, . \, g$ to the rotational side, producing a torque source $-m \, . \, g \, . \, L \cos \theta$. Both graphs are then reduced to this state dependent source working on J_{total}. Generally however such reductions are not possible, as is the case for the next examples. It is also generally preferable to keep all variables in the graph. In Section 4 a more compact bond graph notation, using the multiport-MTF, will be used.

2. Spring connected pendulum

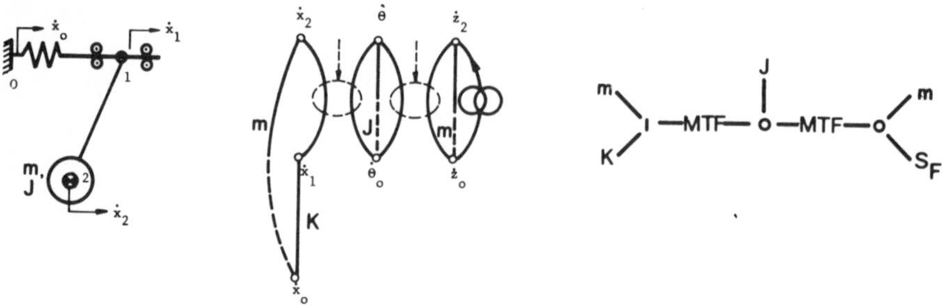

Fig. 6. Network graph and bond graph of spring-connected pendulum.

The left side m still refers to the fixed reference $\dot{x}_0$. The spring K connects $\dot{x}_0$ and $\dot{x}_1$. Drawing a *normal tree* (Fig. 7) of the network graph, to find dependent elements, reveals that the left side m_x is an independent energy storage now.

A tree link formed by the right side m_z reveals that J and m_z are still dependent. So the system has 3 independent storage elements m_x, K and (J, m_z). One integration of $\dot{\theta}$ is necessary to produce the transformation ratio's. So it can be concluded, that the system is of 4th order.

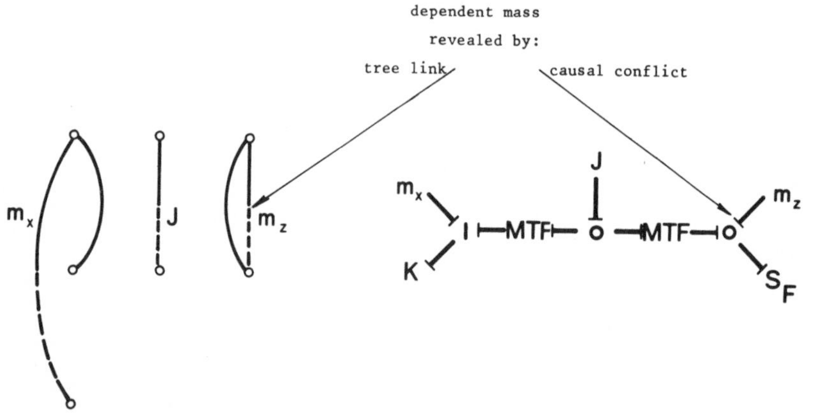

Fig. 7. Normal tree (left) and bond graph, augmented with causal strokes (right), show dependent m_z and J.

In the bond graph dependent elements can easily be detected by augmenting every bond with a *causal stroke* [7]. At the right o-junction 2 causal strokes appear, indicating that 2 bonds determine each an across-variable at a parallel connection. This is impossible and indicates that two mass-type buffers are dependent.

3. Mass-connected pendulum

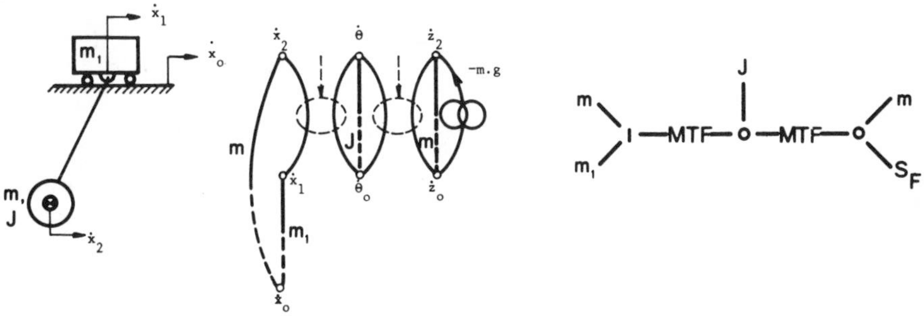

Fig. 8. Models of a mass-connected pendulum.

Left m and m_1 are dependent elements in series and can be replaced by $m_r = \dfrac{m \cdot m_1}{m + m_1}$. Now m_r, m and J are dependent, so the system is of 2nd order. A normal tree or causal stroke analysis gives the same result.

4. Double pendulum

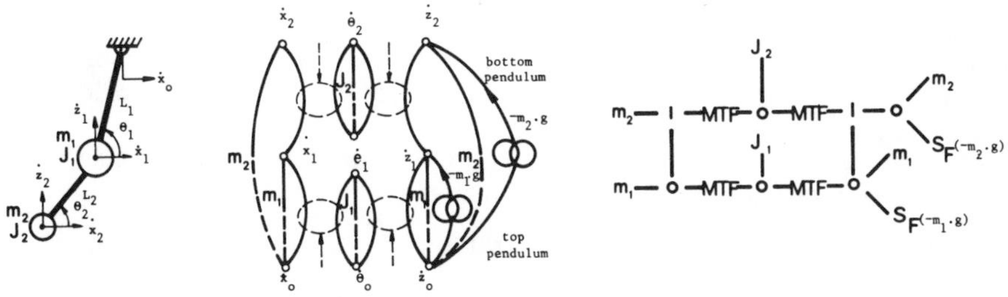

Fig. 9. Models of a double pendulum.

Because we prefer to draw the reference vertices at the bottom, the network graph builds up upside down with respect to the physical model. For comparison the bond graph is drawn the same way.

2.3. Coupling between parallel rotations due to unbalance mass

The double pendulum model may be used to show the "capacitive" coupling, in the electrical sense, which can exist between two rotations, in the presence of an unbalance mass.

Figure 10 shows the typical situation of a movable shaft 1, displaced by the rotation of a fixed shaft 0. The total rotative mass on shaft 1 is assumed to have one principal axis of inertia parallel to both shafts. J_2 is its moment of inertia with respect to the mass centre 2. In the unbalanced situation an eccentricity e exists between shaft 1 and mass centre 2. In that case rotation of shaft 0 produces an unbalance torque on shaft 1.

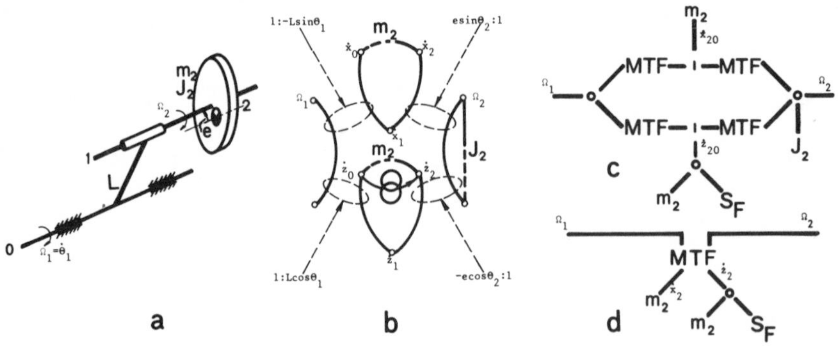

Fig. 10. Parasitic coupling between rotations Ω_1 and Ω_2 due to eccentricity e of mass m_2.

Such situations frequently occur: shaft 1 may be the planet carrier in a planetary gear train, as discussed by Polder in this volume [17]; in a control engineering application 1 may be the axis of rotation of a tracking radar on a ship.

Figure 10 corresponds with Fig. 9: e is identical with L_2, m_1 and J_1 will be negligible. The same graphs are valid for both cases. However to direct the attention to the unbalance coupling between Ω_1 and Ω_2 the graphs of Fig. 9 have been rearranged and rotated a quarter turn to produce the graphs of Fig. 10b, c. These show mass m_2, being analogous (in the across- and through-variable symbolism) to a capacitor, to form a capacitive coupling between Ω_1 and Ω_2 along channels in the x- and z-direction. Such a coupling plays no role in the static state (constant velocities) but may be an important parasitic effect for higher frequencies (high accelerations).

The coupling is non-linear, because all transformers are depending on the angles θ_1 and θ_2. No coupling exists if one transformer in each of the channels is zero. That is the case if $\theta_1 = 0 \pm \pi$ and $\theta_2 = \pm \frac{\pi}{4}$ or if $\theta_1 = \pm \frac{\pi}{4}$ and $\theta_2 = 0 \pm \pi$, corresponding with a square angle between L and e. The coupling is essentially zero if $e = 0$: both right-side transformers disappear, making Ω_2 independent of Ω_1. Furthermore $\dot{x}_2 = \dot{x}_1$ and $\dot{z}_2 = \dot{z}_1$. The graphs illustrate that mass m_2 is still and in fact always connected to Ω_1 by the left side orthogonal transformers. One could reduce m_2, in this case only, to the Ω_1-side resulting in a moment of inertia $m_2 \,.\, L^2$.

As m_2 plays the same role with respect to Ω_2, when $e \neq 0$, one might introduce the concept of mutual mass, like the electrical concept of mutual induction. This has been done by Pawley [11] for a different case. In the author's opinion however this sort of concept does not clarify the model.

The network graph and bond graph can both be used for analysis of the linearized problem, for simulation of the original problem or as a tool for writing down the equations. The bond graph can be drawn in a very compact way, as shown in Fig. 10d, by using the multi-port modulated transformer concept, discussed in Section 4. The general case of coupling by "dynamic" unbalance, occurring when shaft 1 is not parallel to a principal axis of inertia, can be attacked using a multi-port modulated transformer together with the bond graph structure of general rotation (Section 5).

2.4. Some 2-dimensional lever and beam models using 2-port MTF's

A simple massless lever (Fig. 11) performs a transformation from rotation to translations in the x and y direction. The transformation ratios between the across-variables $\dot{x}_{21}, \dot{y}_{21}$ and $\Omega(=\dot{\theta})$ are a function of the angle θ. The relations are modelled by modulated transformers (MTF); only one is shown here.

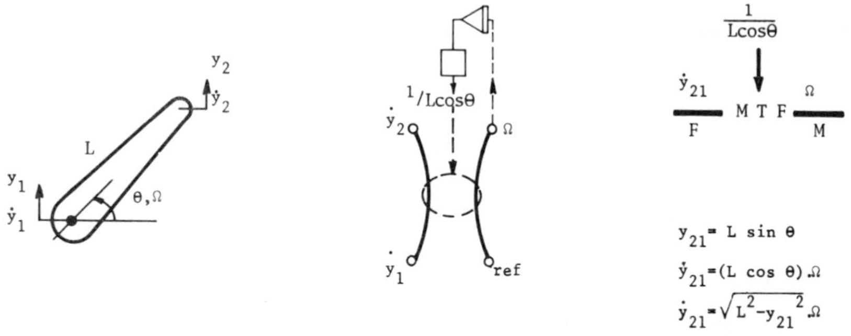

Fig. 11. Massless lever in y-direction.

A double massless lever (Fig. 12) involves 2 modulated transformers in both directions. For small angles θ (nearly horizontal lever) the velocities in the x-direction are neglected.

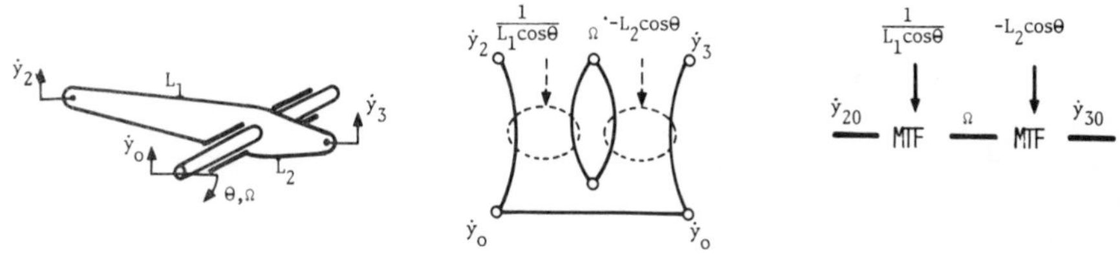

Fig. 12. Double massless lever.

A double lever (Fig. 13) turning about a fixed shaft through its centre of gravity, exhibits a moment of inertia J. A storage element for rotational kinetic energy is added to the graphs.

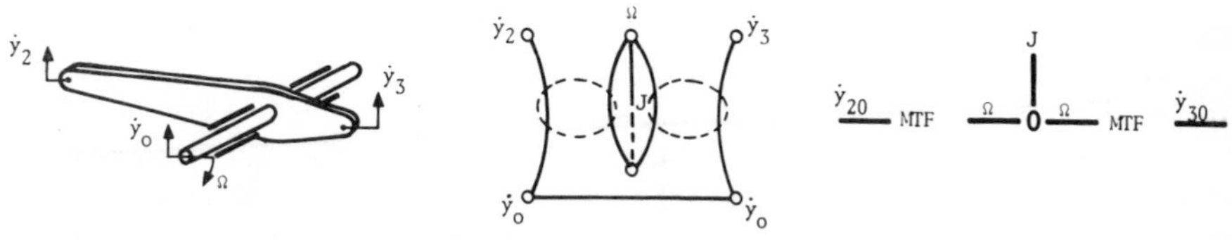

Fig. 13. Double lever exhibiting inertia.

A free beam (Fig. 14) can also be described by the rotation about its centre of gravity, involving a moment of inertia J with respect to this axis. Furthermore, the centre of gravity has a translational velocity $\dot{y}_{10}$. The beam mass m represents an independent energy storage component in this direction. The gravitational force $m \, . \, g$ appears as an external force source working on point y_1 in the negative direction. The network graph is shown in two different ways. For comparison the

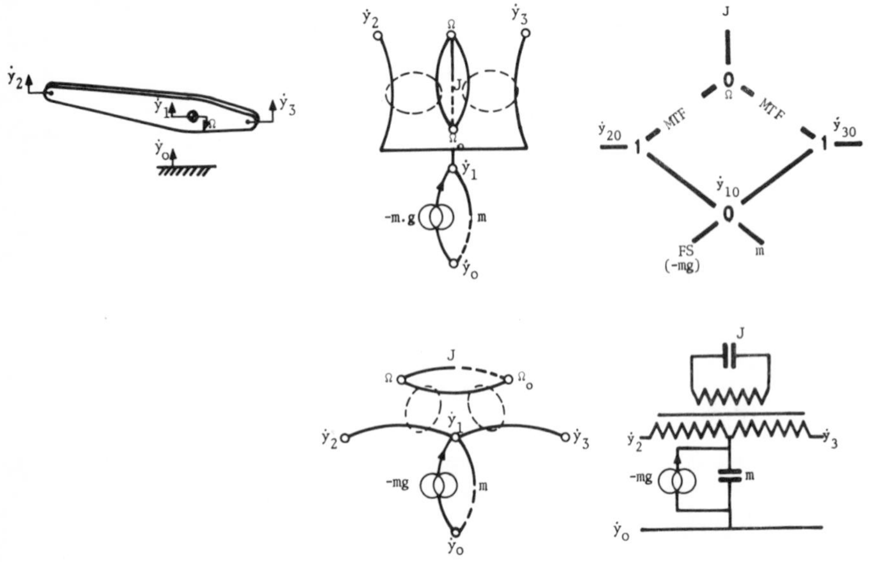

Fig. 14. Free beam models and electric analogue.

electric analogue is drawn, showing the well-known balance transformer. It should be kept in mind, however, that the transformer is of a modulated type (a function of θ) and thus non-linear; furthermore its relation is valid for the static state (d.c.) as well.

2.5. Modelling plane motion of a motor-car using 2-port MTF's

In the first step to an idealized physical model it is decided to consider vertical motions only. Furthermore the car body is represented by a free beam, having mass m and moment of inertia J. K_3 and K_4 are to represent the main springs, R_1 and R_2 the shock absorbers, m_1 and m_2 the wheel masses, K_1 and K_2 the tyre springs. Points 1 and 2 are considered to be moved by the road.

In the second modelling step it is very simple to extend the network graph and bond graph of a free beam (Fig. 14) with respect to across-variables v_1 to v_4 and with respect to the mentioned mass, spring and damper components. Across-sources (V-Sources) are introduced to represent the movements by the road profile, through-sources (F-Sources) for the gravity forces acting on the three masses.

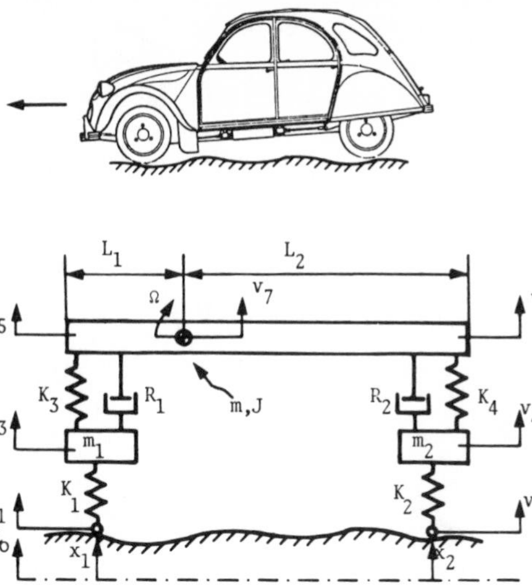

Fig. 15. Idealized physical model of plane motor-car.

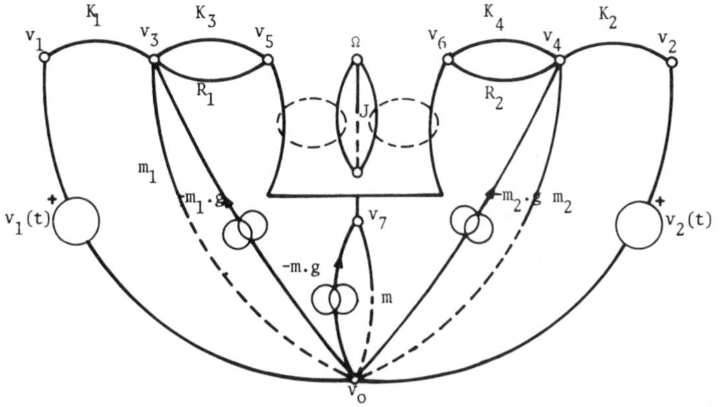

Fig. 16. Network graph of plane motor-car.

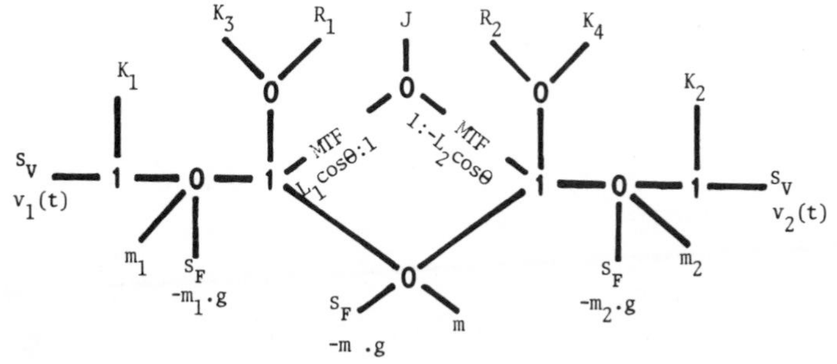

Fig. 17. Bond graph of plane motor-car.

```
C
C        ECAP SIMULATION PLANE MOTORCAR
C        ==============================
C
      AC ANALYSIS
C
C     NETWORK SPECIFICATION
C
C        ONE PORT ELEMENTS      (NODE 8 = OMEGA)
B1    N(1,0),R=1E-9,E=.1E-1
B2    N(1,3),L=4E-6
B3    N(3,0),C=40
B4    N(3,5),L=56E-6
B5    N(3,5),R=4E-4
B6    N(7,0),C=900
B7    N(8,0),C=200
B8    N(2,0),R=1E-9
B9    N(2,4),L=4E-6
B10   N(4,0),C=40
B11   N(4,6),L=56E-6
B12   N(4,6),R=4E-4
C
C        FRONT-SIDE TRANSFORMER
B13   N(5,7),R=1E-9
B14   N(7,8),R=6.95E-10
T1    B(13,14),BETA=1.2
T2    B(14,13),BETA=0.833333
C
C        REAR-SIDE TRANSFORMER
B15   N(6,7),R=1E-9
B16   N(7,8),R=1.56E-9
T3    B(15,16),BETA=-0.8
T4    B(16,15),BETA=-1.25
C
C     SIMULATION RUN CONTROL
      FREQ=1E-1
      PRINT,NV,CA
      EXECUTE


      MODIFY
      FREQ=.2(1.0718)20
      PRINT,NV
      EXECUTE
```

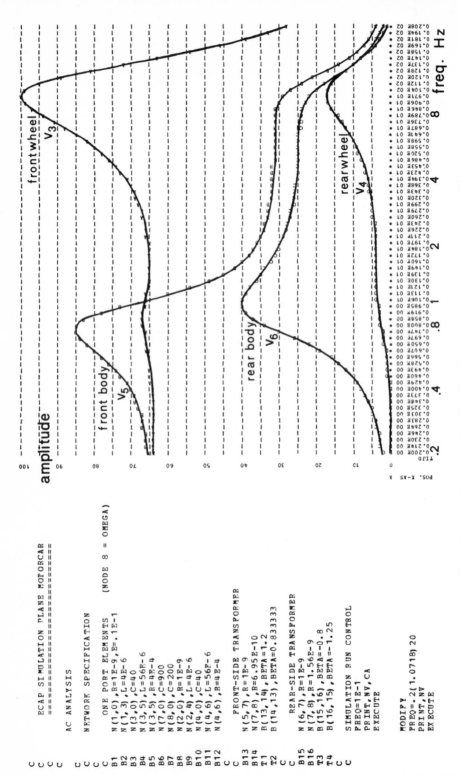

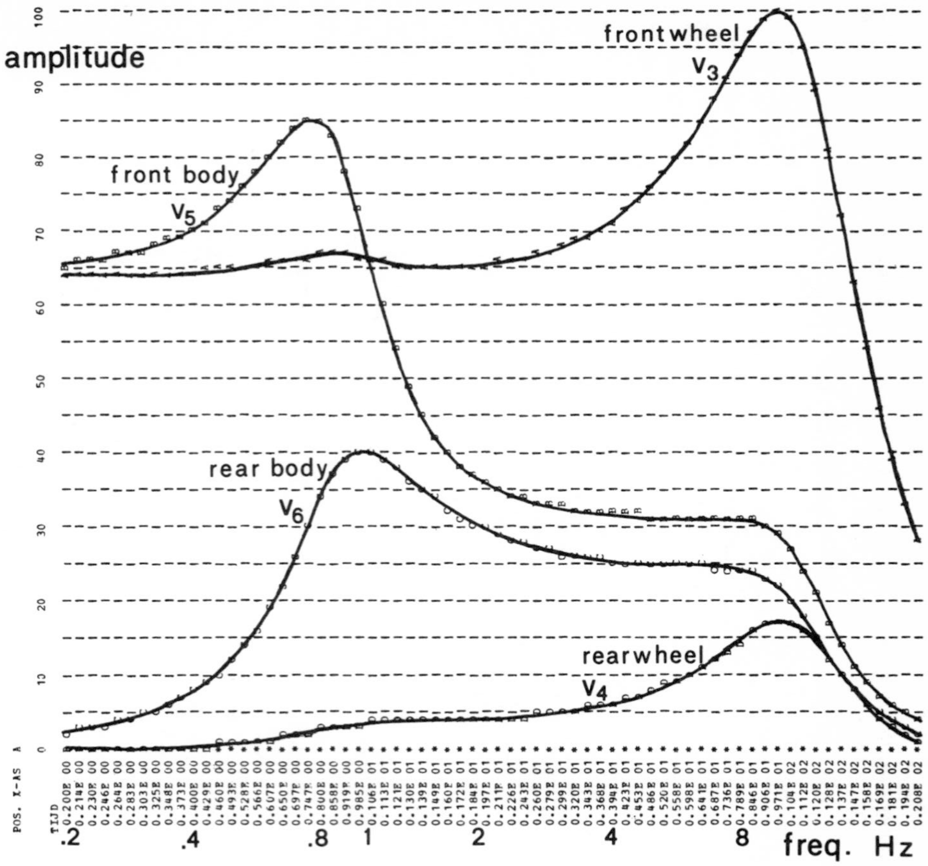

Fig. 18. ECAP-input specification and resulting frequency response of plane motor-car model.

3. ANALYSIS AND SIMULATION OF THE MOTOR-CAR EXAMPLE

Both the network graph and the bond graph present a complete model of the problem in which values can be assigned to the parameters of springs, masses and dampers. The tyre springs (K_1, K_2) have a highly non-linear broken characteristic, because pull forces cannot occur; the compression range is non-linear by virtue of the gas law. Realistic shock absorbers (R_1, R_2) are non-linear. The MTF's are non-linear too, but their transformation ratio's can still be considered constant for fairly big movements.

In the next step one generally wants to investigate the properties and the dynamic behaviour of the model for given disturbances, in this case for road movements $v_1(t)$ and $v_2(t)$ on front and back wheel. The dynamic behaviour of non-linear systems generally can not be calculated analytically. It will be necessary to use digital (numerical) or analogue *simulation.* For both network graphs and bond graphs straightforward ways exist for digital simulation. Standard electric network analysis programs, such as ECAP, [12], accept a network structure as its input. In its "d.c. (direct current) mode" the program calculates the steady state values of forces and displacements, such as the spring compressions. In the "transient mode" the responses to arbitrary input time functions are found. In the "a.c. mode" the *frequency response* of the model in a given working point can be plotted, showing resonances and damped parts of the spectrum. For bond graphs a conversational program ENPORT exists, accepting linear and non-linear bond graphs [16].

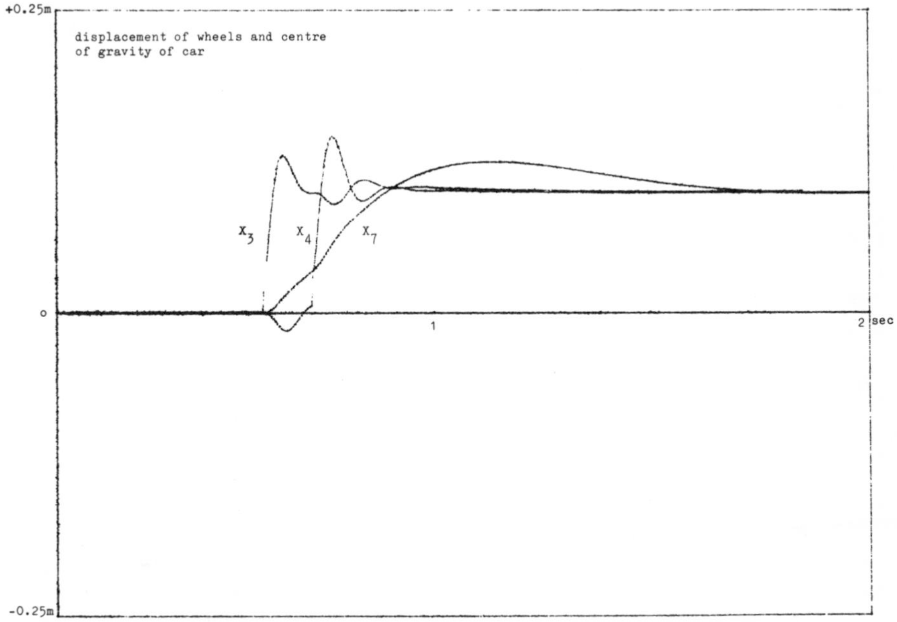

Fig. 19. Step response of plane motor-car.

In Fig. 18 the network graph of the plane motor-car (Fig. 16) is written in the format for the ECAP-program: numbered branches (*B*) are oriented and specified between their "from" and their "to" nodes, *N* (from, to), followed by the element type and parameter. According to the across- and through-variable approach spring-compliances are represented by *L*'s, masses and moments of inertia by *C*'s and slippage by *R*. A structure of controlled current sources and resistances is necessary to specify a transformer. The resulting *AC*-analysis shows the frequency responses of the velocities of both wheels (v_3, v_4) and both body ends (v_5, v_6) if the front wheel (v_1) is subjected to a sinusoidal velocity source. The four resonance frequencies of this asymmetric 8th-order model can be well distinguished.

Figure 19 shows the result of a digital simulation on a small computer of the car model. A sudden step in the road reaches the front wheel first and 0.12 s later the rear wheel. The plot shows the quick displacements of both wheels (x_3, x_4) and the slower vertical motion of the centre of gravity (x_7).

3.1. Block diagram

In control and systems engineering it is often preferred to bring the model in the *block diagram* form. For linear systems the corresponding *signal flow graph* can be used. All interactions between the across-variables and through-variables, that were *implicit* in the network graph and bond graph, are shown in these diagrams in an *explicit* way. A change from power variables to oriented signals, a change from the power domain to the information domain has taken place.

Using simple algorithms the network graph and bond graph of the example can be converted to a block diagram, Fig. 20.

Signals for across-variables are drawn on a level above the level of through-variable signals. All blocks perform the indicated operations; small circles represent summing points, the standard triangular block represents integration with respect to time. The negative clippers in the block diagram assure the correct characteristic relation for the tyre springs. They prevent negative (pull) forces to appear in the model at the places where these springs touch the road.

4. MULTI-PORT MODULATED TRANSFORMERS

In the foregoing part is has been shown that non-linear kinematic relations in mechanical systems can be modelled by ideal (power conserving, non-storing) modulated transformers. Some examples were given of systems in plane motion. Both the network graph and the bond graph symbolism proved to be well suited to represent the abstracted structures. The network graph is slightly nearer to the physical geometrical reality than the bond graph is. This is an advantage, in education especially. The examples show however that the network graph is more elaborate than the bond graph. For many technical systems, for instance for those having a chain structure, like systems built up of electrical to rectilinear mechanical to acoustical transducers, this is no disadvantage.

For mechanical systems in more general than plane motion however, the relations between a restricted number of kinematic across-variables are so numerous that the network graph becomes impractical. Because of the ingenious simplicity of Paynter's power bond notation, the bond graph remains workable. The major breakthrough in representing general mechanical systems by bond graphs has become possible by:

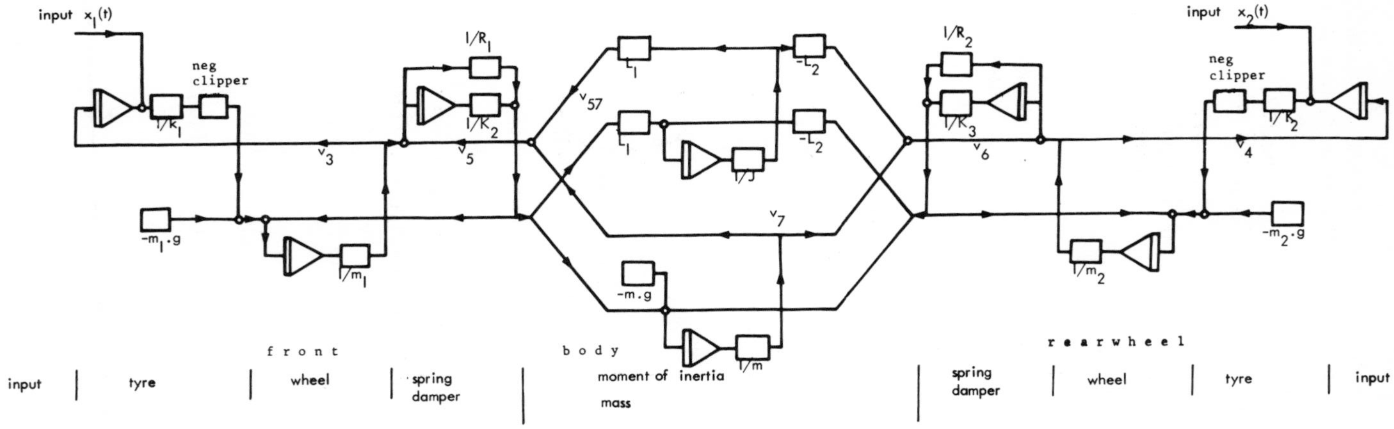

Fig. 20. Block diagram of plane motor-car.

(*a*) The representation of Euler's equations for the rate of change of the angular momentum of a rigid body by a ring structure of modulated gyrators [7], Section 5.

(*b*) The multi-port modulated transformer concept [13]. Some applications of both will be given here, others can be found in [14].

In all previous examples a number of coupled two-port MTF's were used. In 1969 Karnopp [13] introduced the ideal multi-port modulated transformer, a symbolic bond graph element for modelling all kinematic couplings together. Adams discusses in this volume [15] the basic idea of such non-energic *n*-ports.

4.1. Defining the multi-port MTF

Taking the earlier example of a rigid beam in plane motion, the kinematic constraints between vertical velocities $\dot{y}_{10}, \dot{y}_{20}, \dot{y}_{30}$ (measured with respect to a

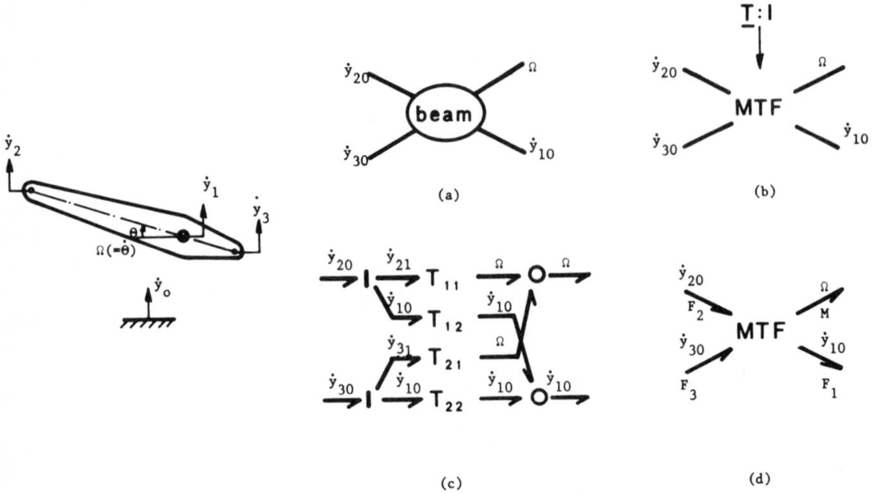

Fig. 21. Models of some kinematic constraints in a beam.

constant reference velocity $\dot{y}_0$) and angular velocity Ω $(=\dot{\theta})$ are, in different notations:

$$\dot{y}_{20} = (L_1 \cos\theta) \,.\, \Omega + \dot{y}_{10} \tag{9}$$

$$\dot{y}_{30} = (-L_2 \cos\theta) \,.\, \Omega + \dot{y}_{10} \tag{10}$$

or

$$\begin{vmatrix} \dot{y}_{20} \\ \dot{y}_{30} \end{vmatrix} = \begin{vmatrix} L_1 \cos\theta & 1 \\ -L_2 \cos\theta & 1 \end{vmatrix} \begin{vmatrix} \Omega \\ \dot{y}_{10} \end{vmatrix} \tag{11}$$

By defining

$$\begin{vmatrix} \dot{y}_{20} \\ \dot{y}_{30} \end{vmatrix} \triangleq \dot{\mathbf{y}}, \qquad \begin{vmatrix} \Omega \\ \dot{y}_{10} \end{vmatrix} \triangleq \mathbf{\Omega}$$

$$\begin{vmatrix} L_1 \cos\theta & 1 \\ -L_2 \cos\theta & 1 \end{vmatrix} \triangleq \mathbf{T}$$

the elemental relation of the multi-port transformer can be written:

$$\dot{\mathbf{y}} = \mathbf{T}\boldsymbol{\Omega} \tag{12}$$

T is a matrix, having state- and thus time-dependent elements.

The kinematic constraints imposed by the beam can be modelled by a word bond graph (Fig. 21a); by a multi-port transformer, having a transformation ratio **T** (Fig. 21b); by 4 two-port transformers, having transformation ratios corresponding with the matrix elements T_{ij} (Fig. 21c). In this figure the 1-junctions represent series-junctions, having a common through-variable, the across-variables adding up according to the half-arrows; the o-junctions have the indicated common across-variable. In order to state the power conserving or non-energic property, through-variables and a positive direction for power and thus for the through-variable have to be indicated at each bond (Fig. 21d), leading to

$$\dot{y}_{20} \,.\, F_2 + \dot{y}_{30} \,.\, F_3 = \Omega \,.\, M + \dot{y}_{10} \,.\, F_1 \tag{13}$$

In matrix notation, defining

$$\mathbf{F}^T \triangleq |F_2 \quad F_3|$$

$$\mathbf{M}^T \triangleq |M \quad F_1|$$

power conservation is stated by

$$\mathbf{F}^T\dot{\mathbf{y}} = \mathbf{M}^T\boldsymbol{\Omega} \tag{14}$$

The transformer relation was

$$\dot{\mathbf{y}} = \mathbf{T}\boldsymbol{\Omega} \tag{12}$$

and so

$$\mathbf{M} = \mathbf{T}^T\mathbf{F} \tag{15}$$

Like in the two-port MTF case (eqn. 7) the same state dependent transformation ratio, though transposed, holds between opposite through-variables.

In the beam example this leads to

$$\begin{vmatrix} M \\ F_1 \end{vmatrix} = \begin{vmatrix} L_1 \cos\theta & -L_2 \cos\theta \\ 1 & 1 \end{vmatrix} \begin{vmatrix} F_2 \\ F_3 \end{vmatrix} \tag{16}$$

Bearing in mind the power directions, adopted in Fig. 21d, this correctly states that F_1 (exerted on the *environment*) equals $F_2 + F_3$ (both being exerted on the *beam*); also $M = L_1 \cos\theta \,.\, F_2 - L_2 \cos\theta \,.\, F_3$ is correct.

The *free beam* of Fig. 14 can now be modelled in a compact way:

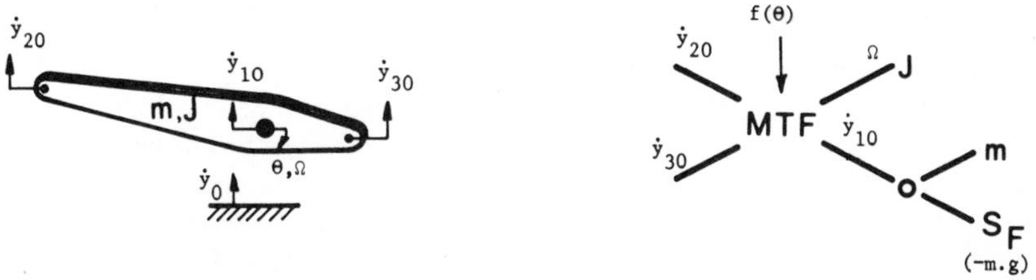

Fig. 22. Compact free beam model (plane motion, vertical velocities only).

4.2. Modelling the plane motor-car using multi-port MTF's

Figure 23 shows the steps in modelling the plane motor-car. The word bond graph (Fig. 23b) relates the kinematic constraints of the body to both inertia elements of the body and to both wheel models. Figure 23c is partly more detailed:

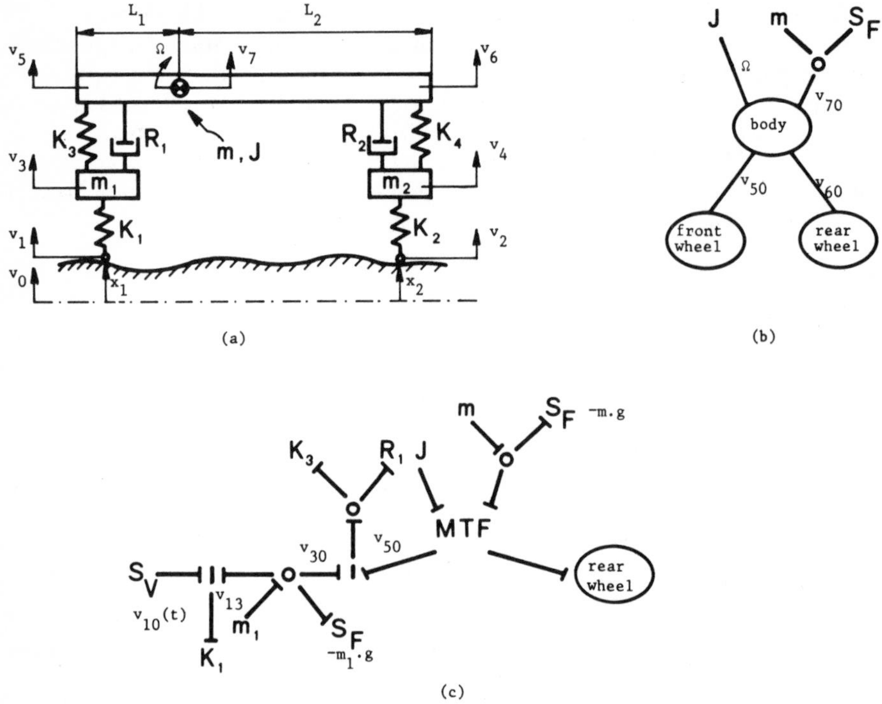

Fig. 23. Recommended bond graph modelling of plane motor-car.

the bond graph is augmented with causal strokes, showing that no dependent storage elements exist. The MTF in the model can of course be split up further, as effected in Fig. 21c.

4.3. Modelling a simplified 3-dimensional motor-car using multi-port MTF's

The above treatment will be extended to a four-wheel motor-car body, moving in 3 dimensions, but restricted to 3 degrees of freedom. It is assumed that the body can rotate about both "horizontal" principal axis of inertia, having principal moments of inertia J_1 and J_2 in these directions. Rotation about the "vertical" principal axis is constrained to zero.

A reference frame u, v, w is fixed to the body having its origin in the body mass centre 5. An inertial reference frame x, y, z is fixed to the earth. The angular velocity vector $\mathbf{\Omega}$, describing the rotation of the body with respect to u, v, w has components Ω_u and $\Omega_v (\Omega_w = 0)$ along the body frame. The angular position of this frame with respect to the x, y, z frame is described by θ_u and θ_v. For commonly encountered values of θ_u and θ_v the approximation $\dot{\theta}_u = \Omega_u$ and $\dot{\theta}_v = \Omega_v$ is valid.

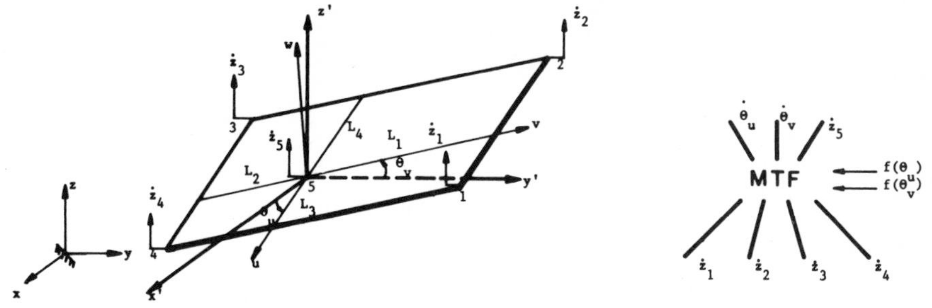

$$\begin{vmatrix} \dot z_1 \\ \dot z_2 \\ \dot z_3 \\ \dot z_4 \end{vmatrix} = \begin{vmatrix} -L_3\cos\theta_u\cos\theta_v & (L_1\cos\theta_v+L_3\sin\theta_u\sin\theta_v) & 1 \\ L_4\cos\theta_u\cos\theta_v & (L_1\cos\theta_v-L_4\sin\theta_u\sin\theta_v) & 1 \\ L_4\cos\theta_u\cos\theta_v & (-L_2\cos\theta_v-L_4\sin\theta_u\sin\theta_v) & 1 \\ -L_3\cos\theta_u\cos\theta_v & (-L_2\cos\theta_v+L_3\sin\theta_u\sin\theta_v) & 1 \end{vmatrix} \begin{vmatrix} \dot\theta_u \\ \dot\theta_v \\ \dot z_5 \end{vmatrix}$$

Fig. 24. Kinematic constraints of body having 3 degrees of freedom.

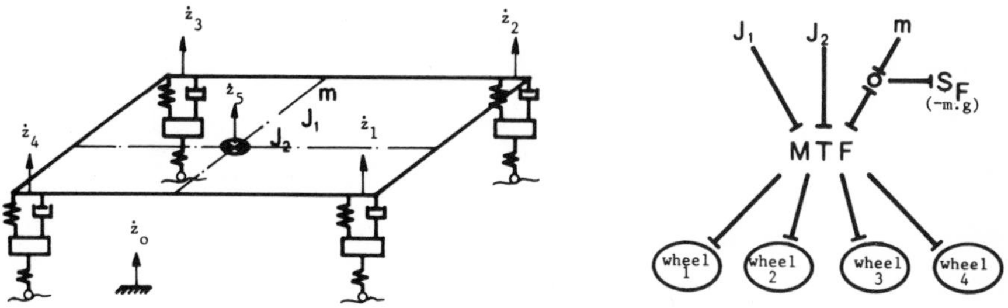

Fig. 25. Idealized 3-dimensional physical model and compact bond graph.

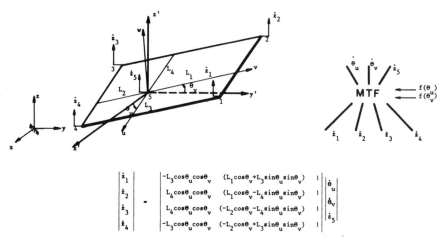

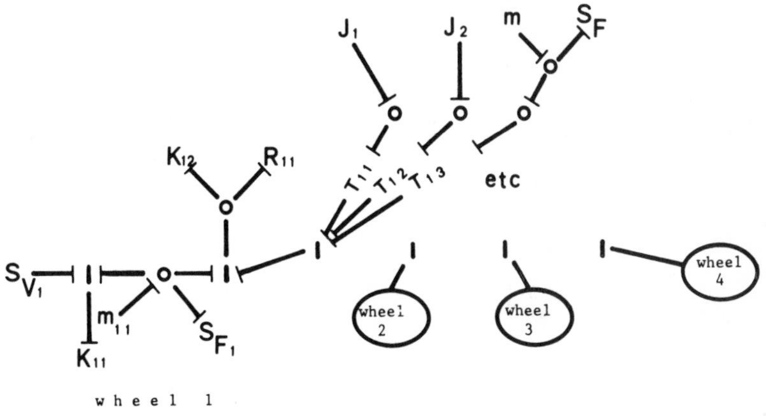

Fig. 26. Detailed and augmented bond graph of 3-dimensional motor-car.

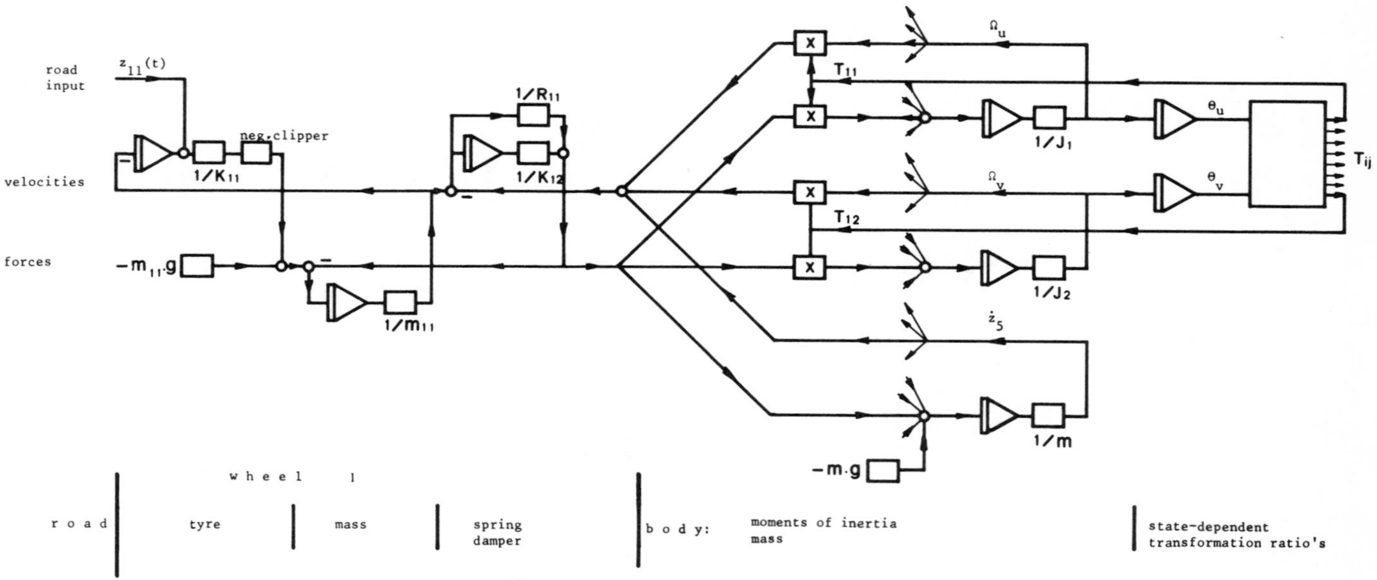

Fig. 27. Block diagram of 3-dimensional motor-car.

Both rotations and the vertical translation of the body mass centre are related to the vertical velocities of points 1 to 4 (z_{10} to z_{40}) by a multi-port MTF having state-dependent transformation matrix **T** (Fig. 24).

It is easy to extend this MTF with respect to the four, independently suspended, wheels and with respect to the three inertial storage elements m, J_1 and J_2 (Fig. 25). The wheel models are identical to those used in the 2-dimensional case. Remembering the "computational causality" of a wheel (Fig. 23) the compact bond graph of Fig. 25 can in this stage already be augmented with causal strokes. These show that no conflict, for instance by dependent masses, occurs.

It may be desirable, for further analysis or for deduction of a block diagram, to expand the multi-port MTF in 12 two-port MTF's, corresponding with matrix elements T_{ij} (Fig. 26).

Changing from the power domain to the information domain by drawing the *block diagram* (Fig. 27) doubles the amount of relations: instead of 12 2-port MTF's 24 multipliers show up, etc.

The efficiency of the bond graph, symbolizing exactly the same relations, is striking.

Figure 28 shows results of a simulation with this model, in which case left-side wheels 2, 3 suddenly start riding on a higher part of the road.

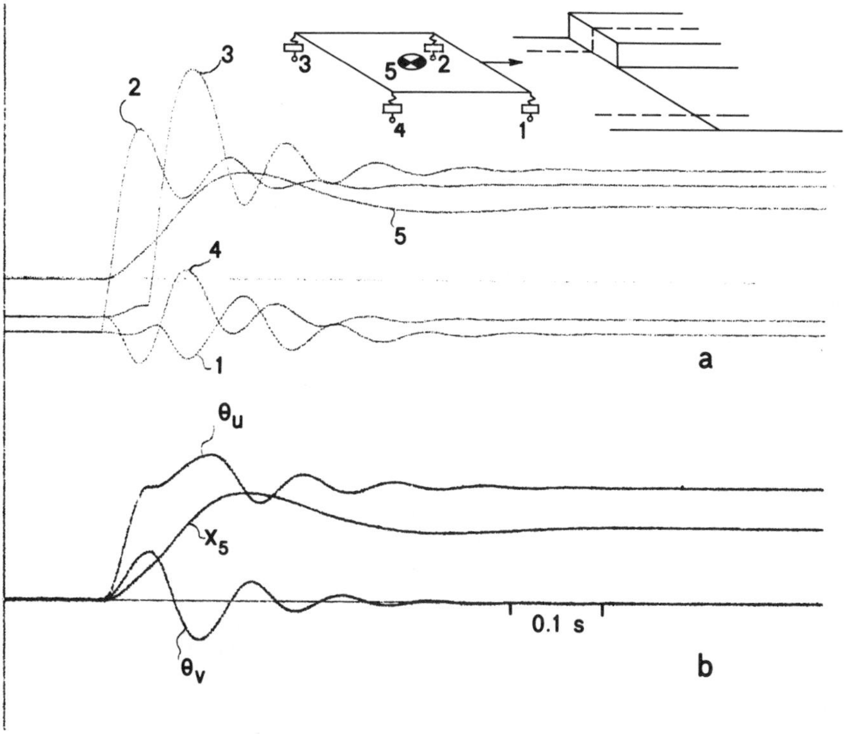

Fig. 28. Simulation of 3-dimensional motor-car. a, vertical displacements of wheels 1 to 4 and of centre of gravity 5; b, vertical displacement and rotations of body frame.

5. BOND GRAPH STRUCTURE OF GENERAL ROTATION

In the previous example the body was assumed to be constrained to 2 independent rotations about its principal axes. Rotation about 3 principal axes, identical with general rotation, is well-known to be a more complicated phenomenon. To the angular momentum **H** of a free rotating body a conservation law applies; the vector **H** has a fixed position with respect to an inertial reference frame x, y, z. With respect to a reference frame u, v, w, however, attached in the body masscentre along its principal axes of inertia, vector **H** has a changing position. So in this frame its components H_u, H_v, H_w are time variables. The angular velocity components $\Omega_u, \Omega_v, \Omega_w$ are also time variables, because of eqn. (17). In this equation the inertia tensor has its canonical form, due to the chosen body frame.

$$\begin{vmatrix} H_u \\ H_v \\ H_w \end{vmatrix} = \begin{vmatrix} J_1 & 0 & 0 \\ 0 & J_2 & 0 \\ 0 & 0 & J_3 \end{vmatrix} \begin{vmatrix} \Omega_u \\ \Omega_v \\ \Omega_w \end{vmatrix} = \begin{vmatrix} J_1 \,.\, \Omega_u \\ J_2 \,.\, \Omega_v \\ J_3 \,.\, \Omega_w \end{vmatrix} \tag{17}$$

The torques $J_1 \,.\, \dot{\Omega}_u$, etc., producing these changes, correspond to the rates of change of angular momenta in the principal directions:

$$\begin{vmatrix} \dfrac{dH_u}{dt} \\ \dfrac{dH_v}{dt} \\ \dfrac{dH_w}{dt} \end{vmatrix} = \begin{vmatrix} J_1 \,.\, \dot{\Omega}_u \\ J_2 \,.\, \dot{\Omega}_v \\ J_3 \,.\, \dot{\Omega}_w \end{vmatrix} \tag{18}$$

These changes of the components of angular momentum **H** are a result of the movement $\mathbf{\Omega}$ of the body frame. They are related to this movement by the skew symmetric square matrix **R** of the components of Ω

$$\frac{d\mathbf{H}}{dt} = \mathbf{RH}$$

or

$$\begin{vmatrix} \dfrac{dH_u}{dt} \\ \dfrac{dH_v}{dt} \\ \dfrac{dH_w}{dt} \end{vmatrix} = \begin{vmatrix} 0 & \Omega_w & -\Omega_v \\ -\Omega_w & 0 & \Omega_u \\ \Omega_v & -\Omega_u & 0 \end{vmatrix} \begin{vmatrix} H_u \\ H_v \\ H_w \end{vmatrix} \tag{19}$$

Equating (19) and (18)

$$(J_2 - J_3)\,\Omega_v\Omega_w = J_1 \,.\, \dot{\Omega}_u$$

$$(J_3 - J_1)\,\Omega_u \Omega_w = J_2\,.\,\dot{\Omega}_v$$

$$(J_1 - J_2)\,\Omega_u \Omega_v = J_3\,.\,\dot{\Omega}_w \tag{20}$$

In 1968 Karnopp and Rosenberg [7] presented the ring-like bond graph structure of Fig. 29b, modelling equations (20). Ideal modulated gyrators MGY, as defined in Fig. 29a, are needed to represent the coupling between the three motions. The structure clearly shows that the exchange of kinetic energy between two

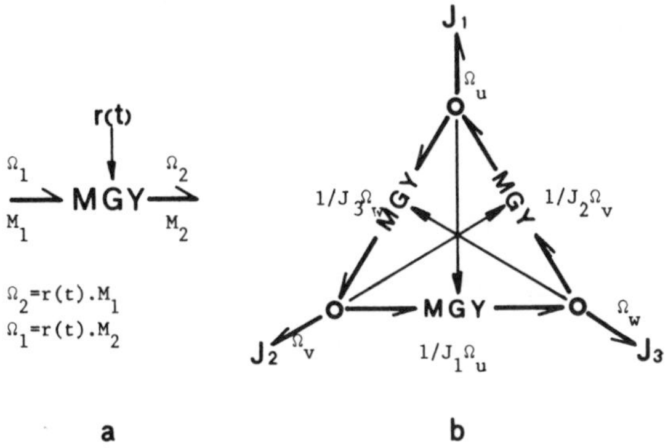

Fig. 29. a, modulated gyrator; b, MGY coupled rotations of a free rigid body.

moments of inertia is governed by the velocity of the third. If for instance Ω_u and Ω_v are both positive, then $\Omega_w > 0$ implies power flow from u to v, $\Omega_w < 0$ implies power flow from v to u and $\Omega_w = 0$ implies no coupling between u and v.

If a rigid body is subjected to an external torque **M**, having components M_u, M_v, M_w with respect to the body frame, these torques are added to the left sides of (20); the result is called *Euler's equations.* In the bond graph these external components are represented by power bonds to the three o-junctions.

5.1. Bond graph of general motor-car suspension

A motor-car body having 6 degrees of freedom is represented by the upper part of the bond graph of Fig. 30. The three angular velocities with respect to the principal direction body frame u, v, w are coupled by the MGY-ring. The origin of this frame is in the mass centre, 5, and exhibits translational velocities $\dot{x}_5, \dot{y}_5, \dot{z}_5$ with respect to the inertial frame x, y, z. The lower part of the bond graph shows the bonds between the connection points 1 to 4 and the wheels. It will be clear that the wheels in this model necessarily have x and y velocity components. The wheel models have to be more sophisticated then previously used. The MTF represents the kinematic constraints relating the 6 body velocities to the 12 velocities of points 1, 2, 3 and 4.

Systematic procedures for finding the elements of the MTF can be followed, like:

1. The coordinates of points $i = 1, 2, 3, 4$ in the u, v, w (body) frame u_i, v_i, w_i, are determined.

2. The rotation $\mathbf{\Omega}$ of the body frame, which causes translational velocities of points i with respect to the body frame, is described by the relation

$$\begin{vmatrix} \dot{u}_i \\ \dot{v}_i \\ \dot{w}_i \end{vmatrix} = \begin{vmatrix} 0 & \Omega_w & -\Omega_v \\ -\Omega_w & 0 & \Omega_u \\ \Omega_v & -\Omega_u & 0 \end{vmatrix} \begin{vmatrix} u_i \\ v_i \\ w_i \end{vmatrix} = \mathbf{R} \begin{vmatrix} u_i \\ v_i \\ w_i \end{vmatrix} \qquad (21)$$

in which figurizes the previously used skew symmetric matrix $\mathbf{R}$.

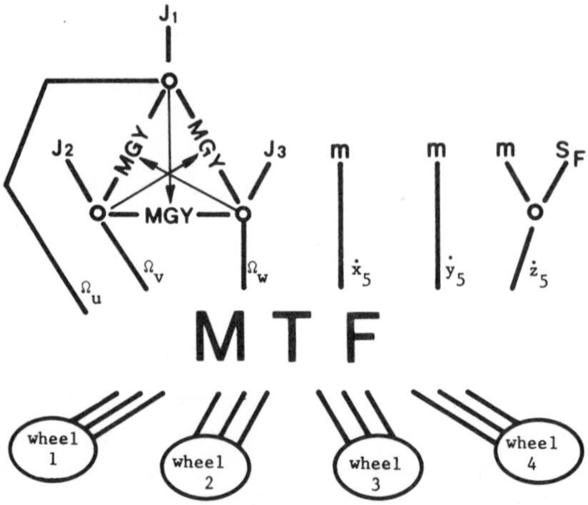

Fig. 30. Compact bond graph of general motor-car suspension, having a 6 degrees of freedom body.

3. The position of the u, v, w frame with respect to a x', y', z' frame having the same origin in point 5 is described by means of the rotational transformation matrix $\mathbf{C}$, whose elements are the direction cosines between both frames:

$$\begin{vmatrix} 1_{x'} \\ 1_{y'} \\ 1_{z'} \end{vmatrix} = \mathbf{C} \begin{vmatrix} 1_u \\ 1_v \\ 1_w \end{vmatrix} \qquad (22)$$

4. The velocities of point i with respect to the x, y, z frame now follow from equations (21) and (22):

$$\begin{vmatrix} \dot{x}_i \\ \dot{y}_i \\ \dot{z}_i \end{vmatrix} = \mathbf{CR} \begin{vmatrix} u_i \\ v_i \\ w_i \end{vmatrix} + \begin{vmatrix} \dot{x}_5 \\ \dot{y}_5 \\ \dot{z}_5 \end{vmatrix} \qquad (23)$$

5. As u_i, v_i, w_i are constants, the above expression can be written in another form

$$\begin{vmatrix} \dot{x}_i \\ \dot{y}_i \\ \dot{z}_i \end{vmatrix} = \mathbf{T} \begin{vmatrix} \Omega_u \\ \Omega_v \\ \Omega_w \\ \dot{x}_5 \\ \dot{y}_5 \\ \dot{z}_5 \end{vmatrix} \tag{24}$$

T is the desired state-dependent (sub)matrix of the multi-port MTF relating the body velocities to the velocities of point $i = 1, 2, 3, 4$.

6. CONCLUSION

It has been shown that network modelling of mechanical systems is not necessarily restricted to one-dimensional systems, as is generally found in dynamic systems literature. Kinematic constraints in multi-dimensional systems can be modelled by displacement modulated (non-energic) transformers, introduced in the bond graph literature by Karnopp. For modelling one- or two-dimensional systems, either network graphs or bond graphs, incorporating such MTF's, can be used. For three-dimensional systems the network graph becomes too elaborate, in contrast to the extra compact bond graph, which can be obtained using the multi-port modulated transformer concept (this applies also to the 2-dimensional models of Section 2). Another bond graph concept, not used in this paper, is the ideal multi-port energy storage element [7], or so-called "field", which can also be defined for non-linear networks. In the motor-car examples they might be added to the model to account for the bending phenomena of the car body, combining rotational and translational elasticity.

The main difference between the network and bond graph approach and the conventional analysis in applied mechanics is that velocities and displacements are the only explicit variables. Accelerations and forces (centripetal, tangential, Coriolis) which are a result of the non-linear transformers in the model, remain hidden variables, like electric currents in a circuit diagram. They are not necessary for setting up the model, though they can easily be made explicit.

It is not claimed that this topological approach to modelling is better or even always possible for mechanical systems. However, the fact that the same procedure can be used as the first step in the process of modelling various kinds of physical systems, mechanical systems not excluded, makes further investigation worthwhile.

REFERENCES

1. Shearer, J. L., Murphy, A. T. and Richardson, H. H. *Introduction to System Dynamics.* Addison-Wesley, Reading, Mass., 1967.
2. Martens, H. R. and Allen, D. R. *Introduction to systems theory*. Merrill Publ. Co., 1969.

3. Perkins, W. R. and Cruz, J. B. *Engineering of dynamic systems.* Wiley, New York, 1969.
4. Roe, P. H. O. N. *Networks and Systems.* Addison-Wesley, Reading, Mass., 1966.
5. MacFarlane, A. G. J. *Dynamical System Models.* Harrap, London, 1970.
6. Koenig, H. E., Tokad, Y. and Kesavan, H. K. *Analysis of discrete physical systems.* McGraw-Hill, New York, 1967.
7. Karnopp, D. and Rosenberg, R. C. *Analysis and simulation of multiport systems.* M.I.T. Press, Cambridge, Mass., 1968.
8. Paynter, H. M. *Analysis and design of engineering systems.* M.I.T. Press, Cambridge, Mass., 1960.
9. "Special issue on bond graphs". *J. Dyn. Syst., Meas., Control, Trans ASME, Series G, 94*, Sept. 1972.
10. Karplus, W. J. and Soroka, W. W. *Analog Methods.* McGraw-Hill, New York, 1959.
11. Pawley, M. "The design of a mechanical analogy for the general linear electrical network with lumped parameters". *J. Franklin Inst. 223,* 1937, 179–198.
12. Jensen, R. W. and Liebermann, M. D. *IBM Electronic Circuit Analysis Program. Techniques and Applications.* Prentice Hall, 1968.
13. Karnopp, D. "Power-conserving transformations: Physical interpretations and applications using bond graphs". *J. Franklin Inst. 288,* 1969, 175–201.
14. Rosenberg, R. C. "Multiport models in mechanics". *J. Dyn. Syst., Meas., Control, Trans. ASME, Series G, 94,* 1972, 206–212.
15. Adams, K. M. "Applicability of basic concepts in classical network theory to engineering systems", in Dixhoorn, J. J. van and Evans, F. J. (eds.) *Physical Structure in Systems Theory*, Academic Press, London, 1974 (this volume).
16. *ENPORT.* Users manual and program available from Prof. R. C. Rosenberg, Dept. Mech. Eng., Michigan State Univ., U.S.A.
17. Polder, J. W. "Model of a mechanical transmission", in Dixhoorn, J. J. van and Evans, F. J. (eds.) *Physical Structure in Systems Theory*, Academic Press, London, 1974 (this volume).

NOMENCLATURE

x, y, z	spatial coordinates in inertial frame	Ω	$\dot{\theta}$, rotational velocity
		F	force
u, v, w	spatial coordinates in body frame	M	torque
		$\mathbf{H}$	angular momentum vector
v	translational velocity (used in plane motor-car model only)	m	mass
		J	moment of inertia
		K	spring compliance
$\mathbf{1}_x, \mathbf{1}_y, \mathbf{1}_z$	unit vectors in x, y, z directions	R	dissipator slippage (across ÷ through)
$\dot{x}_{21}$	$\dot{x}_2 - \dot{x}_1$	S_v	across-type source (velocity, voltage)
θ	angle of rotation		

S_F	through-type source (force, torque, current)
MTF	modulated transformer (2-port or multi-port)
MGY	modulated gyrator
$m(t)$	2-port transformation ratio
$r(t)$	gyration ratio (across ÷ through)
$\mathbf{T}(t)$	multi-port transformation matrix (across ÷ across)
$\mathbf{T}^T(t)$	transposed multi-port transformation matrix (through ÷ through)
g	acceleration of free fall
a.c.	alternating current
d.c.	direct current
$\triangleq$	is defined by
o	in bond graphs: junction of ports having a common across-variable (parallel connection)
1	in bond graphs: junction of ports having a common through-variable (series connection)

APPLICABILITY OF BASIC CONCEPTS IN CLASSICAL NETWORK THEORY TO ENGINEERING SYSTEMS

K. M. Adams
Department of Electrical Engineering
Delft University of Technology
Delft, The Netherlands

SUMMARY

Many basic concepts that are well known in classical electrical network theory have been widely employed in modelling and analysing physical and engineering systems. Other network theoretical concepts that are less well known can also be employed to advantage. This paper presents some of these concepts in a fairly general manner and discusses their relevance.

1. INTRODUCTION

Many concepts that are employed in analysing and modelling physical, economic and biological systems have had a long history of development in the classical theory of electrical networks. As network theory combines both dynamical and structural theories, it provides a convenient frame of reference within which new and strange phenomena can be modelled and analysed. Since the standard of rigour employed in network theory is both uniform and high, many difficulties and inconsistencies in other fields can often be easily resolved once the problem has been formulated in network-theoretical terms.

For these reasons it is worth-while to present some of the basic concepts of classical network theory together with some of the background material that has led to their present state of evolution. Many of these concepts are quite well-known to workers in other branches of engineering; others, however, will probably be new to most people who have not followed closely the recent developments in the subject.

With regard to the material selected, no attempt has been made to be complete or to develop a systematic account of what is now available. Instead topics have been chosen that are connected with the discussions held during the Twente Symposium and which appear to need further clarification and exposition.

Finally a word of explanation of what we mean by classical (electrical) network theory is called for. In the first place, the words "electrical network" refer to a theory which includes Kirchhoff's laws as essential ingredients. Secondly, "classical" refers to a body of theory which has been well developed, tried, tested for consistency and found to satisfy high standards of rigour as well as to be relevant and "useful" in the theoretical sense. Some parts of classical network theory are quite old whereas other parts have appeared only recently. Viewed in this way, the term "modern" says nothing about the quality of a theory. The relevance of the term is in any case subject to rapid devaluation with time. The term "modern network theory" is for these reasons to be deprecated when anything more than a rapidly changing fashion is meant. If the theory is growing healthily, a sizable body of modern theory will graduate after due testing to classical theory; the rest will be rejected.

2. NETWORK ELEMENTS AND A GENERAL PRINCIPLE

The familiar linear elements: resistor, capacitor, inductor are passive one-ports. They have been abstracted from real physical devices by a process of idealization and constitute the simplest one-port networks (or systems) of their type. Thus the resistor is the simplest (idealized) device that exhibits the property of dissipation, the capacitor the simplest (idealized) device that exhibts the property of electric energy storage. Any interconnexion of these elements is an acceptable electrical network. In such an electrical network, if the energy distribution at time $t = 0$ is known, then the current and voltage distribution throughout the network are uniquely determined for all $t > 0$. We then say that the network has a unique, computable behaviour.

If now we add to this collection the ideal voltage source and the ideal current source, which are not passive, and if the intensities of these sources are known functions of time, then *nearly* every interconnexion of these five elements forms an acceptable electrical network with a unique, computable behaviour. The exceptional cases occur when some of the voltage sources form a loop or when some of the current sources form a cut-set. Then, depending on the values of the source intensities, either one or both of Kirchhoff's laws is violated, in which case either the interconnexion is not an acceptable electrical network, or the network does not have a unique computable behaviour. This second situation will normally also be excluded from any workable theory.

The above result constitutes one of the basic theorems of network analysis. For the purposes of this paper, however, the first important point to note is that we can construct a workable theory with these five elements. The exceptional situations which must be excluded from the set of acceptable networks, both on theoretical and physical grounds, can be precisely formulated, are readily recognizable and have a negligible measure compared to the measure of the normal situations in which the theory does work.

The second important point is that, when one is confronted with a new situation, such as arises when proposing to introduce new elements or sub-systems, any network formed by interconnecting the new elements and the already accepted elements, must have a unique and computable behaviour except in situations which can be precisely formulated, are easily recognizable and constitute a negligible proportion of all possible situations arising from any network consisting of the same kinds of elements. This statement is a modified version of a general principle formulated by Tellegen in 1954 [1], for judging the acceptability of new (idealized) devices.

Once these ideas have been established, it is not difficult to consider two-ports exhibiting magnetic coupling and to extract from such two-ports by a series of idealizations the concept of the ideal transformer as network element which can be added to the collection of elements already accepted. The ideal transformer exhibits the property of voltage or current amplification or attenuation, and is the simplest passive two-port displaying this property. In networks composed of all six kinds of these elements, the exceptional situations that have to be excluded cannot in general be specified by a simple graph-theoretical statement, but only in algebraic terms. Nevertheless they satisfy the criteria of the acceptability principle.

Next, one can introduce the gyrator as the simplest passive network violating the reciprocity principle. Again judged by the above criterion, the gyrator is a perfectly acceptable element, and the exceptional situations although somewhat more complicated than in the previous cases are quite definite and can be formulated quite precisely in algebraic terms.

The five types of passive element thus obtained form a complete set in the sense that any passive n-port which can be characterized by rational functions of the complex frequency is equivalent to an n-port constructed from these elements. As a bonus we obtain the result that any interconnexion of the five types of passive element is an acceptable network.

Finally we note that each of the elements we have so far mentioned is characterized by a single parameter, such as resistance, or voltage (of a source) or transformer ratio. There are good reasons for adopting this property as a necessary condition for any entity which is to be called a network element.

3. NON-ENERGIC SYSTEMS AND NON-ENERGIC CONSTRAINTS

The term non-energic was introduced into mechanics by Birkhoff [2, 3], who used it to describe mechanical systems with the following property. Consider an n-port such that two functions of the time $\xi_i(t)$ and $\eta_i(t)$ are associated with port i $(i = 1, \ldots, \eta)$. For concreteness we can think of $\xi_i(t)$ as a voltage and $\eta_i(t)$ as a current, subject to the polarity and direction conventions shown in Fig. 1.

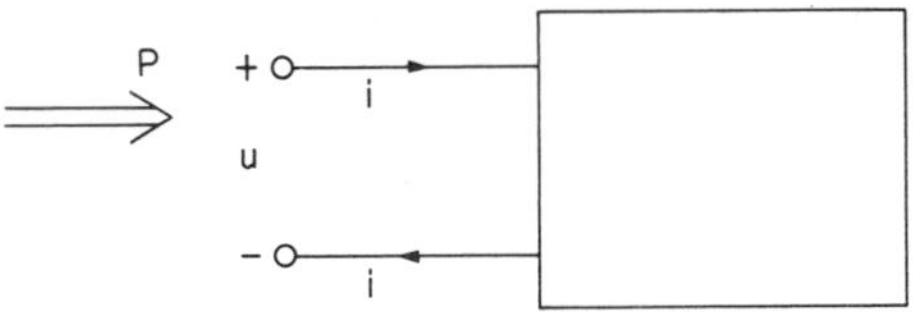

Fig. 1. Reference voltage polarities, current directions and direction of power-flow in a one-port.

If now

$$\sum_{i=1}^{\eta} \xi_i(t)\,\eta_i(t) = 0 \quad \text{for all } t,$$

then the n-port is called *non-energic*. Alternatively, we can say that the vector $\boldsymbol{\xi}(t)$

with ith element $\xi_i(t)$ and the vector $\boldsymbol{\eta}(t)$ with ith element $\eta_i(t)$ belong to complementary vector sub-spaces if the n-port is non-energic. That is to say any vector $\boldsymbol{\xi}(t)$ admitted by the n-port is always orthogonal to the corresponding vector $\boldsymbol{\eta}(t)$. This means that such an n-port is incapable of dissipating or storing energy in an available form. The n-port can only transmit power instantaneously from some ports of its environment to other ports.

One must be careful to distinguish this concept from the concepts of losslessness (non-dissipativeness) and non-reactivity which have sometimes (regrettably) also been termed non-energic. Non-energicness is a property that involves much more stringent conditions than these other properties do.

In mechanics the concept of workless constraint has been employed since the early eighteenth century. But the point of view and the emphasis differ in mechanical contexts (with the exception of Birkhoff's work) from that of a large body of network theory. Indeed one of the reasons for Lagrange to develop his analytical dynamics was to eliminate the need to include workless forces in the equations of motion explicitly. Although the corresponding situation also arises in many branches of network theory, there are other parts of the subject, particularly where synthesis and modelling are relevant, where the opposite occurs. Namely, one deliberately focuses attention on the non-energic sub-systems of a system and studies their separate properties. We shall now go into these properties in more detail.

Any non-energic n-port is characterized by the equation

$$\boldsymbol{\xi}^T(t)\,\boldsymbol{\eta}(t) = \mathbf{0} \quad \forall\, t. \tag{1}$$

Thus the port currents and the port voltages are not all independent. Thus $\boldsymbol{\xi}(t)$ belongs to a sub-space X, which is orthogonal to the sub-space Y to which $\boldsymbol{\eta}(t)$ belongs, and neither of these sub-spaces is empty.

Let the ranks of X and Y be r and s respectively. Let A be a matrix of rank r whose row vectors belong to X.

Then

$$\mathbf{A}\boldsymbol{\eta}(t) = \mathbf{0} \quad \forall\, t. \tag{2}$$

It is important to note that the elements of $\mathbf{A}$ can be functions of time or can depend on voltages, currents, charges, fluxes etc. None of these possibilities is barred by the non-energic condition. All that we know is that each row of $\mathbf{A}$ is a possible (transposed) $\boldsymbol{\xi}(t)$ vector. Henceforth we shall drop the explicit mention of time dependence.

Since $\mathbf{A}$ is of rank r it contains a non-singular sub-matrix of order r, so that r of the elements of $\boldsymbol{\eta}$ can be expressed in terms of the remaining $n-r$ elements.

Thus

$$\boldsymbol{\eta}_1 = \mathbf{M}\boldsymbol{\eta}_2 \tag{3}$$

where

$$\begin{bmatrix} \boldsymbol{\eta}_1 \\ \boldsymbol{\eta}_2 \end{bmatrix} = P\boldsymbol{\eta}$$

and P is some permutation operator acting on the sequence η_i $(i = 1, \ldots n)$. If we now partition $\boldsymbol{\xi}$ conformally in $\begin{bmatrix} \boldsymbol{\xi}_1 \\ \boldsymbol{\xi}_2 \end{bmatrix}$, then we obtain from (1) and (3)

$$(\boldsymbol{\xi}_1^T \mathbf{M} + \boldsymbol{\xi}_2^T)\, \boldsymbol{\eta}_2 = \mathbf{0}. \tag{4}$$

Since $\boldsymbol{\eta}_2$ consists of linearly independent elements, it may be chosen arbitrarily and thus

$$\boldsymbol{\xi}_2 = -\mathbf{M}^T \boldsymbol{\xi}_1. \tag{5}$$

In the same way one can start from an equation of the form

$$\mathbf{C}\boldsymbol{\xi} = \mathbf{0}, \tag{6}$$

where $\mathbf{C}$ is of rank s and consists of row vectors belonging to Y, and successively deduce (5) and (3).

Obviously $\mathbf{CA}^T = \mathbf{0}$.

Equations (3) and (5) are the canonical characterization of non-energic n-ports. It is important to note that not only can $\mathbf{M}$ be time dependent, but also the partitioning of $\boldsymbol{\eta}$ into $\boldsymbol{\eta}_1$ and $\boldsymbol{\eta}_2$, since this partitioning depends on the choice of the non-singular sub-matrix of $\mathbf{A}$ and this sub-matrix may be singular for certain values of t or other variables. The other important point is that the rank s of Y is actually $n-r$. Thus from (3), X is spanned by the vectors of $[\mathbf{1}_r - \mathbf{M}]^T$, where $\mathbf{1}_r$ is the unit matrix of order r, and from (5) Y is spanned by the vectors of $[\mathbf{M}^T \mathbf{1}_{n-r}]^T$ so that $U_n = X \oplus Y$, where U_n is the n-dimensional vector space. In other words a non-energic n-port is characterized by at least n linearly-independent equations in n of the port voltages and currents even when the n-port is non-linear. The qualification "at least" arises from the fact that the n-port itself may impose conditions in addition to non-energicness.

It may seem strange that we do not take it for granted that an n-port should be characterized by n equations which are independent in some sense. It is true that this property does hold for nearly every n-port and that it can be deduced from the basic axiom of the same type as in section 2. However, n-ports that do not have this property, and thus violate the axiom, are known although they have very limited theoretical use.

An example is the nullator, a one-port characterized by two equations, $u = 0$, $i = 0$, and thus also non-energic. We shall return to this subject in section 4.

In mechanics, workless constraints which are linear relations in the virtual displacements, and thus the virtual velocities, are a well-known ingredient of the Lagrangian theory. The forces of constraint which belong to the orthogonal complement of the sub-space of virtual velocities are proportional to linear functions of the Lagrange multipliers. The coefficients in the equations of constraint can for example be functions of the displacements and the velocities [4].

4. SOME EXAMPLES

4.1. The simplest type of non-energic network is an n-port formed by interconnecting the various terminals of the ports.

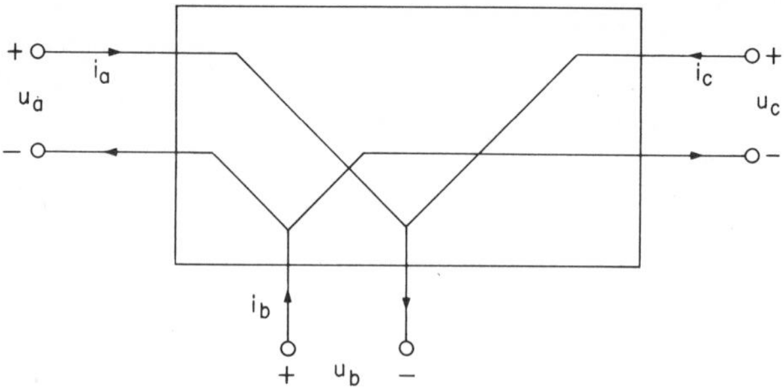

Fig. 2. A non-energic three-port subject only to Kirchhoffian constraints.

The port voltages and currents are then subject only to Kirchhoff's laws and the result is that

$$\mathbf{u}_1 = \mathbf{B}\mathbf{u}_2, \tag{7}$$

$$\mathbf{i}_2 = -\mathbf{B}^T\mathbf{i}_1, \tag{8}$$

where $\mathbf{u}_2$ and $\mathbf{i}_2$ are the vectors of branch voltages and currents of some tree, $\mathbf{u}_1$ and $\mathbf{i}_1$ are the corresponding link voltages and currents, and $\mathbf{B}$ is the basic-loop-link-incidence matrix of the graph formed by the port branches and their interconnexions.

4.2. If the interconnexions contain ideal (lossless) switches, then both the matrix $\mathbf{B}$ and the partitioning and the order of the port vectors $\mathbf{u}_1, \mathbf{i}_1, \mathbf{u}_2, \mathbf{i}_2$ will be in general time dependent.

4.3. The next type of non-energic network results when we consider an n-port formed by an interconnexion of multi-winding ideal transformers [5].

The ideal transformer can be thought of as derived from a system of magnetically coupled coils in various steps of idealization. First the dissipation and capacitive coupling are supposed to be zero. Secondly, the leakage flux is supposed to tend to zero and thirdly the magnetizing inductance is supposed to tend to infinity. If, for example, the windings of a multi-winding transformer are wound on some magnetic structure with relative permeability tending to infinity, then the last two steps of the idealization will be approximately achieved. Thus if the magnetic flux density is finite then the magnetic field strength in the material is zero, so that the line integrals of the magnetic field strength around any loop within the magnetic material is zero, and thus (homogeneous) linear constraints on the port currents apply. Furthermore, the magnetic energy stored in the material is zero. However, the magnetic field strength in the free space surrounding the material cannot be zero everywhere and still satisfy Maxwell's equations. The system is thus only approximately non-energic but the total stored energy averaged over some time interval is in general small compared to the energy transferred between ports over the same time interval.

Neglecting the stored energy we then obtain equations of the form (7) and (8) where $\mathbf{B}$ now is no longer a simple incidence matrix of ones and zeros but is composed of elements from the field of real numbers.

The corresponding n-port in the mechanical context is a system of intermeshing gear wheels, which are massless and capable of frictionless rotation, mounted in massless frames which are also capable of frictionless rotation. See in this connexion the paper by Polder in these symposium proceedings.

In network theory a further generalization has been made by allowing **u**, **i**, **B** to be taken from the field of complex numbers [5].

However, $\mathbf{B}^T$ in (8) is then replaced by $\tilde{\mathbf{B}} = \mathbf{B}^{*T}$, the Hermitian conjugate of B, and (1) is replaced by $\tilde{\boldsymbol{\xi}}\boldsymbol{\eta} = \mathbf{0}$.

4.4. Our next example is an interconnexion of gyrators and ideal transformers. We then obtain equations of the form

$$\begin{bmatrix} 1 & R_{12} & 0 & \cdots & & & & & & & 0 \\ 0 & \cdots & 0 & 1 & 0 & \cdots & 0 & R_{jk} & 0 & \cdots & 0 \\ \mathbf{0} & \cdots & & & & & & & & & \mathbf{A}_m \end{bmatrix} \begin{bmatrix} u_1 \\ i_2 \\ \cdot \\ \cdot \\ \cdot \\ u_j \\ \cdot \\ \cdot \\ i_k \\ \cdot \\ \cdot \\ \mathbf{i}_m \end{bmatrix} = \mathbf{0}, \tag{9}$$

corresponding to (2), and the complementary set

$$\begin{bmatrix} -R_{12} & 1 & 0 & \cdots & & & & & 0 \\ 0 & \cdots & 0 & -R_{jk} & 0 & \cdot & 1 & \cdot & 0 \\ \mathbf{0} & \cdots & & & & & & & \mathbf{C}_m \end{bmatrix} \begin{bmatrix} i_1 \\ u_2 \\ \cdot \\ \cdot \\ i_j \\ \cdot \\ u_k \\ \cdot \\ \mathbf{u}_m \end{bmatrix} = \mathbf{0}, \tag{10}$$

corresponding to (6). Then $\mathbf{A}_m \mathbf{C}_m^T = \mathbf{0}$.

In the mechanical context, such an n-port could be realized by workless interconnexions of ideal gear wheels and gyrostats or by combinations of electromechanical and magneto-mechanical transducers. Note, that once the gyrator or gyrostat are accepted as elements that can be freely interconnected, the distinction between voltage and current or force and velocity as system variables disappears. We can of course still measure the voltage or current at some location in the system by inserting a measuring instrument in the form of a one-port with magnitude indication, in which the voltage and current are distinguishable, but the distinction between "through"- and "across"-variables has vanished.

4.5. The previous examples exhaust the possibilities of what one can do with linear passive elements. In other words it can be shown that any non-energic linear n-port is equivalent to a network of one of the types already discussed. If we want a new situation we must consider non-linear non-energic n-ports.

The simplest examples are the non-linear three-ports known as the conjunctor and the traditor [3, 6, 7]. Although at first they may seem to be rather strange devices and it may even seem stranger to regard them as basic network elements, they have in fact passed all the commonly used theoretical tests and there is no known reason for not accepting them as equally well as one accepts the ideal transformer and the gyrator [3].

Their defining equations are:

conjunctor:

$$u_1 = Bi_2 i_3, \quad u_2 = -Bi_1 i_3, \quad u_3 = 0, \tag{11}$$

where B is a constant.

traditor:

$$u_1 = -Aq_2 i_3, \quad u_2 = -Aq_1 i_3, \quad u_3 = A(q_1 i_2 + q_2 i_1), \tag{12}$$

where A is a constant.

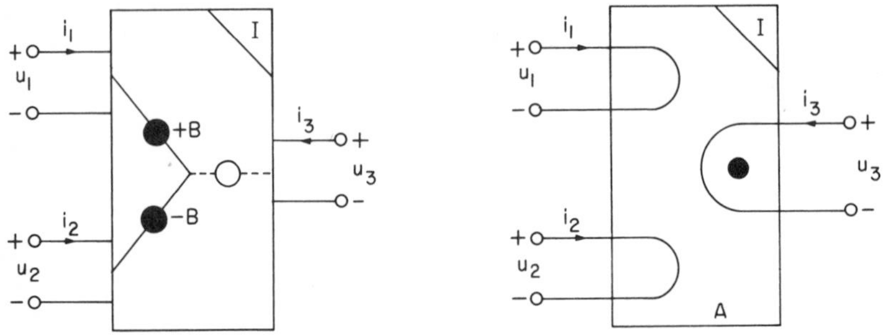

Fig. 3. The conjunctor (left) and traditor (right).

The conjunctor is not only non-energic but also has a non-energic port. We can think of it as a gyrator two-port in which the gyrational resistance can be controlled by a current through a third port. This control does not demand any energy consumption.

Equations (11) actually define a conjunctor of type I. Other types of conjunctor follow by interchanging voltages and the corresponding currents. This can be effected by connecting gyrators to the ports.

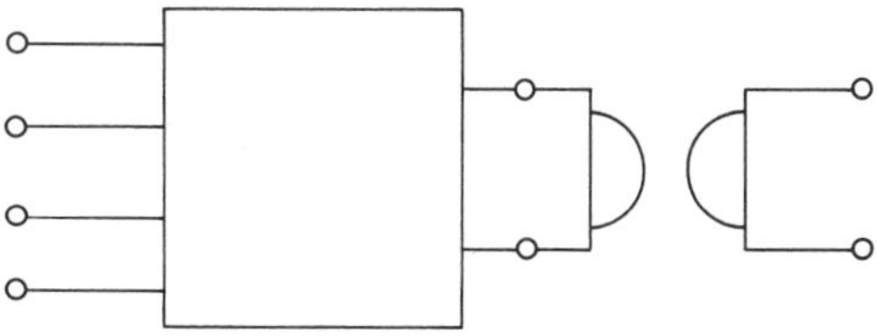

Fig. 4. Interchange of port voltage and current by means of a gyrator.

In order to obtain the equations in the form (3) and (5) we can write

$$\begin{bmatrix} 1 & -Bi_3 & 0 \\ 0 & 0 & 1 \end{bmatrix} \begin{bmatrix} u_1 \\ i_2 \\ u_3 \end{bmatrix} = \mathbf{0}$$

$$\begin{bmatrix} Bi_3 & 1 & 0 \end{bmatrix} \begin{bmatrix} i_1 \\ u_2 \\ i_3 \end{bmatrix} = \mathbf{0},$$

or equivalently

$$\begin{bmatrix} u_1 \\ u_3 \end{bmatrix} = \begin{bmatrix} Bi_3 \\ 0 \end{bmatrix} [i_2] \tag{13}$$

$$[u_2] = [-Bi_3 \quad 0] \begin{bmatrix} i_1 \\ i_3 \end{bmatrix}. \tag{14}$$

We can regard the traditor also as derived from a specially controlled gyrator, namely formed by connecting one port of a gyrator with charge-controlled gyrational resistance in series with a port of a similar gyrator.

Thus

$$u_1 = -Aq_2 i_3, \qquad u_2 = -Aq_1 i_3,$$
$$u_3' = Aq_2 i_1, \qquad u_3'' = Aq_1 i_2,$$
$$u_1 = u_1' + u_3''.$$

The rather curious property that the gyrational resistance of each gyrator should be controlled by the charge (time integral of the current) of one port of the other gyrator makes an effective physical realization of this element rather difficult, although recent technological developments do indicate considerable promise for the future [8].

Equations (12) actually describe a traditor of type I. Other types follow by suitably connecting gyrators to the ports, thereby interchanging the port voltage and current or charge and flux.

In order to obtain the equations in the form (3) and (5) we write

$$\begin{bmatrix} 1 & 0 & Aq_2 \\ 0 & 1 & Aq_1 \end{bmatrix} \begin{bmatrix} u_1 \\ u_2 \\ i_3 \end{bmatrix} = \mathbf{0}.$$

$$[Aq_2 \quad Aq_1 \quad -1] \begin{bmatrix} i_1 \\ i_2 \\ u_3 \end{bmatrix} = \mathbf{0},$$

or equivalently

$$\begin{bmatrix} u_1 \\ u_2 \end{bmatrix} = \begin{bmatrix} -Aq_2 \\ -Aq_1 \end{bmatrix} [i_3], \tag{15}$$

$$[u_3] = [Aq_2 \quad Aq_1] \begin{bmatrix} i_1 \\ i_2 \end{bmatrix}. \tag{16}$$

The traditor and the conjunctor have been proposed as useful elements for generating, in conjunction with the energic elements, various non-linear characteristics. Indeed, there are good reasons for supposing that no other element type is needed for generating n-port characteristics that can be expressed as rational functions of currents, voltages, charges and fluxes [6, 7]. In this sense these two non-linear elements together with the linear elements form a complete set. Furthermore, all the common amplifier characteristics can be generated in this way, as this set of elements, although globally passive, exhibits the property of local activity [6, 7]. The other interesting point to note here is that the traditor is derivable from a Lagrangian, viz. $Aq_1q_2\dot{q}_3$, whereas the conjunctor is not. However, recently a formalism, similar to the Lagrangian, has been given which includes the conjunctor [9].

We conclude this section with an example of modelling a mechanical system in which the traditor can usefully be employed.* Consider the possible motion of a massless, rigid rod which is constrained to move in a plane. If the coordinates of the end points of the rod are (x_1, y_1) and (x_2, y_2), then the constraints are

$$(x_1 - x_2)^2 + (y_1 - y_2)^2 = r^2$$

and

$$(x_1 - x_2)(\dot{x}_1 - \dot{x}_2) + (y_1 - y_2)(\dot{y}_1 - \dot{y}_2) = 0.$$

These are non-linear constraints; furthermore the rod is non-energic. It readily follows that the forces act along the rod and are oppositely directed [4]. The

*The problem of finding passive electrical analogues for a system of massless, rigid rods and massive particles constrained to coplanar motion was proposed by Franksen during the symposium.

corresponding traditor analogue, based on the correspondence force ↔ voltage, velocity ↔ current is shown in Fig. 5.

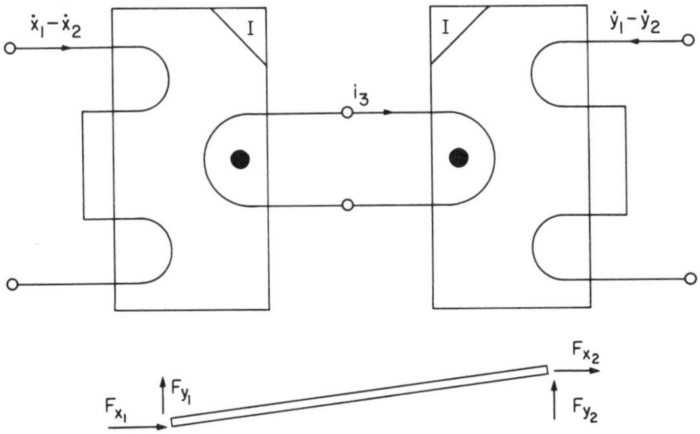

Fig. 5. Electrical network analogue of an ideal massless and infinitely rigid rod in two-dimensional motion. The port voltages are $F_{x_1} = F_{x_2}$ and $F_{y_1} = F_{y_2}$ respectively.

Two evident extensions can be made: first to the problem of three-dimensional motion by including an extra traditor with the "dot port" in series with the "dot ports" of the other two traditors. Secondly, the inclusion of massive particles in the analogue is readily effected as shown in Fig. 6, where the traditor combination is simply represented by the three-port "rod", from which any appropriate mechanical network of such devices, complete with the boundary conditions, can be built up.

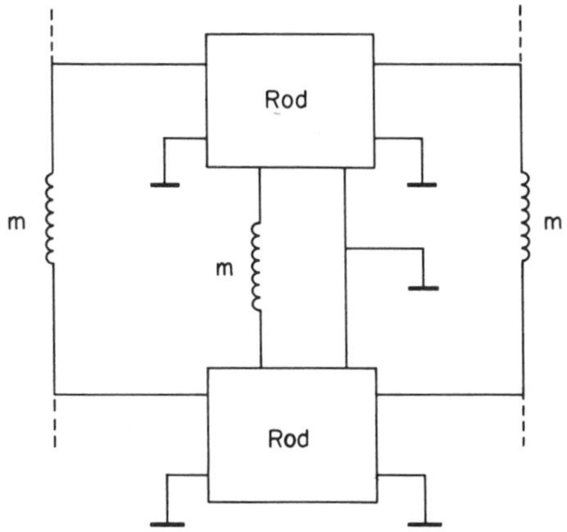

Fig. 6. Network analogue of a system of masses and rigid rods in three-dimensional motion.

This does not mean to say that a practical analogue of this mechanical system should necessarily be built in this way. A practical analogue will depend on the physical building blocks available, and on the combinations permitted which can result in physical subsystems that are good approximations to the idealized systems to be simulated [10].

4.6. The important point to be noted, which all the preceding discussion implies, is that, when analysing a system, it is often profitable to isolate the purely energy storing and dissipation functions – which are associated with one-ports – from the transmission and constraint functions which are purely non-energic. Any passive network is, for example, to be regarded as a non-energic n-port in which the ports are terminated by the elementary energy storage, dissipation or source one-ports discussed in Section 2 [5]. Of course, one is free in analysis to follow the Lagrangian approach in which the non-energic elements and constraints are eliminated automatically by the analysis. In synthesis and modelling, however, such an approach often involves loss of information and insight into the structure.

5. ENERGIC CONSTRAINTS – THE NULLOR

The next problem we have to consider is the inclusion of active elements in the theory. There are various definitions of active elements and devices in vogue which lead to different results. In the case of passive elements, however, there is generally no difficulty arising as a result of the various definitions. The definition we shall adopt is the following:

A passive n-port is an n-port in which the energy that the n-port is capable of delivering through its ports to its environment does not exceed the net energy that the n-port has absorbed from the environment through its ports during its previous history.

Thus

$$\int_{-\infty}^{t} \mathbf{u}(\tau)^T \, \mathbf{i}(\tau) \, d\tau \geqslant 0,$$

where the polarity and direction conventions of Fig. 1 apply for each port, and $\mathbf{u}(t)$ and $\mathbf{i}(t)$ are any voltage and current vectors simultaneously admitted by the n-port.

An active n-port is then an n-port which is capable of delivering more energy through its ports to its environment than it has already received through its ports in its previous history. An important point about these definitions is that what one regards as active or passive is largely a question of how one defines the boundaries of the system and its environment. Thus according to this definition, a lead-acid accumulator is a passive electrical one-port but a dry torch battery is an active electrical one-port.

Another effect is the distinction between local and global activity. Thus a transistor is passive when regarded globally as a non-linear device, but if one considers the relations between small signals superimposed on a fixed operating (bias) point, as defining the device, then it is (locally) active.

With these remarks and their physical significance in mind, we can now go on to considering concrete active elements. It appears that we now have a considerable choice of devices that can be idealized in various ways, leading to a wide range of active elements that are not all independent. We shall not reproduce the arguments that have led to all these elements, the discussion of their properties and relative advantages. Instead we shall concentrate on one linear active element from which all other linear active elements can be derived by combinations with the accepted passive elements.

This element was originally called the ideal amplifier [1], at a later stage the unitor [11], and most recently the nullor [12]. It is a two-port defined by the equations

$$\begin{aligned} u_1 &= 0 \\ i_1 &= 0. \end{aligned} \tag{17}$$

This means that u_2 and i_2 are arbitrary. At first sight this seems a strange sort of concept to be regarded as network element. However, it is in fact closely approximated by the modern differential operational amplifier. In such an amplifier the input impedance is extremely high, so that equating the input current to zero seems reasonable. Secondly the gain is very high, so that with an output voltage within the normal dynamic range the input voltage will be very small and thus equating this voltage to zero seems reasonable. The only problem is how to deal with the arbitrary output current and voltage.

Now one can show on theoretical grounds [13] that the nullor can be used only if it is embedded in some other network that admits a transmission path between port 2 and port 1 or is itself a nullor with port 2 as non-energic port. This is precisely the condition that electronic practice prescribes. The operational amplifier is always used in conjunction with some feedback network; otherwise the gain will not be sufficiently stabilized. Thus we can conclude that the nullor is under normal working conditions approximated very well by an accepted and reliable technological device. The theoretical restriction on its use is not really different in kind from the restriction imposed on the voltage and current sources, and thus it is perfectly acceptable as network element. It is rather special in that it is characterized by the constant zero instead of a parameter. For this reason we prefer to call the nullor an active constraint.

Since there is apparently no connexion between the ports, it is tempting to split the nullor into two one-ports which are called the nullator (non-energic) and the norator (energic) respectively. However, it has been shown [13] that any combination of nullators and norators with the accepted network elements is an acceptable network if and only if the nullators and norators can be grouped so as to form nullors which in turn satisfy the feedback restriction. The only point to be gained by using the nullator and the norator – and it is very marginal – is that from a given realization of a circuit involving more than one nullor, other realizations can be immediately deduced by regrouping the nullator–norator combinations. This can sometimes be helpful in electronic design.

As far as modelling other systems is concerned, the nullor concept can be useful provided one remembers that every active system really portrays a special aspect of a more complete system that is really passive. Activity implies some form of energy conversion, usually via a non-linear characteristic, and if the details of this conversion are not important in the problem under investigation, then the system may be modelled by active elements. One must, however, be very clear about what is to be understood by this term.

6. THE SCATTERING MATRIX [5]

In many problems connected with wave and information transmission the scattering matrix is the most suitable tool for analysis and synthesis of systems. The situation for a linear two-port network is that shown in Fig.7.

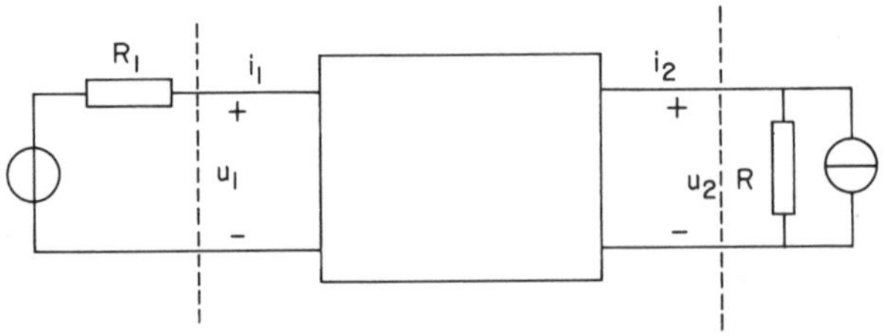

Fig. 7. Configuration for the definition of the scattering matrix.

A linear two-port transmission network is terminated by two active one-ports. Under sinusoidal or d.c. conditions these one-ports have an equivalent Thévenin or Norton circuit as shown. If such a one-port is terminated by a passive load, then the average power absorbed by the load is a maximum if the load impedance is purely resistive and is equal to the internal resistance of the active one-port. This power absorbed by the load is then called the maximum available power deliverable by the one-port.

The scattering matrix is a means of specifying the properties of the transmission two-port in such a way that the maximum available power of the energy sources appears in a very natural manner in the equations.

We write

$$\mathbf{b} = \mathbf{S}\mathbf{a}$$

where **a** is called the incident and **b** the reflected wave vector and **S** is the scattering matrix of the two-port. We define **a** and **b** as

$$\mathbf{a} = \frac{1}{2}\begin{bmatrix} \dfrac{U_1}{\sqrt{R_1}} + \sqrt{R_1}I_1 \\ \dfrac{U_2}{\sqrt{R_2}} + \sqrt{R_2}I_2 \end{bmatrix}$$

$$\mathbf{b} = \frac{1}{2}\begin{bmatrix} \dfrac{U_1}{\sqrt{R_1}} - \sqrt{R_1}I_1 \\ \dfrac{U_2}{\sqrt{R_2}} - \sqrt{R_2}I_2 \end{bmatrix}$$

where $|U_1|$, $|U_2|$ and $|I_1|$, $|I_2|$ etc. are the voltage and current amplitudes (not r.m.s. values).

The average power absorbed by the two-port is then

$$\frac{1}{2}\operatorname{Re}(\tilde{\mathbf{a}}\mathbf{a} - \tilde{\mathbf{b}}\mathbf{b})$$

where ~ denotes the complex conjugate of the transposed matrix.

If port 2 of the two-port is simply terminated by the resistance R_2, then $|S_{21}|^2$ is equal to the ratio of the (average) power transmitted to this resistance to the maximum available (average) power of the source at port 1. This is easily seen if we note that the strength of the voltage source associated with port 1 is simply $2\sqrt{R_1}a_1$. Under the same conditions $|S_{11}|^2$ represents the fraction of the available power that is not absorbed by port 1, i.e. is "reflected" back into the source.

This simple situation characterized by a scattering matrix can be generalized in various ways [5]. First, the extension of the above considerations to n-ports is immediate. Secondly, one can develop a theory based on complex source impedances, provided all currents and voltages are sinusoidal. Thirdly, one can consider signals with a time dependence of the form exp (λt) and still retain all essentials of the theory provided the source impedances are real. Finally, one can develop a time-domain theory with retention of the essential features of the simple model discussed here.

The advantages of using the scattering formalism are many, for example:

(*a*) For passive networks the scattering matrix always exists, in contrast to the impedance, admittance and chain matrices. In the case of active networks the scattering matrix exists nearly always for certain isolated choices of the source resistances. However, the nullor does not possess a scattering matrix. This is not surprising since the nullor terminated by one-ports is not an acceptable network.

(*b*) In the case of passive networks the matrix $1_n - \tilde{\mathbf{S}}\mathbf{S}$ is positive semi-definite, where 1_n is the unit matrix of order n. This means that the elements of **S** are bounded and lie within the unit circle: $|S_{ij}| \leqslant 1, i, j = 1, \ldots, n$. If the network is also lossless, **S** is unitary.

(*c*) Lossless two-ports have a very important sensitivity property, viz, in the neighbourhood of the real-frequency zeros of S_{11}, the characteristic $|S_{21}|^2$ has a very low sensitivity to variations in the element values of the lossless two-port and to variations in the source and load resistances. The sensitivity is actually zero at the real-frequency zeros of S_{11}.

It is as a result of this last property that high-quality filter circuits can be built. The same quality is not attainable when one attempts to use other network configurations that do not consist of a lossless two-port terminated in resistances [14]. It must be noted, however, that not every realization of a desired characteristic by a lossless two-port is suitable. The accuracy with which the zeros of S_{21} can be established also enters into the picture. These properties of classical filter networks have recently been exploited in the theory of digital filter processing by constructing digital networks that imitate structurally the classical filter networks [15].

As far as the author is aware the scattering matrix was originally developed in nuclear physics for studying the collision and scattering of particles moving in the neighbourhood of atomic nuclei. However, some of the basic ideas go back to the early history of wave optics, and indeed the scattering formulism is nearly always relevant when energy transmission by wave propagation is being discussed. Before its introduction into classical network theory, the scattering matrix was applied extensively in electromagnetic problems such as waveguide networks and filters.

Seeing its wide applications to date and the close relation to the central theme of this symposium, one would expect that the scattering concept should find useful applications in other fields such as thermodynamics and economics.

7. POTENTIAL FUNCTIONS

Potential functions such as the Lagrangian, the content, the co-content, the Brayton–Moser mixed potential as well as the energy and co-energy are discussed elsewhere in these proceedings and will consequently only be very briefly mentioned here. There are, however, two simple properties of potential functions of networks which we should specially like to emphasize.

The first is a form of the reciprocity principle which is valid for functions with continuous second partial derivatives.

Thus

$$\frac{\partial^2 P}{\partial x_i \partial x_j} = \frac{\partial^2 P}{\partial x_j \partial x_i} \tag{16}$$

where $P(x_i)$ $(i = 1, \ldots, n)$ is any potential function of the "coordinates" x_i and with continuous partial derivatives. For linear n-ports described by a content or co-content function, this means that the n-port satisfies the reciprocity principle for currents and voltages. Then impedance (content) and admittance (co-content) matrices respectively exist and are symmetric. For non-linear networks the small-signal departures from some equilibrium state satisfy the same principle. Thus neither a gyrator nor a conjunctor of type I possesses a content or co-content function.

In the case of the Brayton–Moser mixed potential applied to linear n-ports, the associated matrix of a linear n-port is a symmetric modified hybrid n-port-matrix. The analysis leads to the same conclusion as above.

Similarly the existence of the energy and co-energy functions implies reciprocity relations for the voltages and charges on the one hand and for the currents and fluxes on the other.

In the case of the Lagrangian the situation is more complicated. The consequences of eqn. (16) are not readily expressible in terms of simple relations but involve terms with both forces and displacements as well as velocities and momenta. We know, however, that no reciprocity of force and displacement or force and velocity is implied, since non-reciprocal systems can be represented by a Lagrangian function.

As far as the author is aware no set of necessary and sufficient conditions for the existence of the Lagrangian, and thus for the validity of eqn. (16), is known. We do know, however, that any conservative system describable by analytic relations has an associated Lagrangian function, which, however, does not completely determine the system equations [2, 3].

The second important property of these potential functions is that of additivity [3]. When a system is constructed by interconnecting various sub-systems subject to Kirchhoff's laws, then except in one special case, the potential function of the whole system is equal to the sum of the potential functions of the sub-systems.

The exceptional case is that the additivity property does not apply to those terms of the Lagrangian function which are linear in the velocities. All other terms of the Lagrangian function possess the additivity property.

8. TELLEGEN'S THEOREM IN NEW CONTEXTS

We have already discussed in Section 2 the important consequences of considering a set of all admissible voltages belonging to a sub-space that is the orthogonal complement of the sub-space which is spanned by all admissible current vectors. In fact Tellegen's theorem is simply another way of stating this result since any network can be regarded as consisting of an n-port of interconnexions (and is thus non-energic) in which the ports are terminated by various one-, two- or multi-ports whose ports can be identified as branches. The n-port of interconnexions then guarantees that Kirchhoff's laws are satisfied by the "branch" voltages and currents. This idea of orthogonal sub-spaces is extremely useful and keeps on appearing in all kinds of forms and places in mathematics. The applications to network theory are very numerous, as has been shown in a recent monograph [16]. Here we simply wish to mention a recent application of the type of result of which Tellegen's theorem is a prototype.

In a signal-flow network in which signals are combined linearly, one can divide the signals into two sets, viz, those flowing into a node and those flowing out of a node. Then there exists a certain simple relation between these variables and the corresponding variables of a signal-flow network with complementary topology. This relation corresponds to the difference form of Tellegen's theorem of classical networks. From this result similar results to the well-known properties of classical networks, such as conservation theorems, relations between sensitivity and energy and inter-reciprocity can be derived.

Considering the discussions on models of economic systems held during the Twente Symposium, it should be possible to apply some of these ideas fruitfully to economic problems.

9. CONCLUSION

In this paper we have attempted to present some network theoretical ideas which have or could be expected to have useful applications in other fields. These ideas constitute only a small part of the basis of classical network theory. That does not mean to say that what we have not mentioned is not relevant in applications, but only that the direct relevance did not manifest itself during the symposium. A general rule is that what has been found especially useful in one field is almost sure to find applications in another.

REFERENCES

1. Tellegen, B. D. H. "La recherche pour une série complète d'éléments de circuit idéaux non-linéaires". *Rendiconti del seminario Matematico e Fisico di Milano, 1953–54, 25*, 1955, 134–144.

2. Birkhoff, G. D. *Dynamical Systems*, ch. 1. Am. Math. Soc., New York, 1927.
3. Adams, K. M. "Some basic concepts in nonlinear network theory", in G. Biorci (ed.) *Network and Switching Theory*. Academic Press, New York, 1968, pp. 361–381.
4. Pars, L. *A Treatise on Analytical Dynamics*, chs. 2, 6. Heinemann, London, 1965.
5. Belevitch, V. *Classical Network Theory*, chs. 1, 4, 6. Holden-Day, San Francisco, 1968.
6. Duinker, S. "Traditors, a new class of non-energic non-linear network elements". *Philips Res. Rep. 14,* 1959, 29–51.
7. Duinker, S. "Conjunctors, another new class of non-energic, non-linear network elements". *Philips Res. Rep. 17,* 1962, 1–19.
8. Deprettere, E. "Electronic realization of non-energic linear and non-linear network elements". *Int. J. Electronics, 34,* 1973, 237–240.
9. Deprettere, E. "On the synthesis and electronic realization of non-energic linear and non-linear network elements". *Digest of Technical Papers, London 1971 I.E.E.E. International Symposium on Electrical Network Theory*, I.E.E.E., New York, 1971.
10. Adams, K. M. and Morris, R. A. "An analogue computer for the simulation of the behaviour of structures during earthquakes". *J. Electr. Control, 12,* 1962, 143–176.
11. Keen, A. W. "A topological non-reciprocal network element". *Proc. I.R.E., 47,* 1959, pp. 1148–1150.
12. Carlin, H. J. and Youla, D. C. "Network synthesis with negative resistors". *Proc. I.R.E., 49,* 1961, 907–920.
13. Tellegen, B. D. H. "On nullators and norators". *I.E.E.E. Trans. Circuit Theory,* CT–13, 1966, 467–469.
14. Neirynck, J. and Thiran, J. P. "Sensitivity analysis of lossless two-ports by uniform variation of the element values". *Electron. Letters, 3,* 1967, 74–76.
15. Sedlmeyer, A. and Fettweis, A. "Digital filters with true ladder configuration". *Int. J. Circuit Theory and Applications, 1,* 1973, 5–10.
16. Penfield, P., Spence, R., and Duinker, S. *Tellegen's Theorem and Applications,* M.I.T. Press, Cambridge, Mass., 1970.
17. Fettweis, A. "Some general properties of signal-flow networks", in J. W. Skwirzynski and J. O. Scanlan (eds.) *Signal and Network Theory*. Peregrinus, London, 1973, pp. 48–59.
18. Oster, G. F. and Desoer, C. A. "Tellegen's Theorem and thermodynamic inequalities". *J. Theor. Biol. 32,* 1971, 219–241.

NOMENCLATURE

u	port voltage
i	port current
$\mathbf{u}$	vector of port voltages
$\mathbf{i}$	vector of port currents
ξ	a voltage or current
$\boldsymbol{\xi}$	vector of voltages or (and) currents
η	a voltage or current complementary to ξ
$\boldsymbol{\eta}$	vector orthogonal to $\boldsymbol{\xi}$

$\mathbf{A}$	matrix with row vectors orthogonal to $\boldsymbol{\eta}$
$\mathbf{C}$	matrix with row vectors orthogonal to $\boldsymbol{\xi}$
$\mathbf{B}$	basic-loop incidence matrix with the unit matrix removed
$\tilde{\mathbf{B}}$	the Hermitian conjugate of $\mathbf{B}$
$\mathbf{B}^T$	the transpose of $\mathbf{B}$
A, B	scalar constants
q	charge, time integral of current
x, y	spatial coordinates
m	mass
F	force
$\mathbf{S}$	scattering matrix
$\mathbf{a}, \mathbf{b}$	wave variables (vectors)
R	normalizing or reference resistance
P	a potential function

KRON'S WAVE AUTOMATON

J. W. Lynn *and* R. A. Russell
Department of Electrical Engineering and Electronics
University of Liverpool
England

SUMMARY

Gabriel Kron developed a wave automaton, energized by multi-dimensional magnetohydrodynamic waves, and envisaged its use in modelling partial differential equations, adaptive control systems oscillating crystals and brain function. This paper presents the background of orthogonal network theory and some of the structural features of the wave automaton.

1. INTRODUCTION

For a decade (1958–68) Gabriel Kron [1], developed and used the self-organizing electromagnetic polyhedral wave automaton. He has published results obtained with this multi-dimensional network model, for the case of surface fitting with discrete data [2]. Additional applications are suggested in his published papers, for example the solution of non-linear partial differential equations, integration theory, adaptive control, crystal studies and models of some of the self-adaptive features of brain function [3].

As with a good deal of Kron's work, the wave model is at first sight a simple device and it is only when the attempt is made to construct and operate such a model, for an actual data system, that the scope of the device becomes clear.

Kron has never published details of his method of making the polyhedron self-organizing, although his published results show that in this state it has some remarkable properties, associated with harmonic integrals on multiply connected spaces.

In the present paper the self-organizing nature of the device is not investigated. The paper describes the background of generalized network theory which led Kron to the construction of the dualistic field system, and goes on to discuss the structural features of the model and the technique of computing some values in different parts of the system.

Basically, electromagnetic wave vectors are here represented on two spatially orthogonal networks (Fig. 1). Together these carry eight quantities, $\mathbf{e}\ \hat{\mathbf{b}}\ \dot{\mathbf{d}}\ \hat{\mathbf{h}}$ and $\hat{\mathbf{E}}\ \dot{\mathbf{B}}\ \hat{\mathbf{D}}\ \mathbf{H}$ where the lower case letters represent conventional wave quantities and the capital letters represent electrical (and magnetic) charges and current. The "capped" quantities flow on the dual network.

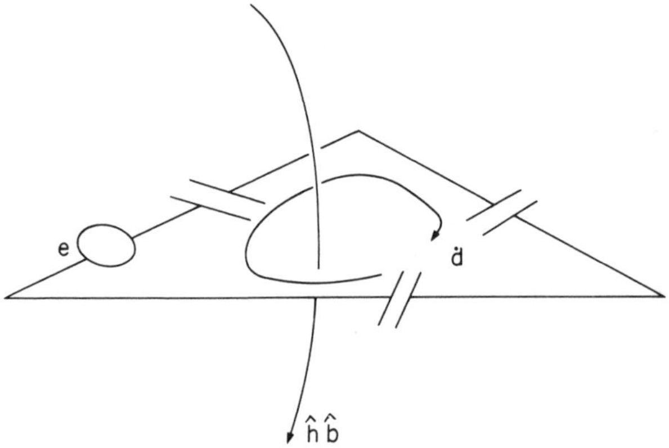

Fig. 1. Electromagnetic orthogonality.

The wave system satisfies Maxwell's equations and the theorems of Stokes, Green, Ampere, etc., for electromagnetic fields. When the charges and fields are coupled by velocity and acceleration effects, self-oscillation can occur and multiple feed-back gives the self-optimizing condition.

2. ORTHOGONAL NETWORKS [4]

In the context of electrical networks, the word "orthogonal" refers to a mathematical relationship between transformation matrices associated with open and closed paths on the network. Thus, one may have current vectors which are orthogonal, on the same network. Later these "orthogonal" currents will be considered to flow in dual circuits which are orthogonal in the conventional sense.

In Kron's network analysis, a circuit of connected branches may be characterized by its singular connection matrices $\mathbf{C}_{cl}$ and $\mathbf{A}^{op}$. These matrices show the relationship between the unconnected, oriented primitive branches and selected closed and open paths traced out along the branches of the connected network.

The matrix $\mathbf{C}_{cl}$ is set up by equating the branch and mesh currents (through-variables) using Kirchhoff's laws. Similarly the matrix $\mathbf{A}^{op}$ relates branch voltages to the selected open-path voltages (across-variables). The singular connection matrices are related by

$$(\mathbf{A}^{op})^T \mathbf{C}_{cl} = \mathbf{O} \tag{1}$$

The use of these singular matrices is restricted, since they have no inverse. The orthogonal theory developed by Kron introduced square non-singular connection matrices. The rectangular matrix $\mathbf{C}_{cl}$ now has extra columns added to include current flowing in open paths, in addition to the closed path currents. The branch current is then

$$(\mathbf{i} + \mathbf{I}) = \mathbf{C}(\mathbf{i}' + \mathbf{I}') \tag{2}$$

where **C** is now square, since

$$\text{no. of branches} = \text{no. of open paths} + \text{no. of closed paths} \tag{3}$$

In terms of singular connection matrices the circuit relations between primitive (branch) and (primed) connected path values can be written

$$\begin{array}{llll} \text{Closed paths} & \mathbf{i} = \mathbf{C}\mathbf{i}' & \text{Open paths} & \mathbf{V} = \mathbf{A}\mathbf{V}' \\ & \mathbf{v}' = \mathbf{C}^T\mathbf{v} & & \mathbf{I}' = \mathbf{A}^T\mathbf{I} \\ & \mathbf{v} = \mathbf{Z}\mathbf{i} & & \mathbf{I} = \mathbf{Y}\mathbf{V} \\ & \mathbf{v}' = \mathbf{Z}'\mathbf{i}' & & \mathbf{I}' = \mathbf{Y}'\mathbf{V}' \\ & \mathbf{Z}' = \mathbf{C}^T\mathbf{Z}\mathbf{C} & & \mathbf{Y}' = \mathbf{A}^T\mathbf{Y}\mathbf{A} \end{array} \tag{4}$$

When the non-singular **C** is written for the network, with currents flowing in both open and closed paths

$$\mathbf{A}^T = \mathbf{C}^{-1}$$

$$\mathbf{A}^T\mathbf{C} = \mathbf{1} \tag{5}$$

also

$$(\mathbf{A}^{cl})^T\mathbf{C}_{cl} = \mathbf{1} \quad (\mathbf{A}^{cl})^T\,\mathbf{C}_{op} = \mathbf{0}$$

$$(\mathbf{A}^{op})^T\mathbf{C}_{cl} = \mathbf{0} \quad (\mathbf{A}^{op})^T\mathbf{C}_{op} = \mathbf{1}$$

hence the designation "orthogonal" given to the relationship between currents in the open and closed paths.

The circuit equations in terms of quantities in both open and closed paths become

$$(\mathbf{v}' + \mathbf{V}') = \mathbf{Z}'(\mathbf{i}' + \mathbf{I}') \tag{6}$$

where

$$\mathbf{Z}' = \mathbf{C}^T\mathbf{Z}\mathbf{C}, \text{ etc.} \tag{7}$$

and

$$(\mathbf{i} + \mathbf{I}) = \mathbf{C}(\mathbf{i}' + \mathbf{I}') \tag{8}$$

Similarly, the equation may be written

$$(\mathbf{i}' + \mathbf{I}') = \mathbf{Y}'(\mathbf{v}' + \mathbf{V}') \tag{9}$$

$$\mathbf{Y}' = \mathbf{A}^T\mathbf{Y}\mathbf{A}, \text{ etc.} \tag{10}$$

3. DUALITY

Having demonstrated the advantages of the use of connection matrices on networks of one-dimensional elements, Kron then showed that exactly the same relationships hold between unconnected branches or cells and connected paths (chains) of *n*-dimensional elements, for example, two-dimensional surfaces connected together into a tetrahedron, or the four zero-dimensional vertices of the tetrahedron. Furthermore, the concepts of open and closed chains now embrace sets of zero-dimensional elements as well as those of higher dimension.

It is now found that the incidence matrices for the structure are given in terms of the **C** and **A** matrices in various dimensions, by relationships such as

$$\mathbf{M}_1^0 = \mathbf{C}_{cl}^0\,(\mathbf{A}_1^{op})^T \tag{11}$$

$$\mathbf{M}_2^1 = \mathbf{C}_{cl}^1\,(\mathbf{A}_2^{op})^T, \text{ etc.} \tag{12}$$

The remaining parts of the square **C** and **A** matrices give similar incidence matrices for a dual structure in which the cells of dimension k are orthogonal to those of dimension $(n-k)$ on the primal network.

Referring to Fig. 2b

$$\mathbf{C}^0 = \begin{array}{c} \\ W \\ X \\ Y \\ Z \end{array}\!\!\begin{array}{c} \begin{array}{cccc} p & q & r & op \end{array} \\ \left[\begin{array}{cccc} 1 & \cdot & \cdot & 1 \\ -1 & -1 & -1 & 1 \\ \cdot & 1 & \cdot & 1 \\ \cdot & \cdot & 1 & 1 \end{array}\right] \end{array} \qquad \mathbf{A}_0 = \begin{array}{c} \\ W \\ X \\ Y \\ Z \end{array}\!\!\begin{array}{c} \begin{array}{cccc} p & q & r & op \end{array} \\ \left[\begin{array}{cccc} 3 & -1 & -1 & 1 \\ -1 & -1 & -1 & 1 \\ -1 & 3 & -1 & 1 \\ -1 & -1 & 3 & 1 \end{array}\right] \end{array} \frac{1}{4}$$

$$\mathbf{C}^1 = \begin{array}{c} \\ 1 \\ 2 \\ 3 \\ 4 \\ 5 \\ 6 \end{array}\!\!\begin{array}{c} \begin{array}{cccccc} a & b & c & p & q & r \end{array} \\ \left[\begin{array}{cccccc} 1 & 1 & \cdot & 1 & 1 & 1 \\ \cdot & -1 & \cdot & \cdot & \cdot & \cdot \\ -1 & \cdot & \cdot & \cdot & \cdot & \cdot \\ \cdot & 1 & -1 & \cdot & 1 & \cdot \\ 1 & \cdot & 1 & \cdot & \cdot & 1 \\ \cdot & \cdot & -1 & \cdot & \cdot & \cdot \end{array}\right] \end{array} \qquad \mathbf{A}_1 = \begin{array}{c} \\ 1 \\ 2 \\ 3 \\ 4 \\ 5 \\ 6 \end{array}\!\!\begin{array}{c} \begin{array}{cccccc} a & b & c & p & q & r \end{array} \\ \left[\begin{array}{cccccc} \cdot & \cdot & \cdot & 1 & \cdot & \cdot \\ \cdot & -1 & \cdot & \cdot & 1 & \cdot \\ -1 & \cdot & \cdot & \cdot & \cdot & 1 \\ \cdot & \cdot & \cdot & -1 & 1 & \cdot \\ \cdot & \cdot & \cdot & -1 & \cdot & 1 \\ \cdot & \cdot & -1 & \cdot & -1 & 1 \end{array}\right] \end{array}$$

and

$$\mathbf{M}_1^0 = \mathbf{C}_{cl}^0\,(\mathbf{A}_1^{op})^T = \begin{array}{c} \\ W \\ X \\ Y \\ Z \end{array}\!\!\begin{array}{c} \begin{array}{cccccc} 1 & 2 & 3 & 4 & 5 & 6 \end{array} \\ \left[\begin{array}{cccccc} 1 & \cdot & \cdot & -1 & -1 & \cdot \\ -1 & -1 & -1 & \cdot & \cdot & \cdot \\ \cdot & 1 & \cdot & 1 & \cdot & -1 \\ \cdot & \cdot & 1 & \cdot & 1 & 1 \end{array}\right] \end{array}$$

Here $\mathbf{C}_{cl}^0$ describes "closed chains" in the zero-dimension. This simply means closed groups of points – here taken as pairs of points. An open chain would be the full set of points.

The dual network has been defined in geometrical terms by Veblen [5] and from this it is seen that

$$\underset{\text{(dual)}}{\left(\widetilde{\mathbf{M}}_k^{k-1}\right)^T} = \underset{\text{(primal)}}{\mathbf{M}_{(n-k+1)}^{n-k}} \tag{13}$$

where the cells are of dimension k on an n-dimensional structure. For example, in a 3-dimensional space the tetrahedron will have incidence matrices $\mathbf{M}_1^0, \mathbf{M}_2^1, \mathbf{M}_3^2$; $k = 1, 2, 3$ and $n = 3$.

In terms of equations (11), (12) and (13)

$$\mathbf{M}^{n-k}_{(n-k+1)} = \mathbf{C}^{n-k}_{cl}\left(\mathbf{A}^{op}_{(n-k+1)}\right)^T \tag{14}$$

$$\left(\tilde{\mathbf{M}}^{k-1}_{k}\right)^T = \left(\tilde{\mathbf{C}}^{k-1}_{cl}\,\tilde{\mathbf{A}}^{op\,T}_{k}\right)^T = \tilde{\mathbf{A}}^{op}_{k}\,\tilde{\mathbf{C}}^{k-1\,T}_{cl} \tag{15}$$

The dual connection matrix may now be written down by inspection, since

$$\begin{array}{ll} \mathbf{C}^{n-k}_{cl} \rightarrow \tilde{\mathbf{A}}^{op}_{k} & \mathbf{A}^{cl}_{n-k+1} \rightarrow \tilde{\mathbf{C}}^{k-1}_{op} \\ \mathbf{C}^{n-k}_{op} \rightarrow \tilde{\mathbf{A}}^{cl}_{k} & \mathbf{A}^{op}_{n-k+1} \rightarrow \tilde{\mathbf{C}}^{k-1}_{cl} \end{array} \tag{16}$$

and so on.

Thus the dual network is completely described in terms of all the components of the primal **C** and **A** matrices, corresponding to open and closed paths, or chains, in every dimension. Figure 2c shows a dual network derived in this way, on which all of the open and closed paths bear a dual relationship to those on the primal network, for example, 1-cells forming a closed chain on the primal are identified with 1-cells forming an open chain on the dual network etc.

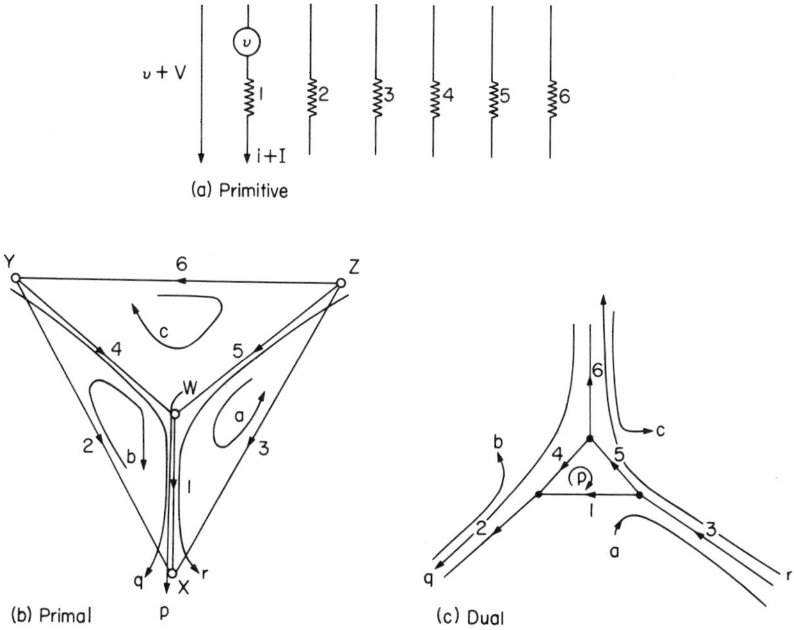

Fig. 2. 1-Dimensional Paths.

The scheme is illustrated by Fig. 3. This shows the connection relationships for the network of Fig. 2. From left to right, the dimensions of the primal chains increase from 0 to 2 and the corresponding dual incidence relationships may be traced out for the dual system, of dimension 0 to 2 from right to left.

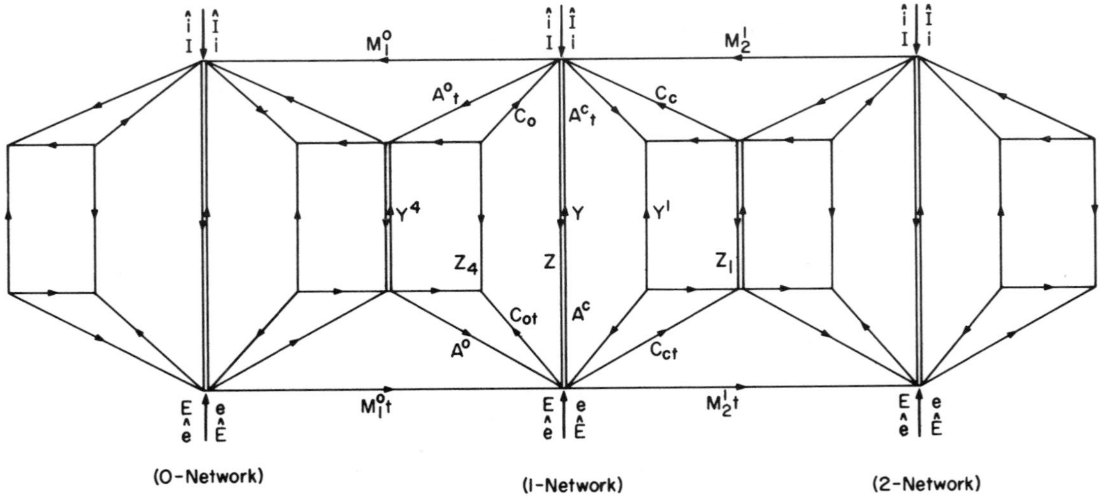

$$e + \hat{E} = Z_1(i + \hat{I}) + Z_2(\hat{i} + I)$$
$$\hat{e} + E = Z_3(i + \hat{I}) + Z_4(\hat{i} + I)$$

Fig. 3. Simplified algebraic diagram.

The algebraic diagram is an adaptation of that suggested for electrical circuits by Roth [10]. By convention all products are taken against the direction of the arrows.

It is now seen that the word "orthogonal" arises in a new context, namely that the currents and voltages divide naturally into groups associated with the spatially orthogonal primal and dual networks. At this stage Kron realized that orthogonal electromagnetic quantities were necessary, to fill the n-dimensional chains, rather than currents and voltages. He also extended the algebraic diagram to cover more than three dimensions.

4. THE ELECTROMAGNETIC POLYHEDRON

In 1944 Kron published the equivalent circuit for Maxwell's equations [7], on which all the field vectors are represented by voltages and currents on 1-cells. However, the matching of circuit quantities to the terms in the differential equations is not straightforward since the co-variant and contra-variant nature of the variables must be preserved [8]. For example, on the electrical network, voltage is co-variant (indicated by a subscript) and current is contra-variant (indicated by a superscript). In the field equation

$$\text{curl}\,(E_\alpha) = -\frac{\partial(B_\alpha)}{\partial t} \tag{17}$$

E and B are co-variant. However, curl E is contra-variant and must be associated with a contra-variant form of B. The equation then becomes

$$(\text{curl}\,E)^\beta = -\frac{\partial B^{\beta'}}{\partial t} \tag{18}$$

where

$$B^{\beta'} = g^{\beta\alpha}\sqrt{g}\,B_\alpha \tag{19}$$

$g^{\beta\alpha}$ is the inverse of the space-metric tensor and g is its determinant.
$B^{\beta'}$ is a contra-variant vector density.

Using the rules of vector and tensor calculus, all of the electromagnetic field vectors can be represented on the network by voltages and currents. One very important point arises however. The quantities on finite meshes of the network represent line, surface and volume integrals of the vector densities in the electromagnetic field.

The same boundary operations and constraints apply to the representation of fields on the polyhedron and its dual. With this proviso, the concepts of field vectors and their integrals along edges and surfaces can be used to energize and couple the dual systems – rather in the same manner as a 2-phase machine. The same considerations apply to the introduction of charge fields. These will then produce electromagnetic effects which interact with the original field quantities, which in turn influence the velocity and acceleration of the charges.

Figure 4 shows the algebraic diagram with some of the electromagnetic quantities applied. It is seen that the incidence matrices now play the role of div and curl operations, in associating quantities of successive dimensions – satisfying Stokes' theorem.

$$\hat{\mathbf{E}}_1 = \text{div } \mathbf{B}_0 = \mathbf{M}_1^0 \mathbf{B}_0 \tag{20}$$

$$\dot{\mathbf{d}}^1 = \text{curl } \hat{\mathbf{h}}^2 = \mathbf{M}_2^1 \hat{\mathbf{h}}^2 \tag{21}$$

It is important to note also that two successive boundary operations give zero, that is

$$\mathbf{M}_2^{1T} \mathbf{M}_1^{0T} \mathbf{B}_0 = \mathbf{0} \tag{22}$$

This corresponds to the topological statement that the boundary of a boundary is zero, or in the notation of the calculus of differential forms

$$d d \boldsymbol{\omega} = 0 \tag{23}$$

where $\boldsymbol{\omega}$ is any p-form.

In the following section the electromagnetic quantities are used to model a small discrete-data table without any self-organizing capability.

5. COMPUTATION

An n-dimensional polyhedron and its dual, with electromagnetic quantities impressed, forms the basis of Kron's wave automaton. Electromagnetic waves can propagate throughout the polyhedral structure, and may assume an "oscillatory", self-adaptive state. The resulting oscillatory polyhedron may be used in several different modes, for multi-dimensional information processing.

When energizing the wave automaton it is necessary to direct the propagation of the electromagnetic waves across the model, according to the needs of the particular problem. A multi-dimensional form of divided differences can be produced by allowing propagation of an electromagnetic wave, without feed-back. The values of stored electrostatic and electromagnetic energies in each dimension represent the square of the divided differences.

As an example, we consider a function of two variables

$$z = x^2 + 2y^2 - 3xy + 5x - 3y + 6 \tag{24}$$

The function is sampled at four points W, X, Y and Z

Node	x	y	z
W	1	1	8
X	1	−2	32
Y	−2	3	27
Z	4	2	20

A polyhedral reference frame is constructed around the four data points.

Six lines (branches) interconnect the nodes, and three planes are bounded by sets of three branches, Fig. 2b. The squares of the branch lengths are assigned as the impedance of the 1-network (ϵ_{11}) and the squares of the areas are assigned as the impedance of the 2-network (μ_{22}). The four values of the dependant variable are impressed as bound magnetic charges $\mathbf{B}_0$ on the 0-network of the polyhedron.

Electromagnetic quantities propagate across the polyhedron by the path, shown in Fig. 4 by a bold line. The values of the electromagnetic quantities are

$\dot{\mathbf{B}}_0$	$\dot{\mathbf{B}}_0'$	$\mathbf{E}_1$	$\dot{\mathbf{D}}^1 + \dot{\mathbf{d}}^1$	$\dot{\mathbf{d}}^{1\prime}$	$\mathbf{h}^2$	$\dot{\mathbf{b}}_2$
8	−24	−24	−2.6666666	0.48000000	0.48000000	9.7200000
32	−5	−5	−0.14705882	0.14705882	0.14705882	2.9779411
27	−12	−12	−0.48000000	0.18918918	0.18918918	3.8310808
20		19	1.4615384			
		12	1.2000000			
		−7	−0.18918918			

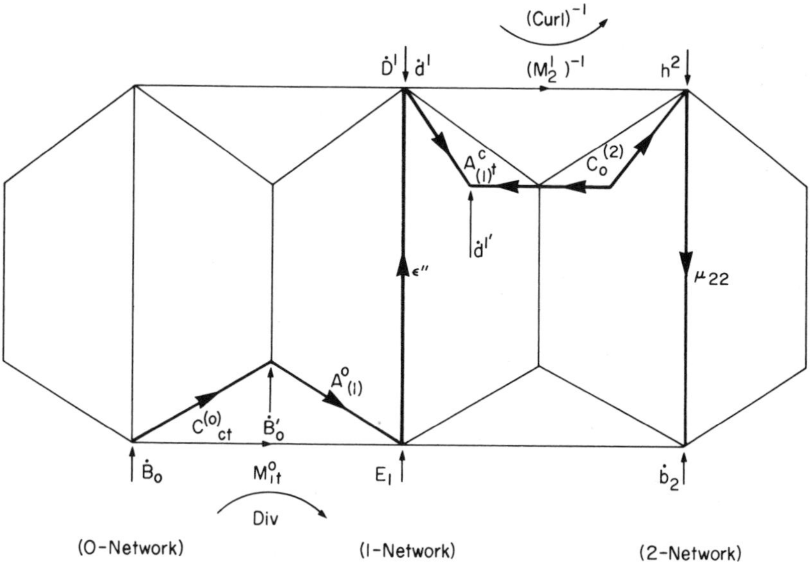

Fig. 4. Field computation.

The 1-network impedance matrix is

$$\boldsymbol{\epsilon}_{11} = (\boldsymbol{\epsilon}^{11})^{-1} = \begin{bmatrix} 9 & \cdot & \cdot & \cdot & \cdot & \cdot \\ \cdot & 34 & \cdot & \cdot & \cdot & \cdot \\ \cdot & \cdot & 25 & \cdot & \cdot & \cdot \\ \cdot & \cdot & \cdot & 13 & \cdot & \cdot \\ \cdot & \cdot & \cdot & \cdot & 10 & \cdot \\ \cdot & \cdot & \cdot & \cdot & \cdot & 37 \end{bmatrix}$$

The 2-network impedance matrix is

$$\boldsymbol{\mu}_{22} = (\boldsymbol{\mu}^{22})^{-1} = \begin{bmatrix} 20.25 & . & . \\ . & 20.25 & . \\ . & . & 20.25 \end{bmatrix}$$

The stored electrostatic and electromagnetic energies on the polyhedron are
electrostatic energy = $(\dot{\mathbf{d}}^1 + \dot{\mathbf{D}}^1)\boldsymbol{\epsilon}_{11}(\dot{\mathbf{d}}^1 + \dot{\mathbf{D}}^1)$ = (1st divided differences)^2
electromagnetic energy = $(\dot{\mathbf{b}}_2 + \dot{\mathbf{B}}_2)\boldsymbol{\mu}^{22}(\dot{\mathbf{b}}_2 + \dot{\mathbf{B}}_2)$ = (2nd divided differences)^2 (25)

The divided differences may be obtained from the field quantities on the polyhedron.

$$\text{1st Divided differences} = (\dot{\mathbf{d}}^1 + \dot{\mathbf{D}}^1)\sqrt{\boldsymbol{\epsilon}_{11}}$$
$$\text{2nd Divided differences} = (\dot{\mathbf{b}}_2 + \dot{\mathbf{B}}_2)\sqrt{\boldsymbol{\mu}^{22}} \quad (26)$$

1st Divided differences	*2nd Divided differences*
−7.9999998	2.1600000
−0.85749292	0.66176468
−2.4000000	0.85133128
5.2696516	
3.7947331	
−1.1507928	

There is a close correspondence between the equations of diakoptics and the equations of least-squares regression. In a measurement process, there is a set of observations $\mathbf{y}$, and a measurement matrix $\tilde{\mathbf{H}}$. It is required to find $\mathbf{x}$ knowing that

$$\mathbf{y} = \tilde{\mathbf{H}}\mathbf{x} \quad (27)$$

but $\mathbf{x}$ is overdetermined. The least-squares solution to this problem [9] is

$$\hat{\mathbf{x}} = (\tilde{\mathbf{H}}^T\mathbf{W}\tilde{\mathbf{H}})^{-1}\tilde{\mathbf{H}}^T\mathbf{W}\mathbf{y} \quad (28)$$

where $\hat{\mathbf{x}}$ is the "best estimate" (by the least-squares criterion) of $\mathbf{x}$ and $\mathbf{W}$ is a weighting matrix. The basic equation of diakoptics, where the removed impedance is zero [6], is

$$\begin{bmatrix} \mathbf{E} \\ \mathbf{o} \end{bmatrix} = \begin{bmatrix} \mathbf{Z} & \mathbf{ZC} \\ \mathbf{C}^T\mathbf{Z} & \mathbf{C}^T\mathbf{ZC} \end{bmatrix} \begin{bmatrix} \mathbf{I} \\ \mathbf{i} \end{bmatrix} \quad (29)$$

From eqn. (29) the intersection network current is given by

$$\mathbf{i} = -(\mathbf{C}^T\mathbf{ZC})^{-1}\mathbf{C}^T\mathbf{ZI} \quad (30)$$

If the quantities used in diakoptics are compared with those of regression theory, the following correspondences are found.

Symbol	Diakoptic quantity	Symbol	Regression quantity
I	current impressed across a junction pair	**y**	observed dependent variable
E	junction potentials	**e**	error $\mathbf{y} - \hat{\mathbf{y}}$ where $\hat{\mathbf{y}}$ is the best estimate of **y**
C	connection matrix branch-mesh	$\tilde{\mathbf{H}}$	measurement matrix
Z	branch impedance matrix	**W**	weighting matrix
i	current in the intersection network	$\hat{\mathbf{x}}$	best estimate of **x**

It is seen that the diakoptic equation, for the case where the removed impedance is zero, bears a close relationship to the equation for least-squares regression.

Kron combined the concepts of the polyhedral model for generating divided differences and the use of the diakoptic equations for regression analysis.

The resulting automaton could produce an extremely close fit to data, while still retaining the original degrees of freedom of the data. The precise way in which this was achieved has yet to be fully investigated.

6. CONCLUSION

The simple description of the non-oscillatory wave model given above is sufficient to show that this system contains the necessary physical and mathematical techniques for studies of data smoothing with several variables and for signal estimation, prediction and filtering. The connection between the one-dimensional electrical network problem and least-squares estimation has in fact now been discussed in detail by Nicholson [9].

The multi-dimensional polyhedron combines (explicity or implicity) orthogonal functions, generalized Taylor's series, concepts of differential forms and boundary operators, duality and the Hodge star operator, deRham's theorem and harmonic integrals.

Little has been published in the literature about the self-organizing oscillatory state of Kron's electromagnetic polyhedron. This condition arises when the charges **E**, **B**, **D** and **H** are considered to flow across the k-cells, setting up their own voltages and fluxes which provide feed-back. In this (almost) oscillatory state the complementary orthogonal primal k-cells and dual $(n-k)$-cells act like generalized machines. The circuits are, however, coupled by ideal transformers instead of the usual rotational effects.

REFERENCES

1. Kron, G. "Basic concepts of multi-dimensional space filters". *Trans. A.I.E.E. 1,* 78, 1959, 554–561.
2. Kron, G. "Multi-dimensional curve-fitting with self-organizing automata". *J. Math. Anal. and Appl. 5,* 1962, 46–69.
3. Kron, G. "Power-system Type Self-organizing Automata". *R.A.A.G., Memoirs, 3,* G–IV, 1962, 392–417.
4. Kron, G. *Tensor Analysis of Networks*, Wiley, New York, 1939, and MacDonald, London, 1965.
5. Veblen, O. *Analysis Situs*, American Math. Soc., New York, 1931.
6. Kron, G. *Diakoptics – the piecewise solution of large-scale systems.* MacDonald, London, 1963.
7. Kron, G. "Equivalent circuits of the field equations of Maxwell–1". *Proc. I.R.E., 32,* 1944, 289–299.
8. Lynn, J. W. *Tensors in electrical engineering*, Edward Arnold, London, 1963.
9. Nicholson, H. "Sequential least-squares estimation and the electrical-network problem". *Proc. I.E.E., 118,* 1971, 1635–1641.
10. Roth, J. P. "An application of algebraic topology to numerical analysis. On the existence of a solution of the network problem". *Proc. Nat. Ac. Sci. 41,* 1955, 518–521.
11. Balasulnamanian, N. V., Lynn, J. W. and Sen Gupta, D. P. "Differential Forms on Electromagnetic Networks." Butterworth, London, 1970.

NOMENCLATURE

A	branch – mesh connection matrix
B	longitudinal magnetic flux
b	transverse magnetic flux
C	branch-open-path connection matrix
D	longitudinal electrostatic flux
d	transverse electrostatic flux
E	longitudinal electric field intensity
e	transverse electric field intensity
$g_{\alpha\beta}$	space-metric tensor
g	determinant of $g_{\alpha\beta}$
H	longitudinal magnetic field intensity
$\tilde{\mathbf{H}}$	measurement matrix
h	transverse magnetic field intensity
I	vector of open path currents
i	vector of closed path currents
M	incidence matrix
t	time
V	vector of open path voltages
v	vector of closed path voltages
W	least-squares weighting matrix
x	state vector
$\hat{\mathbf{x}}$	"best" estimate of **x**
Y	admittance matrix
y	vector of observed values
Z	impedance matrix
$\boldsymbol{\epsilon}$	electrostatic permittivity
$\boldsymbol{\mu}$	electromagnetic permeability
$\boldsymbol{\omega}$	a p-form

THE CLASSIFICATION OF PHYSICAL VARIABLES IN NETWORK THEORY AND MECHANICS WITH APPLICATIONS TO VARIATIONAL ANALYSIS

D. L. Jones
A.W.R.E.
Aldermaston
Berkshire
England

D. J. Holding *and* F. J. Evans
Department of Electric and Electronic Engineering
Queen Mary College
University of London
England

SUMMARY

A classification of physical variables is given that is based on aspects of measurement, and on the topological properties of linear graphs. These criteria are quite general to any set of variables that can be associated with network models of physical systems.

It is shown that this classification can be imbedded in a generalized algebraic structure (derived by Sewell in another chapter) which governs the variational formulation for a general class of dissipative system. The classification also provides a systematic method of choosing proper sets of independent coordinates. The relative merits of classifications based on intensive and extensive variables (defined in a thermodynamic sense) are discussed, and an extended form of the intensive formulation is given that overcomes the difficulties associated with the singularity of certain topological matrices. The network-derived functions of content and co-content are shown to be of special significance in the generalized variational analysis.

Finally it is shown that the basic classification and generalized algebraic structure can be extended to spatially distributed field systems.

1. INTRODUCTION

Electrical network theory and its extension, by analogies, into other classes of physical system can provide a useful basis from which to embark on a classification of physical variables. The reason for this rests in the "lumped" nature of networks which are easily definable from a topological point of view [1]. No difficulty exists in network theory of establishing the necessary independent variables, unlike classical variational dynamics which can provide no systematic way of selecting them.

A network can be viewed simply as a system in which there is only one "external" independent variable, namely time. It is shown here that it can also be considered as a special case resulting naturally from a far more general class of system, the description of which has, in addition, independent spatial coordinates. Hence a common formalism is possible in which the fundamental tensor concepts of co- and contra-variance are central, and in which it is also possible to supplement the generalized variational analysis proposed by Sewell in this volume [10].

Classical variational mechanics rests firmly on the proper identification of certain functional quantities, and these in turn are inseparable from an adequate classification of physical variables. It is hoped that some benefit might result from a joint consideration of these two topics together with network theory. This hope is fulfilled to an encouraging degree, but it is also interesting that certain questions related to classical variational methods, which are not usually discussed, arise naturally from this approach. Not all of these are satisfactorily answered here, and in some cases it is even suggested that a more highly structured network approach to these problems highlights the limitations of the more classical point of view.

Network theory provides substantial reasons for proposing that a "power" description of a system, in terms of voltages and currents, should be considered as the most general form from which any other, more restricted case, may be derived. From a classification of lumped variables a power – or intensive – formulation will be given that can be used, unambiguously, to generate all the functions appearing in Sewell's variational algebra. The term "intensive" is used here in a thermodynamic sense, (i.e. variables that are independent of the system extent, and are normally measurable). The classification can also be used to generate other, more general, functions.

It is then shown that an extended form of the network description, similar to that employed by Kron [2], enables network equations containing non-singular matrices to be developed for the intensive variables, and for the basic scalar functions to be the subject of reversible Legendre transformations. In this way a greatly increased flexibility is brought to the formulation which is completely complementary to the approach of Sewell [3] and includes the earlier concepts of Jones and Evans [4].

The implications of reducing the general power formulations to an "extensive" formulation is discussed, and it can be seen that severe restrictions appear to exist to the general application of the conventional conservative Lagrangian/Hamiltonian approach. Again the term "extensive" is used in a thermodynamic sense (i.e. a variable that is dependent on the system extent, and the storage of which is associated with the accumulation of internal energy).

Those descriptions related to the state-space formulation of modern control and systems theory are compared to those, previously generated, that arise from the gradient operation on scalar potential functions. A clear distinction must be made between variable transformations and topological transformations in this context.

The concepts discussed in this chapter are complementary to those of Sewell and to the application of generalized tensor fields to dynamic system analysis by Korn *et al.* [18]. This, and its thermodynamic basis, are discussed elsewhere in this volume, and only serve to re-emphasize that the content and co-content functions are fundamental to this integrated approach. They possess the essential property that they can be sensibly interpreted for any class of system element and therefore provide an analytic link between dissipative (entropic) power and the instantaneous power associated with the storage devices.

2. SIMPLE NETWORK THEORY

A network can be described topologically by the connectivity of the associated linear graph [1]. To each branch is allocated an arbitrary direction and the graph is then said to be an "orientated" linear graph. It is now possible to describe its structure algebraically by dividing the graph into a tree and associated co-tree. For the choice of any particular tree there is a minimal description in terms of the fundamental mesh matrix, $\mathbf{B}_f$, in which the sign of the elements is governed by the direction allocated to the members of the co-tree, normally termed links or chords. In each mesh, or loop, the positive direction is that of the associated link.

The connectivity of the network can also be described in a minimal form by the fundamental cut-set matrix, $\mathbf{C}_f$, in which the sign of the elements is governed by the members of the tree; these provide the positive direction for other members of the cut-set. The mesh and cut-set formulations are not independent of each other and this is shown in the simple example of Fig. 1.

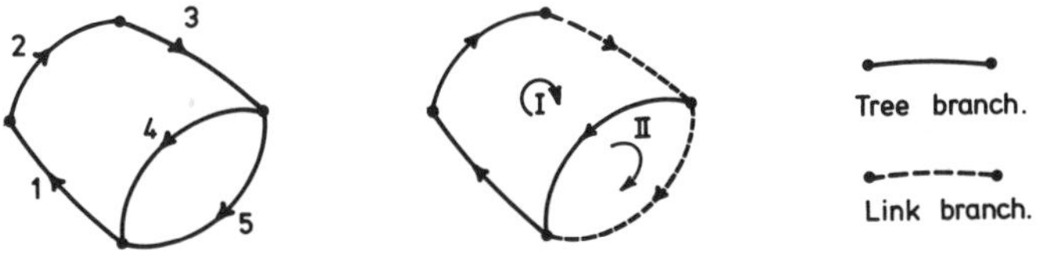

Fig. 1.

Matrix $\mathbf{B}_f$ = $[\mathbf{B}_t \,\vdots\, \mathbf{I}]$ (1)

	1	2	4	3	5
Mesh I	1	1	1	1	0
Mesh II	0	0	−1	0	1

Matrix $\mathbf{C}_f$ = $[\mathbf{I} \,\vdots\, \mathbf{C}_l]$ (2)

	1	2	4	3	5
Cut-set 1	1	0	0	−1	0
Cut-set 2	0	1	0	−1	0
Cut-set 4	0	0	1	−1	1

where

$$\mathbf{B}_t^T = -\mathbf{C}_l = \mathbf{D} \tag{3}$$

and **D** is the fundamental dynamical transformation matrix; the suffices t and l refer to tree and link branches respectively.

If every branch of the network is now allocated two variables: one considered as flowing *through* and the other as being measured *across* the branch [5, 6]; certain fundamental relationships can be stated. In electrical terms the variables are, of course, current and voltage respectively, and the relationships are known as Kirchhoff's laws. In terms of the matrices above these laws can be stated as

$$\mathbf{B}_f \mathbf{v}_b = \mathbf{0} \quad \text{or} \quad \mathbf{v}_l = -\mathbf{B}_t \mathbf{v}_t \tag{4}$$

$$\mathbf{C}_f \mathbf{i}_b = \mathbf{0} \quad \text{or} \quad \mathbf{i}_t = -\mathbf{C}_l \mathbf{i}_l = \mathbf{B}_t^T \mathbf{i}_l \tag{5}$$

It is a direct result of these laws that a further, most important, statement can be made, known as Tellegen's theorem [7].

$$\langle \mathbf{i}_t, \mathbf{v}_t \rangle + (\mathbf{i}_l, \mathbf{v}_l) = 0 \tag{6}$$

The net instantaneous power of the network is zero.

3. CLASSIFICATION OF LUMPED VARIABLES

The classification of variables into the categories of "throughness" and "acrossness", as above, is one that clearly rests on measurement. In essence the topological approach adds to this another based on the division of the network into two; a "system" and an "environment" corresponding to the tree and co-tree. It is possible to transform the variables of one sub-system into those of the other by means of the fundamental mesh matrix, or other linear operators, as in Fig. 2.

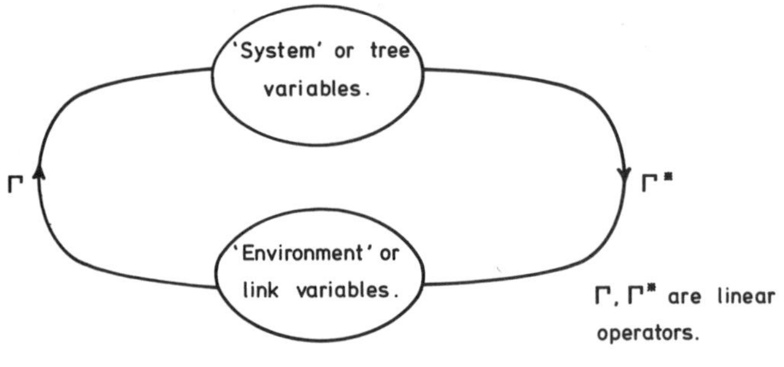

Fig. 2.

It is now possible to combine both the measurement and topological classification into one as shown in Fig. 3.

The overlay of the measurement structure with that of the topology results in a set of four variables, but as given so far there are only two equations relating them. It is therefore necessary to provide two more relationships to complete the combined structure. It will be suggested that two further equations of the form

$$\frac{\partial A}{\partial x_l} = y_l \tag{9}$$

$$\frac{\partial A}{\partial u_t} = v_t \tag{10}$$

would suffice, where A is some scalar function. So, finally, we now have a complete structure as shown in Fig. 4.

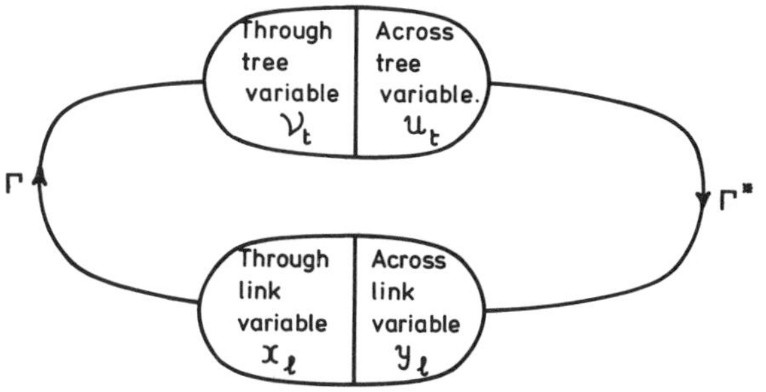

Fig. 3.

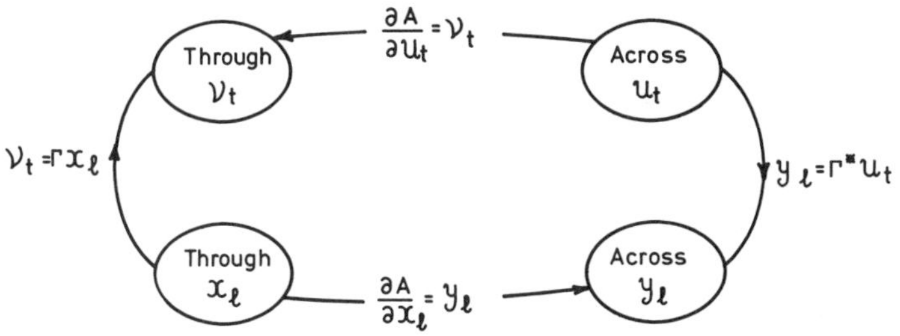

Fig. 4.

Apart from the dependency implied in eqns. (1) and (2) above, (i.e. link voltages and tree currents are derived from the independent sets of tree voltages and link currents respectively) there will exist other dependencies governed by the nature of the independent network elements. The simplest of these are the basic constitutive relationships

$$\text{Resistor: } v_R = Ri_R \qquad \text{Inductor: } \left.\begin{array}{l}\psi_L = Li_L \\ v_L = L\dot{i}_L\end{array}\right\} \qquad \text{Capacitor: } \left.\begin{array}{l}q_C = Cv_C \\ i_C = C\dot{v}_C\end{array}\right\}$$

These can be summarized diagramatically as in Fig. 5, together with the definitions of the important scalar functions [8, 9] that will be used in the subsequent analysis. From classical mechanics it is well known that the following scalar functions can be defined

$$\begin{array}{ll}\text{Lagrangian } L = T^* - U & \text{Co-Lagrangian } L^* = U^* - T \\ \text{Hamiltonian } H = T + U & \text{Co-Hamiltonian } H^* = T^* - U^*\end{array}$$

and also that they can be related by the Legendre dual transformation in the following manner

$$L(q, \dot{q}) = \langle p, \dot{q} \rangle - H(q, p) \tag{11}$$

$$L^*(p, \dot{p}) = H^*(\dot{q}, \dot{p}) - \langle p, \dot{q} \rangle \tag{12}$$

in which q and p are the generalized coordinate and momentum respectively, and refer to the mass of the particle.

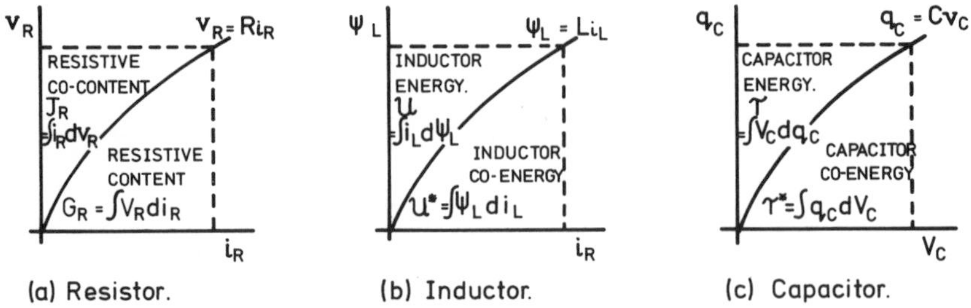

Fig. 5.

It will be seen that this structure, exhibited by the classical formulation is reflected in the topological derivation to follow. Certain aspects of this should, however, at this stage be emphasized. Classical dynamics, arising as it does from particle dynamics, was not faced with certain problems of formulation that only arise in the more general application. These can be illustrated by referring to the conventional forms of the Euler–Lagrange equations and the associated Hamiltonian equations.

$$\frac{\partial L}{\partial q} - \frac{d}{dt}\left(\frac{\partial L}{\partial \dot{q}}\right) = 0 \tag{13}$$

$$\dot{p} = -\frac{\partial H}{\partial q}, \quad \dot{q} = \frac{\partial H}{\partial p} \tag{14}$$

In particle mechanics there normally can exist no ambiguity concerning the definition of the variables in these equations, as they all relate to the particle in which resides the ability to store both kinetic *and* potential energy. In modern mechanical systems analysis the concept of a "spring" as an element that provides the environmental aspects of potential energy to the mass has been introduced. The mass-spring duality in mechanical systems is, of course, reflected in the capacitor-inductor duality of electrical networks.

So in passing from the more restricted particle formulation to the more general systems description, it is imperative that care be taken to ensure the existence of a proper reference frame. At the same time it is also necessary that no unsupportable assumptions be made concerning the reduction of the more general "power" formulation to that normally associated with the classical conservative problem. These aspects will be discussed in greater detail after the power – or intensive – description has been presented.

4. THE USE OF VARIABLE CLASSIFICATION TO DERIVE VARIATIONAL PRINCIPLES

From Fig. 4 it is seen that, in order to proceed, it is necessary to allocate to the variables x_l and u_t some recognizable interpretation, and then to also identify some suitable scalar function A. The "energetic" approach of classical mechanics immediately suggests that if x_l and u_t are considered to be the "extensities" – that is variables that are dependent on the extent or size of the system – then A will be defined in terms of the energy of the system. This is in keeping with the thermodynamic concepts that energy is possessed by a system by virtue of the storage of the extensities. In the electrical network case these extensities are the capacitor charge and inductor flux, and Fig. 6 shows how Fig. 4 is made more specific to the electrical network.

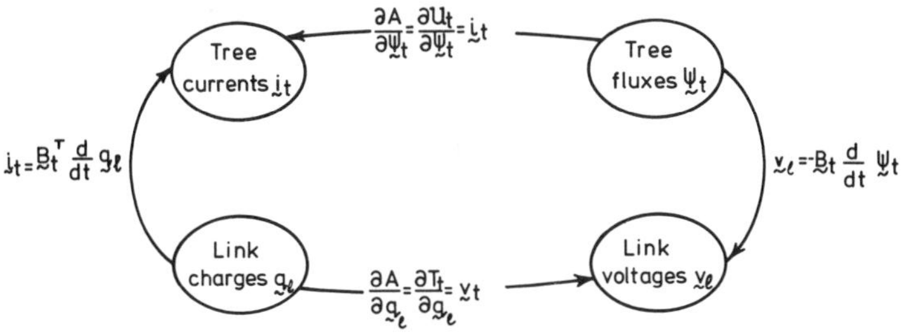

Fig. 6.

Alternatively it is possible to choose the variables so that x_l and u_t be considered as "intensities" – that is variables that are independent of the system extent – and in the electrical system these will be voltages and currents. The scalar A will then have to be redefined as a "power" type function in terms of the content and co-content (Fig. 5, [8, 9, 18]) and Fig. 4 becomes a "power" description as shown in Fig. 7.

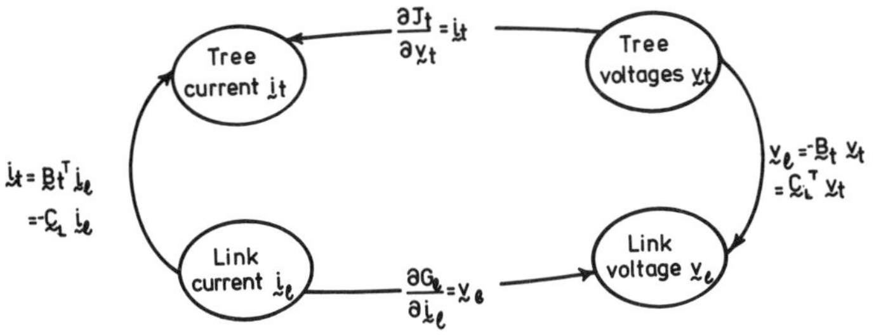

Fig. 7.

The classification of variables can also be used to generate other more general functions.

The differential forms

$$dK = \langle \mathbf{v}_t, d\mathbf{i}_t \rangle - (\mathbf{v}_l, d\mathbf{i}_l) \tag{15}$$

and

$$dK^* = (\mathbf{i}_l, d\mathbf{v}_l) - \langle \mathbf{i}_t, d\mathbf{v}_t \rangle \tag{16}$$

can be shown to be perfect differential forms for all network configurations since

$$\frac{d\mathbf{v}_l}{d\mathbf{i}_t} = \frac{d(-\mathbf{D}^T\mathbf{v}_t)}{d(\mathbf{D}\mathbf{i}_l)} = \frac{\partial(-\mathbf{D}^T\mathbf{v}_t)}{\mathbf{D}^T\partial\mathbf{i}_l} = \frac{-\partial\mathbf{v}_t}{\partial\mathbf{i}_l} \tag{17}$$

Integration of the differential forms gives the functions

$$K = 2\int^{\Gamma} \langle \mathbf{v}_t, d\mathbf{i}_t \rangle = -2\int^{\Gamma} (\mathbf{v}_l, d\mathbf{i}_l) \tag{18}$$

and

$$K^* = 2\int^{\Gamma} (\mathbf{i}_l, d\mathbf{v}_l) = -2\int^{\Gamma} \langle \mathbf{i}_t, d\mathbf{v}_t \rangle \tag{19}$$

These functions have been used to describe network behaviour [13].
Two other functions, N and N^* can also be shown to exist, where

$$N = \langle \mathbf{v}_t, \mathbf{i}_t \rangle - K \tag{20}$$

$$N^* = (\mathbf{i}_l, \mathbf{v}_l) - K^* \tag{21}$$

It is now possible to adopt the basic theoretical framework proposed by Sewell [3] for both these cases, and to clothe this framework almost completely by simple "inspection".

Sewell [10] has proposed a closed chain of saddle-shaped and convex functionals linked by Legendre-type transformations that encompass many theories in applied mathematics and possesses a well-defined scope for generalization. It is summarized in Fig. 8.

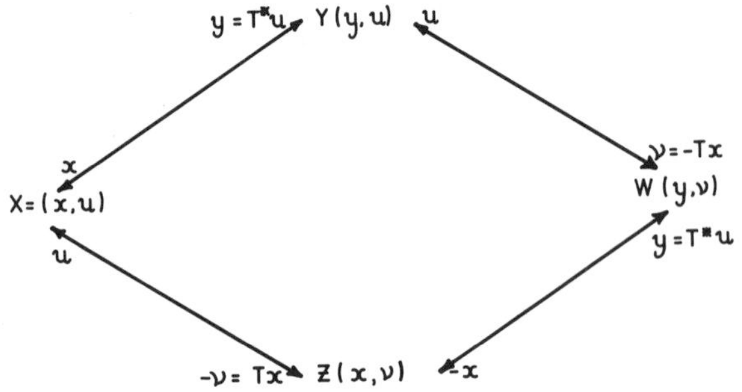

Fig. 8.

where

$X(x, u)$ is convex in x concave in u	$Y(y, u)$ is jointly convex
$Z(x, v)$ is jointly concave	$W(-x, y)$ is concave in y convex in x

$$\begin{aligned} X + Y &= (x, y) \qquad & Y + W &= \langle u, v \rangle \\ Y + W &= \langle u, v \rangle \qquad & W + Z &= (-x, y) \end{aligned} \tag{22}$$

Partial derivatives being

$$\frac{\partial X}{\partial x} = y; \quad \frac{\partial X}{\partial u} = -v; \qquad \frac{\partial W}{\partial v} = u; \quad \frac{\partial W}{\partial y} = -x;$$
$$\frac{\partial Z}{\partial v} = -u; \quad \frac{\partial Z}{\partial x} = -y; \qquad \frac{\partial Y}{\partial y} = x; \quad \frac{\partial Y}{\partial u} = v; \tag{23}$$

Lagrangian formulation:

$$x = \frac{\partial Y}{\partial(T^*u)} \qquad -Tx = \frac{\partial Y}{\partial u}$$

Hence

$$T\left(\frac{\partial Y}{\partial(T^*u)}\right) + \frac{\partial Y}{\partial u} = 0 \tag{24}$$

Hamiltonian formulation:

$$T^*u = \frac{\partial X}{\partial x}; \quad Tx = \frac{\partial X}{\partial u} \tag{25}$$

In order to make the problem completely determinate, Sewell demands the additional conditions that T and T^* are linear operators adjoint in the sense that $\langle x, T^*u\rangle = \langle u, Tx\rangle$. It will therefore be necessary to ensure that any linear operators, such as those already adopted in Figs 2 and 3 do, in fact, satisfy such conditions. (This is achieved if, in the intensive formulation of Section 3 and Section 4, $\Gamma = -T$; $\Gamma^* = T^*$, and in the extensive formulation of Section 4, $\Gamma = T$; $\Gamma^* = T^*$.) It is useful to point out at this stage that the operators $\frac{d}{dt}$ and $\frac{-d}{dt}$ are formally adjoint in the sense required, and also that any matrix and its transpose are adjoint in the same sense.

4.1 Intensive classification

We shall assume that the only truly independent variables are the set of tree voltages $\mathbf{v}_t$ and the set of link currents $\mathbf{i}_l$. Also we select those scalar functions that are naturally associated with these variables, the tree co-content $J_t(\mathbf{v}_t)$ and the link content $G_l(\mathbf{i}_l)$. But as shown in Fig. 7 a slight adjustment is required in order to accommodate the proper adjoint operators which are

$$T = -\mathbf{B}_t^T$$
$$T^* = -\mathbf{B}_t \tag{26}$$

It should be noted that the following useful relationships exist between certain content and co-content functions.

$$dJ_t(\mathbf{v}_t) = \mathbf{i}_t,\ d\mathbf{v}_t = \mathbf{B}_t^T\mathbf{i}_l,\ d\mathbf{v}_t = \mathbf{i}_l,\ d(\mathbf{B}_t^T\mathbf{v}_t) = -\mathbf{i}_l,\ d\mathbf{v}_l = -dJ_l(\mathbf{v}_l) = -\,dJ_l(\mathbf{v}_l) \tag{27}$$

$$dG_1(\mathbf{i}_l) = \mathbf{v}_l,\ d\mathbf{i}_l = -\mathbf{B}_t\mathbf{v}_t,\ d\mathbf{i}_l = -\mathbf{v}_t,\ d(\mathbf{B}_t^T\mathbf{i}_l) = -\mathbf{v}_t,\ d\mathbf{i}_t = -\,dG_t(\mathbf{i}_t) = -\,dG_t(\mathbf{i}_l) \tag{28}$$

Using these we can now proceed to establish the basic functional relationships for the intensive formulation of the Sewell type, Fig. 9:

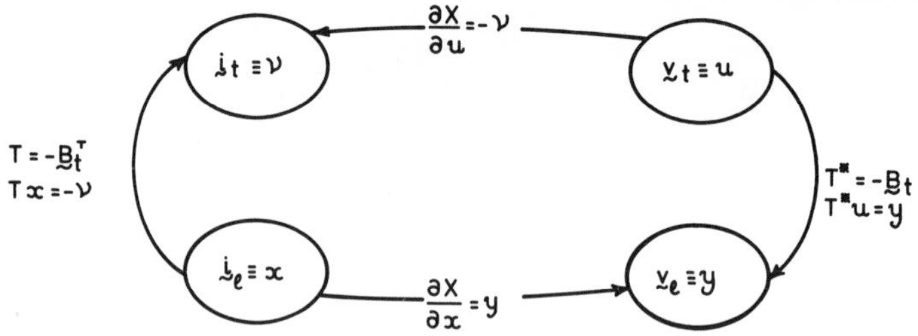

Fig. 9.

in which:

$$T^* = -\mathbf{B}_t; \quad T = -\mathbf{B}_t$$

$$\nu = \mathbf{i}_t; \; u = \mathbf{v}_t; \; x = \mathbf{i}_l; \; y = \mathbf{v}_l$$

$$\begin{aligned}
&X = G_l(\mathbf{i}_l) - J_t(\mathbf{v}_t); && \frac{\partial X}{\partial \mathbf{i}_l} = \mathbf{v}_l = y; && \frac{\partial X}{\partial \mathbf{v}_t} = -\mathbf{i}_t = -\nu \\
&Y = J_l(\mathbf{v}_l) + J_t(\mathbf{v}_t) = 0; && \frac{\partial Y}{\partial \mathbf{v}_l} = \mathbf{i}_l; && \frac{\partial Y}{\partial \mathbf{v}_t} = \mathbf{i}_t \\
&Z = -G_t(\mathbf{i}_t) - G_l(\mathbf{i}_l) = 0; && \frac{\partial Z}{\partial \mathbf{i}_t} = -\mathbf{v}_t; && \frac{\partial Z}{\partial \mathbf{i}_l} = -\mathbf{v}_l \\
&W = -J_l(\mathbf{v}_l) + G_t(\mathbf{i}_t); && \frac{\partial W}{\partial \mathbf{i}_t} = \mathbf{v}_t; && \frac{\partial W}{\partial \mathbf{v}_l} = -\mathbf{i}_l
\end{aligned} \tag{29}$$

Note also

$$\begin{aligned}
X &= G_l(\mathbf{i}_l) - J_t(\mathbf{v}_t) = G_l(\mathbf{i}_l) + J_t(\mathbf{v}_l) = -G_t(\mathbf{i}_t) - J_t(\mathbf{v}_t) \\
W &= J_t(\mathbf{v}_t) + G_t(\mathbf{i}_t) = -J_l(\mathbf{v}_l) - G_l(\mathbf{i}_l)
\end{aligned} \tag{30}$$

It is easy to establish that the proper conditions for the integrability of these functions are satisfied.

For X: $\frac{\partial}{\partial \mathbf{i}_l}\mathbf{i}_t = \frac{-\partial}{\partial \mathbf{v}_t}\mathbf{v}_l$ because $\mathbf{i}_t = \mathbf{B}_t^T\mathbf{i}_l$ and $\mathbf{v}_l = -\mathbf{B}_t\mathbf{v}_t$ (see appendix)

For Y: Since $Y = \frac{K^*}{2} - \frac{K^*}{2}$ (31)

For Z: Since $Z = \frac{K}{2} - \frac{K}{2}$

For W: $\frac{\partial}{\partial \mathbf{i}_t}\mathbf{i}_l = -\frac{\partial}{\partial \mathbf{v}_l}\mathbf{v}_t$ provided no. of links = no. of tree branches

where X is the topologically extended form of the Brayton–Moser mixed potential functions and W is its dual [19, 12].

The corresponding Lagrangian formulation is

$$T\left(\frac{\partial Y}{\partial(T^*u)}\right) + \frac{\partial Y}{\partial u} = 0; \quad -\mathbf{B}_t^T\left(\frac{\partial Y}{\partial \mathbf{v}}\right) + \frac{\partial Y}{\partial \mathbf{v}} = 0 \tag{32}$$

which reduces to

$$\delta Y = 0; \quad \text{similarly } \delta Z = 0 \tag{33}$$

and the Hamiltonian form

$$\begin{aligned} T^*u &= \frac{\partial X}{\partial x}; \qquad Tx = \frac{\partial X}{\partial u} \\ -\mathbf{B}_t \mathbf{v}_t &= \frac{\partial X}{\partial \mathbf{i}_l}; \quad -\mathbf{B}_t^T \mathbf{i}_l = \frac{\partial X}{\partial \mathbf{v}_t} \end{aligned} \tag{34}$$

In the intensive formulation the Y and Z functions are both invariant and both identically equal to zero [11]. This would appear to result from the fact that both of these functions are virtually functions of a single variable because of the dependency of $\mathbf{v}_l$ on $\mathbf{v}_t$ and the dependency of $\mathbf{i}_t$ on $\mathbf{i}_l$ due to the topological constraints. Hence a minimization principle associated with a linear function of a single variable can only be satisfied by the invariance of that function; the invariant value in these cases being zero, as established previously [8].

The intensive formulation presented here agrees with that given elsewhere by Jones and Evans [12]. It would appear that the properties of convexity and concavity demanded by Sewell are reflected here in extreme forms inasmuch that invariant functions can be considered as limiting cases that constitute a class of function that lie on the boundary of those that are doubly convex (or concave) and those that are convex in one variable and concave in another.

There is one important aspect that is not made explicit in this type of formulation and that is central to useful application. Although mathematically there is no restriction on the choice of tree so far, from a physical point of view there is considerable advantage in making a selection that provides the basic functions of tree co-content and link content with a truly resistive significance, compatible with Fig. 5a, by selecting a resistive tree or co-tree. Obviously it is not possible to do this simultaneously, as they are mutually exclusive, and so the whole complexity of the formulation as condensed in Fig. 8 is greatly increased. For every selection of a tree there will be a set of related functions. It may not be possible to proceed through the whole set of transformations, (this point is discussed in greater detail later), but there can be considerable advantage in selecting a preferred tree in order that particular functions are endowed with special physical significance, regardless of any possibility of it being transformed into others.

Furthermore the ability to relate this approach to the underlying fundamental thermodynamic mechanism of all systems, discussed at length by Korn *et al.* in this volume [18], rests on selecting the independent variables (i.e. the tree) to furnish the content and co-content functions with the real resistive properties of dissipation and entropy production. This topic has been introduced earlier by Flower and Evans [13] and is developed in the chapter by Korn *et al.*

4.2. The possibility of an extensive classification

Historically the variational description of conservative particle systems, couched in terms of energy functions, has been the reference on which any other variational formulation has been developed. At first sight it appears that minor changes in the choice of the x, u variables, to make them extensive (e.g. in the electrical network to let $x = \mathbf{q}_l$ and $u = \boldsymbol{\psi}_t$), might achieve the desired objective. We shall now explore this possibility and any limitations that it may have. It will be postulated, for the present, that $T^* = -\mathbf{B}_t \frac{d}{dt}$ and $T = \mathbf{B}_t^T \frac{d}{dt}$, and this will require a slight modification in the sign of one variable.

Consideration of Figs 4 and 6 together suggests that an appropriate choice for the scalar A within the extensive classification would be the sum $T_l(\mathbf{q}_l) + U_t(\boldsymbol{\psi}_t)$ from which we obtain the equations

$$y = \frac{\partial A}{\partial \mathbf{q}_l} = \mathbf{v}_l \equiv \frac{\partial X}{\partial x}; \quad \frac{\partial A}{\partial \boldsymbol{\psi}_t} = \mathbf{i}_t \equiv \frac{\partial X}{\partial u} = -\nu \tag{35}$$

and these correspond to eqns (9) and (10). As we have chosen a function of $\mathbf{q}_l$ and $\boldsymbol{\psi}_t$ which are equivalent to x and u of Figs 3 and 4, then scalar A above can be considered to be equivalent to the function $X(x, u)$ of Fig. 8, and also to the classical Hamiltonian H.

It can easily be demonstrated that to complete the entire structure of Fig. 8 for the extensive classification it is necessary to adopt the set of variables shown in Fig. 10.

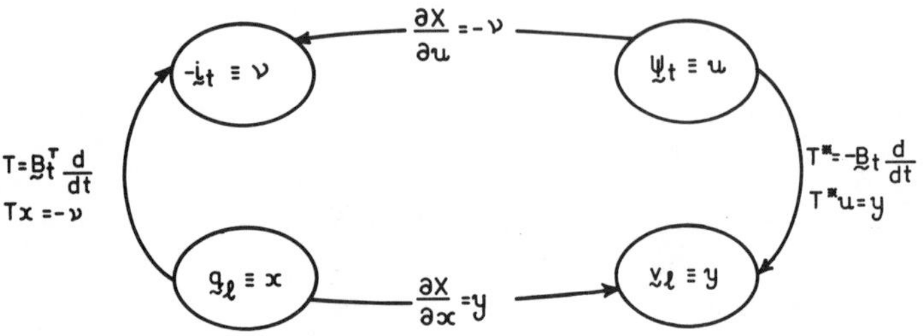

Fig. 10.

From these it follows that the following functions can be derived by the Legendre transformations.

$$
\begin{aligned}
dY &\equiv dL = (\mathbf{q}_l, d\mathbf{v}_l) - \langle \mathbf{i}_t, d\boldsymbol{\psi}_t \rangle; & \frac{\partial Y}{\partial \mathbf{v}_l} &= \mathbf{q}_l; & \frac{\partial Y}{\partial \boldsymbol{\psi}_t} &= -\mathbf{i}_t \\
dZ &\equiv dL^* = \langle \boldsymbol{\psi}_t, d\mathbf{i}_t \rangle - (\mathbf{v}_l, d\mathbf{q}_l); & \frac{\partial Z}{\partial(-\mathbf{i}_t)} &= -\boldsymbol{\psi}_t; & \frac{\partial Z}{\partial \mathbf{q}_l} &= -\mathbf{v}_l \\
dW &\equiv dH^* = -\langle \boldsymbol{\psi}_t, d\mathbf{i}_t \rangle - (\mathbf{q}_l, d\mathbf{v}_l); & \frac{\partial W}{\partial(-\mathbf{i}_t)} &= \boldsymbol{\psi}_t; & \frac{\partial W}{\partial \mathbf{v}_l} &= -\mathbf{q}_l \\
dX &\equiv dH = \langle \mathbf{i}_t, d\boldsymbol{\psi}_t \rangle + (\mathbf{v}_l, d\mathbf{q}_l); & \frac{\partial X}{\partial \mathbf{q}_l} &= \mathbf{v}_l; & \frac{\partial X}{\partial \boldsymbol{\psi}_t} &= \mathbf{i}_t
\end{aligned} \tag{36}
$$

The Lagrangian formulation then becomes

$$T \frac{\partial Y}{\partial (T^*u)} + \frac{\partial Y}{\partial u} = 0; \quad \mathbf{B}_t^T \frac{d}{dt} \frac{\partial Y}{\partial(-\mathbf{B}_t\boldsymbol{\psi}_t)} + \frac{\partial Y}{\partial \dot{\boldsymbol{\psi}}_t} = 0$$

or
$$\mathbf{B}_t^T \frac{d}{dt}\frac{\partial Y}{\partial \mathbf{v}_l} + \frac{\partial Y}{\partial \boldsymbol{\psi}_t} = 0 \tag{37}$$

and the Hamiltonian

$$T^*u = \frac{\partial X}{\partial x}, \quad Tx = \frac{\partial X}{\partial u}; \quad -\mathbf{B}_t \frac{d}{dt}\boldsymbol{\psi}_t = \frac{\partial X}{\partial \mathbf{q}_l} = \mathbf{v}_l$$

$$\mathbf{B}_t^T \frac{d}{dt}\mathbf{q}_l = \frac{\partial X}{\partial \boldsymbol{\psi}_t} = \mathbf{i}_t \tag{38}$$

Both of these agree with those given earlier by Jones and Evans [12].

Clearly the functions that have been derived are generalizations, in a topological form, of the classical Hamiltonian and Lagrangian functions (and their co-forms). If the special case of a network containing only capacitors and inductors (i.e. a purely conservative system) is considered in which all the inductors are assumed to constitute the links and all the capacitors the tree branches, we have a generalization of multivariable classical dynamics. If other types of element are present (e.g. resistors), or such a partitioning of the network as just suggested is not possible, then the formulation is purely a mathematical statement that hinges on the definition of two variables $\boldsymbol{\psi}_t$ and $\mathbf{q}_l$ which are defined by

$$\dot{\boldsymbol{\psi}}_t = \mathbf{v}_t, \quad \dot{\mathbf{q}}_l = \mathbf{i}_l \tag{39}$$

In a restricted class of the problem (a conservative system properly partitioned into tree and links), these variables assume a positive reality.

A comparative study of the extensive formulation and the intensive formulation gives rise to certain, so far unresolved, questions.

If
$$dH = \langle \mathbf{i}_t, d\boldsymbol{\psi}_t \rangle + (\mathbf{v}_l, d\mathbf{q}_l) \tag{40}$$

and
$$\frac{dH}{dt} = \langle \mathbf{i}_t, \mathbf{v}_t \rangle + (\mathbf{v}_l, \mathbf{i}_l) = 0$$

therefore
$$dH = 0 \tag{41}$$

and we can say that H is constant; the usual property distinguishing conservative systems.

If we consider the conditions for integrability for H we obtain

$$\frac{\partial \mathbf{i}_t}{\partial \mathbf{q}_l} = \frac{\partial \mathbf{v}_l}{\partial \boldsymbol{\psi}_t} \tag{42}$$

and reference to the conditions related to the function $X = G_l(\mathbf{i}_l) - J_t(\mathbf{v}_t)$ in the previous intensive classification, and the note in the Appendix, shows that this can only be satisfied by

$$\frac{\partial}{\partial \mathbf{q}_l}\left(\mathbf{B}_t^T \frac{d}{dt}\mathbf{q}_l\right) = 0 = \frac{\partial}{\partial \boldsymbol{\psi}_t}\left(-\mathbf{B}_t \frac{d}{dt}\boldsymbol{\psi}_t\right) \tag{43}$$

It is difficult to understand how the conditions for convexity and concavity demanded by Sewell can be met without resort to rather devious manipulation, if at all. These points may be related to those raised by Seabrook and Evans [14] elsewhere, and be of some interest in future work.

5. AN EXTENDED GENERALIZED FORMULATION

A complete intensive specification for an electrical network has been shown in Section 4.1, to be the complete sets of branch voltages, $\mathbf{v}_b$, and branch currents, $\mathbf{i}_b$. This set of generalized across- and through-variables can be reduced to the independent sets of tree across-variables and link through-variables. In the case of an electrical network the branch voltages, $\mathbf{v}_b = \begin{bmatrix} \mathbf{v}_t \\ \hline \mathbf{v}_l \end{bmatrix}$, and branch currents, $\mathbf{i}_b = \begin{bmatrix} \mathbf{i}_t \\ \hline \mathbf{i}_l \end{bmatrix}$, are dependent on the independent sets $\mathbf{v}_{to} = \begin{bmatrix} \mathbf{v}_t \\ \hline \mathbf{0} \end{bmatrix}$ and $\mathbf{i}_{ol} = \begin{bmatrix} \mathbf{0} \\ \hline \mathbf{i}_l \end{bmatrix}$. This dependency is expressed by the across- and through-variable constraints (eqns (4) and (5)). These equations can be extended and written in partitioned form as

$$\mathbf{v}_b = \mathbf{\Omega}^* \mathbf{v}_{to} \tag{44}$$

$$\mathbf{i}_b = \mathbf{\Omega} \mathbf{i}_{ol} \tag{45}$$

where

$$\mathbf{\Omega}^* = \begin{bmatrix} \mathbf{I} & \mathbf{0} \\ -\mathbf{D}^T & -\mathbf{I} \end{bmatrix} \qquad \mathbf{\Omega} = \begin{bmatrix} -\mathbf{I} & \mathbf{D} \\ \mathbf{0} & \mathbf{I} \end{bmatrix}$$

These equations, similar in form to those used by Kron [2], have the advantage of being non-singular and are therefore reversible (compared to the singular form of eqns (4) and (5)).

For a dynamical network which has a natural motion governed by Tellegen's theorem, the non-singular across- and through-variable mappings (eqns (37) and (38)) can be considered as adjoint transformations between the tree and link partitions of the network [3]. These transformations can be used to construct a complete set of Legendre transformations between the primal and dual (or co-) Hamiltonian and Lagrangian formulations of the network problem. The allocation of the variables and the role played by the content and co-content functions are clearly shown in Fig. 11 which can be regarded as a more generalized form of Fig. 9, the non-singular nature of eqns (37) and (38) adding considerably to the flexibility of the approach and is complementary to the rigorous approach of Sewell.

where

$$(\mathbf{\Omega}^*)^{-1} = \mathbf{\Omega}^* \quad \text{and} \quad (\mathbf{\Omega})^{-1} = \mathbf{\Omega} \quad \text{and} \quad -\mathbf{\Omega}^T = \mathbf{\Omega}^*$$

The Hamiltonian functional associated with this formulation is the explicit bilinear functional, P_T, given by

$$P_T(\mathbf{i}_{ol}, \mathbf{v}_{to}) = (\mathbf{\Omega}^* \mathbf{v}_{to}, \mathbf{i}_{ol}) = -\langle \mathbf{\Omega} \mathbf{i}_{ol}, \mathbf{v}_{to} \rangle \tag{46}$$

where $\mathbf{i}_{ol}^T \mathbf{\Omega}^T \mathbf{v}_{to} = P_t$, the tree power; and $\mathbf{v}_{to}^T \mathbf{\Omega}^{*T} \mathbf{i}_{ol} = P_l$, the link power.

The co-Hamiltonian functional, Q_T, is the explicit bi-linear functional

$$Q_T(\mathbf{v}_b, \mathbf{i}_b) = \langle \mathbf{\Omega}^* \mathbf{v}_b, \mathbf{i}_b \rangle = -(\mathbf{\Omega} \mathbf{i}_b, \mathbf{v}_b) \tag{47}$$

where

$$\mathbf{v}_b^T \mathbf{\Omega}^{*T} \mathbf{i}_b = P_t \quad \text{and} \quad \mathbf{i}_b^T \mathbf{\Omega}^T \mathbf{v}_b = P_l.$$

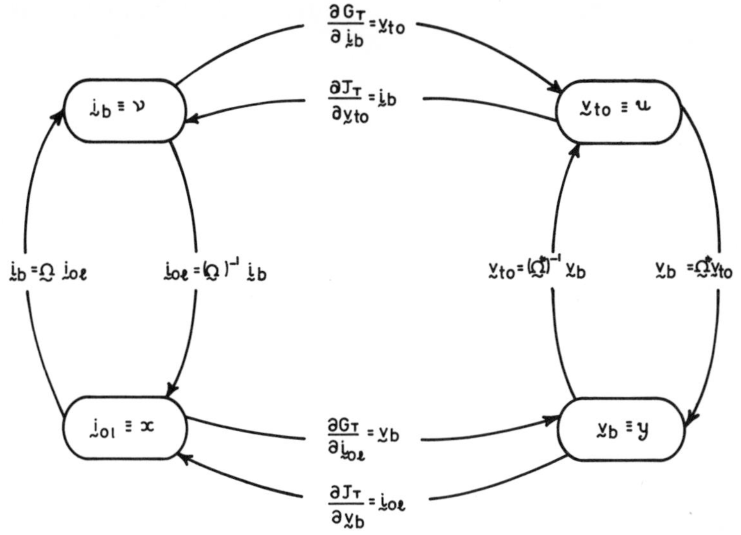

Fig. 11.

The Lagrangian and co-Lagrangian functionals, J_T, and, G_T, are the total network co-content and content respectively, both of which have been shown to be zero.

The complete formulation, and the implicit algebraic and partial differential equations, can be summarized in diagrammatic form in the manner of Sewell, as shown in Fig. 12.

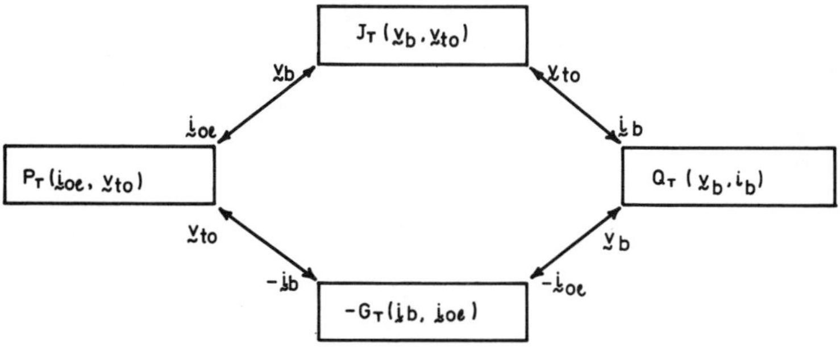

Fig. 12.

The variational principle implicit in this formulation is essentially that derived before, namely

$$\delta J_T = 0; \qquad \delta G_T = 0. \tag{48}$$

The adjoint transformations between the variables used in this formulation can be represented as mappings between inner product spaces, as used by Sewell [10]. These spaces can be identified with the tree and link partitions of the network; the inner products of the space being the instantaneous tree and link power respectively. These products are embodied in the bi-linear functionals P_T and Q_T. The reversibility of the spacial mappings clearly shows the way in which the network tree and links act as system and environment, with a duality of role, as shown in Fig. 13.

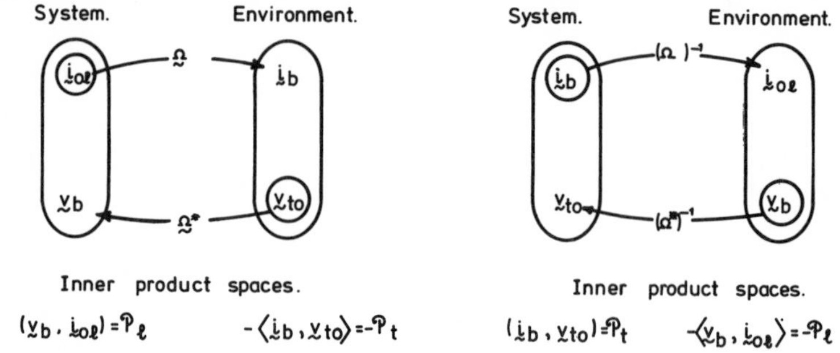

Fig. 13.

The variables of this formulation can be classified from an examination of the role of the passive variables in the complete set of Legendre transformations. The across-variables are passive in transformations from the mixed-variable power functionals P_T and Q_T to the co-content functional J_T; the through-variables being passive in transformations from the mixed variable functionals to the content functional G_T.

The N-independent across-variables and M-independent through-variables (in this extended formulation $N = M$), can be considered as the coordinates of a $(N + M)$-dimensional space; the phase-space of the system. This enables the independent passive variables of the Legendre transformations to be classified as phase-space variables; the complete set of $(N + M)$ branch variables forming the coordinates of the dual, or co-phase space. This classification, shown in Fig. 14, includes as a reduced case the earlier classification of Jones and Evans [4].

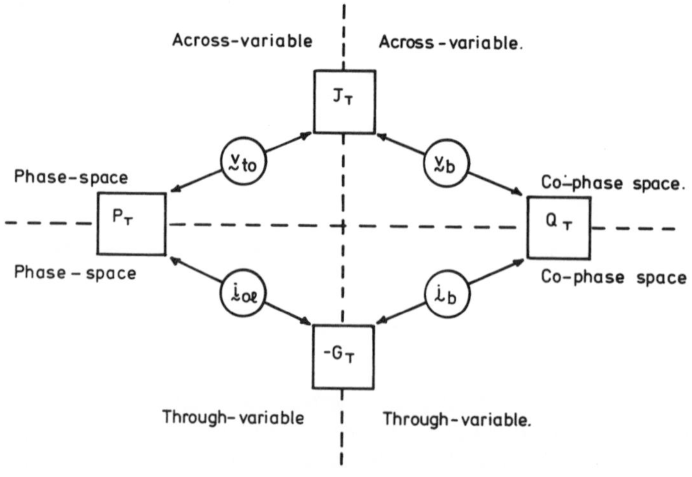

Fig. 14.

The complete set of Legendre transformations can be expressed in terms of operations on the set of independent variables $\mathbf{v}_{to}, \mathbf{i}_{ol}$ (Fig. 15). This enables the derivatives of the primal and dual Lagrangian functionals to be combined to obtain across-variable and through-variable forms of the Euler–Lagrange equations for the intensive network description.

Similarly the derivatives of the primal and dual Hamiltonian functionals provide dual forms of Hamiltons equations. These equations are the phase-space and co-phase space equations of motion for the intensive network description. These equations can be reduced to the state-space equations for the network and have been considered elsewhere [15].

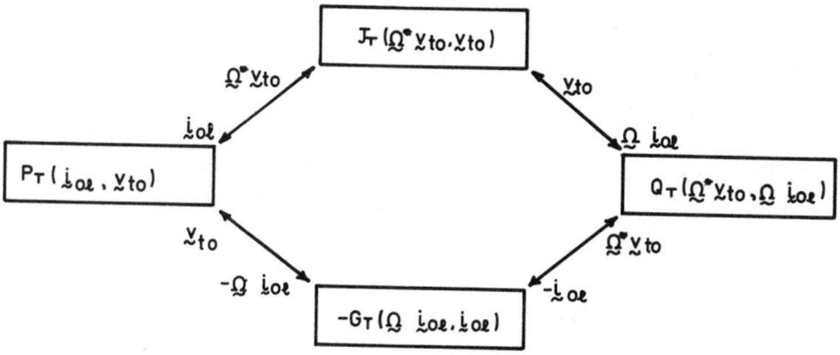

Fig. 15.

Euler–Lagrange equations

$$\frac{\partial J_T}{\partial \mathbf{v}_{to}} - \mathbf{\Omega}\frac{\partial J_T}{\partial(\mathbf{\Omega}^*\mathbf{v}_{to})} = \mathbf{0} \qquad \text{Across-variable form} \qquad (49)$$

$$\frac{\partial(-G_T)}{\partial \mathbf{i}_{ol}} - \mathbf{\Omega}^*\frac{\partial(-G_T)}{\partial(\mathbf{\Omega}\mathbf{i}_{ol})} = \mathbf{0} \qquad \text{Through-variable form} \qquad (50)$$

Hamiltons equations

$$\frac{\partial P_T}{\partial \mathbf{v}_{to}} = -\mathbf{\Omega}\mathbf{i}_{ol}$$

$$\frac{\partial P}{\partial \mathbf{i}_{ol}} = \mathbf{\Omega}^*\mathbf{v}_{to} \qquad \text{Phase space} \qquad (51)$$

$$\frac{\partial Q_T}{\partial(\mathbf{\Omega}^*\mathbf{v}_{to})} = -\mathbf{i}_{ol}$$

$$\frac{\partial Q_T}{\partial(\mathbf{\Omega}\mathbf{i}_{ol})} = \mathbf{v}_{to} \qquad \text{Co-phase space.} \qquad (52)$$

6. CLASSIFICATION OF FIELD VARIABLES

Whereas time was the independent variable in network theory, in distributed field problems one must express the spacial distribution, and in order to do this we can introduce a vector, x^k, of independent space-time variables,

where
$$x^k = [x, y, z, ict]$$

in which i and c are conventionally defined [4]. The raised suffix, k, implies that the vector x^k is of a contravariant nature.

The operators T and T^* introduced earlier in the network classification must now become four-dimensional differential operators of the form

$$\frac{\partial}{\partial x^k} \quad \text{and} \quad -\frac{\partial}{\partial x^k}.$$

It is now possible to separate the variables associated with fields into those belonging to first order tensors (i.e. vectors) and those belonging to zero-order tensors (i.e. scalars), and relate them to the operators shown above. This is similar to the "system-environment" approach used in the networks earlier, and we now have the more general concepts of co-variance and contra-variance corresponding to the "throughness" and "acrossness" used previously. Finally, if we propose a scalar β we can suggest a structure, similar to that of Fig. 4, for the classification of variables for simple fields, as shown in Fig. 16.

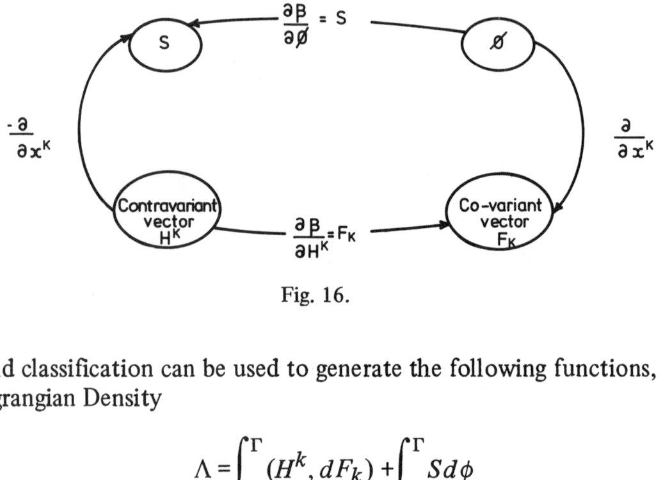

Fig. 16.

This field classification can be used to generate the following functions,
The Lagrangian Density

$$\Lambda = \int^{\Gamma} (H^k, dF_k) + \int^{\Gamma} S\, d\phi \tag{53}$$

(It should be noted that the summation convention should be used in the first integral).

The Co-Lagrangian Density

$$\Lambda^* = \int^{\Gamma} \phi\, dS - \int^{\Gamma} (F_k, dH^k) \tag{54}$$

The Hamiltonian Density

$$\beta = \int^{\Gamma} (F_k, dH^k) + \int^{\Gamma} S\, d\phi \tag{55}$$

and the Co-Hamiltonian Density.

$$\beta^* = \int^{\Gamma} (H^k, dF_k) + \int^{\Gamma} \phi\, dS \tag{56}$$

Note that it is customary to call the time component of the energy-momentum [16, 17] tensor, T_{ik}, the Hamiltonian density, i.e.

$$T_{44} = \int^{\Gamma} (H^4, dF_4) - \int \phi\, dS$$

TABLE 1

	Network description	Field description
Independent variables	t	$x^k = [x, y, z, ict]$
Generalized coordinates	u	ϕ
Generalized momenta	$x = \dfrac{\partial L}{\partial y}$; $\left(y = \dfrac{du}{dt}\right)$	$H^k = \dfrac{\partial \Lambda}{\partial F_k}$; $\left(F_k = \dfrac{\partial \phi}{\partial x^k}\right)$
Lagrangian	$L = \int^{\Gamma} (x, dy) - \int^{\Gamma} \langle v, du\rangle$	$\Lambda = \int^{\Gamma} (H^k, dF_k) - \int^{\Gamma} S\, d\phi$
Co-Lagrangian	$L^* = \int^{\Gamma} \langle u, dv\rangle - \int^{\Gamma} (y, dx)$	$\Lambda^* = \int^{\Gamma} \phi\, dS - \int^{\Gamma} (F_k, dH^k)$
Hamiltonian	$H = \int^{\Gamma} (y, dx) + \int^{\Gamma} \langle v, du\rangle$	$\beta = \int^{\Gamma} (F_k, dH^k) + \int^{\Gamma} S\, d\phi$
Co-Hamiltonian	$H^* = \int^{\Gamma} (x, dy) + \int^{\Gamma} \langle u, dv\rangle$	$\beta^* = \int^{\Gamma} (H^k, dF_k) + \int^{\Gamma} \phi\, dS$
Euler–Lagrange equations	$\dfrac{d}{dt}\dfrac{\partial L}{\partial y} - \dfrac{\partial L}{\partial u} = 0$	$\dfrac{\partial}{\partial x^k}\dfrac{\partial \Lambda}{\partial F_k} - \dfrac{\partial \Lambda}{\partial \phi} = 0$
	$-\dfrac{d}{dt}\dfrac{\partial L^*}{\partial v} - \dfrac{\partial L^*}{\partial x} = 0$; $\left(v = -\dfrac{dx}{dt}\right)$	$-\dfrac{\partial}{\partial x^k}\dfrac{\partial \Lambda^*}{\partial S} - \dfrac{\partial \Lambda^*}{\partial H^k} = 0$; $\left(S = -\dfrac{\partial H^k}{\partial x^k}\right)$
Legendre transformations	$H = \left(\dfrac{\partial L}{\partial y}, y\right) - L$	$\beta = \left(\dfrac{\partial \Lambda}{\partial F_k}, F_k\right) - \Lambda$
	$H^* = L^* - \left(\dfrac{\partial L^*}{\partial x}, x\right)$	$\beta^* = \Lambda^* - \left(\dfrac{\partial \Lambda^*}{\partial H^k}, H^k\right)$
Canonical equations	$y = \dfrac{\partial H}{\partial x}$, $v = \dfrac{\partial H}{\partial u}$	$F_k = \dfrac{\partial \beta}{\partial H^k}$, $S = \dfrac{\partial \beta}{\partial \phi}$
	$x = \dfrac{\partial H^*}{\partial y}$, $u = \dfrac{\partial H^*}{\partial v}$	$H^k = \dfrac{\partial \beta^*}{\partial F_k}$, $= \dfrac{\partial \beta^*}{\partial S}$
Conservation equations	$\dfrac{dH}{dt} = 0$	$\dfrac{\partial \beta}{\partial x^k} = 0$

However this definition is inconsistent as it implies that the canonical momentum is only the time component of H^k and not the space *and* time component of H^k as would normally be expected. In other terms, the canonical momentum derived from T_{44} would not be a Lorentz invariant [16]. The choice of β and β^* as Hamiltonian Densities is entirely consistent with the formalism and allows the field description to be reduced to the network or lumped description in the limiting case of time being the only independent variable.

Before turning to an example of the field classification, it is illuminating to compare the mathematical structure of analytical mechanics in the field description with that of the more familiar structure associated with the network description. This is done in Table 1.

7. EULERIAN FLUID FIELD

The behaviour of an unsteady, irrotational, compressible, inviscid fluid flow may be described by the Eulerian variables of velocity potential, ψ, velocity, v, momentum density, ρv, and sources and sinks, Q. The classification of Fig. 17 gives for this problem

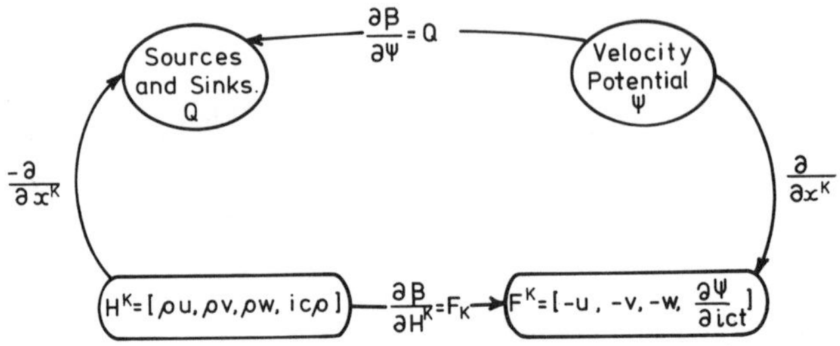

Fig. 17.

The Lagrangian Density can be shown to be

$$\Lambda = \int^{\Gamma} (H^k, dF_k) - \int^{\Gamma} Q\, d\psi \tag{57}$$

and the Co-Lagrangian Density

$$\Lambda^* = \int^{\Gamma} \psi\, dQ - \int^{\Gamma} (F_k\, dH^k) \tag{58}$$

By introducing the specific enthalpy h where

$$h = \frac{\partial\psi}{\partial t} - \frac{1}{2} v^2 \tag{59}$$

the pressure p where

$$p = \int^{\Gamma} \rho\, dh \tag{60}$$

and the internal energy E where

$$\rho E = \int^{\Gamma} h \, d\rho \tag{61}$$

the Lagrangian and Co-Lagrangian densities can be shown to be (ignoring the sources and sinks)

$$\Lambda = p \tag{62}$$

and

$$\Lambda^* = \rho E - \frac{1}{2}\rho v^2 \tag{63}$$

The Hamiltonian Density

$$\beta = (F_k, dH^k) + \int^{\Gamma} Q d\psi \tag{64}$$

can be shown to be equivalent to

$$\beta = \frac{1}{2}\rho v^2 - \rho E \tag{65}$$

when the sinks and sources are ignored.

Finally it is worth noting that T_{44}, as defined earlier, can be shown to be

$$T_{44} = \frac{1}{2}\rho v^2 + \rho E$$

or

$$T_{44} = \rho v^2 - \beta$$

The classification given in Fig. 17 is applicable only to simple fields such as the Eulerian Fluid Field. However, it is possible to develop other classifications for more complex fields, such as the electromagnetic, where the variables may be allocated to the spaces of first and second order tensors, and the operators are of the form $\frac{\partial}{\partial x^k}$ and $\left(\frac{\partial}{\partial x^i} - \frac{\partial}{\partial x^k}\right)$. This has been considered elsewhere [4].

8. CONCLUSIONS

Historically it has been the classical conservative variational formulations which have acted as the "pattern" for other variational problems (e.g. optimal control, approximate solutions, etc).

It has been shown in this paper that a generalized variational formulation can be extended to the general dissipative problem. The functionals characterizing the problem, and the conditions for the natural behaviour of the dissipative system have been identified, and are significant physical functions and well recognized physical laws. The classical Hamiltonian and Lagrangian formulations, and that of Brayton–Moser, have been shown to be reduced cases of the more general formulation.

A common framework is provided for both the lumped and distributed systems and is a valuable unifying concept. It is possible that it could form the basis for useful criteria associated with the numerical solution of partial differential equations.

This work has unified a number of significant physical laws and concepts and should provide new insight into the general structure, and behaviour, of systems. The full implications of this work have yet to be realized.

APPENDIX

The integrability of multi-variable functions

Consider a function of the form

$$F=\int \mathbf{v}_l^T d\mathbf{i}_l - \int \mathbf{i}_t^T d\mathbf{v}_t$$

together with the relationships

$$\mathbf{v}_l = -\mathbf{B}_t \mathbf{v}_t; \quad \mathbf{i}_t = \mathbf{B}_t^T \mathbf{i}_l$$

The conditions for integrability could be written in the form

$$\left(\frac{\partial \mathbf{v}_l^T}{\partial \mathbf{v}_t}\right)^T = -\frac{\partial \mathbf{i}_t^T}{\partial \mathbf{i}_l}$$

But a notation is required to represent the expanded form of these conditions unambiguously. This expanded form is:

$$\begin{bmatrix} \frac{\partial \mathbf{v}_{l1}}{\partial \mathbf{v}_{t1}} & \cdots & \frac{\partial \mathbf{v}_{l1}}{\partial \mathbf{v}_{tt}} \\ & & \\ \frac{\partial \mathbf{v}_{ll}}{\partial \mathbf{v}_{t1}} & & \frac{\partial \mathbf{v}_{ll}}{\partial \mathbf{v}_{tt}} \end{bmatrix}^T = \begin{bmatrix} -\frac{\partial \mathbf{i}_{t1}}{\partial \mathbf{i}_{l1}} & -\frac{\partial \mathbf{i}_{t1}}{\partial \mathbf{i}_{ll}} \\ & \\ -\frac{\partial \mathbf{i}_{t1}}{\partial \mathbf{i}_{ll}} & -\frac{\partial \mathbf{i}_{tt}}{\partial \mathbf{i}_{ll}} \end{bmatrix}$$

So we can interpret $\frac{\partial x_i}{\partial y_j}$ as a $j \times j$ matrix.

REFERENCES

1. Bryant, P. R. *The algebra and topology of electrical networks.* I.E.E. Monograph 414E, Nov. 1960.
2. Kron, G. *Diakoptics.* MacDonald, London, 1965.
3. Noble, B. and Sewell, M. J. "On dual extremum principles in applied mathematics". *J. Inst. Maths. Applics. 9,* 1972, 123–193.
4. Jones, D. L. and Evans, F. J. "A classification of physical variables and its application in variational methods". *J. Franklin Inst. 291,* 1971, 449–467.
5. Trent, H. M. "Isomorphisms between orientated linear graphs and lumped physical systems". *J. Acoust. Soc. Am. 27,* 1955, 500–527.

6. Paynter, H. M. *Analysis and design of engineering systems.* M.I.T. Press, Cambridge, Mass., 1960.
7. Tellegen, B. D. H. "A general network theorem with application". *Philips Res. Rep. 7,* 1952, 259–269.
8. Millar, W. "Some general theorems for non-linear networks possessing resistance". *Phil. Mag. 42,* 1951, p. 1150.
9. Cherry, C. "Some general theorems for non-linear networks possessing reactance". *Phil. Mag. 42,* 1951, p. 1161.
10. Sewell, M. J. "On applications of saddleshaped and convex generating functionals", in Dixhoorn, J. J. van and Evans, F. J. (eds.) *Physical Structure in Systems Theory*. Academic Press, London, 1974 (this volume).
11. Macfarlane, A. G. J. "An integral invariant formulation of a canonical equation". *Int. J. Control, II,* No. 3, 1970.
12. Jones, D. L. and Evans, F. J. "Variational analysis of Electrical Networks". *J. Franklin Inst. 295*, No. 1., 1973.
13. Flower, J. O. and Evans, F. J. "Irreversible thermodynamics and stability considerations of electrical networks". *J. Franklin Inst. 291,* 1971, 121–136.
14. Seabrook, R. and Evans, F. J. "Some questions related to the general variational principle". *J. Franklin Inst. 296*, No. 4, 1973.
15. Holding, D. J. and Evans, F. J. "Tensor Field Interpretations of the dynamic behaviour of Networks". *Electronic Letters 9,* 1973.
16. Leech, J. W. *Classical mechanics*, chs. 9, 11. Methuen, London, 1965.
17. Lanczos, C. *The variational principles of mechanics.* Toronto Press, 1966.
18. Korn, J., Evans, F. J. and Holding, D. J. "Thermodynamics and Field Theory as a basis for System Dynamics" in Dixhoorn, J. J. van and Evans, F. J. (eds.) *Physical Structure in Systems Theory*. Academic Press, London, 1974 (this volume).
19. Brayton, R. K. and Moser, J. K. "A theory of non-linear networks." *Quart. Appl. Maths. 22*, 1964, 1.

NOMENCLATURE

General

A	scalar function
$\mathbf{B}_f$	fundamental mesh matrix
$\mathbf{B}_t$	tree component of $\mathbf{B}_f$
b	suffix, branch
$\mathbf{C}_f$	fundamental cut-set matrix
$\mathbf{C}_1$	link component of C_f
C	suffix, Capacitance
$\mathbf{D}$	fundamental transformation matrix
G	content
H, H^*	Hamiltonian and co-Hamiltonian
i	current
J	co-content
K, K^*	scalar functions
L	suffix, inductance
l	suffix, link
L, L^*	Lagrangian and co-Lagrangian
N, N^*	scalar functions
ol	suffix, extended independent link variables
p	generalized momentum
P_T	Hamiltonian bi-linear functional
P_t	tree power

P_1	link power
q	generalized coordinate
q_C	electrostatic charge
Q_T	co-Hamiltonian bi-linear functional
R	suffix, resistance
t	suffix, tree
T, T*	capacitive energy and co-energy
T, T^*	linear operators
to	suffix, extended independent tree variables
u_t	across tree variable
U, U^*	inductor energy and co-energy
v_t	through tree variable
v	voltage
W	co-Hamiltonian functional
X	Hamiltonian functional
x_l	through link variable
Y	Lagrangian functional
y_l	across link variable
Z	co-Lagrangian functional
Γ, Γ^*	linear operators
ψ	electromagnetic flux
Ω, Ω^*	extended circuit matrices

Field symbols

$\mathbf{F}_k$	co-variant vector
$\mathbf{H}^k$	contra-variant vector
S	source variables
ϕ	scalar potential
β, β^*	Hamiltonian and co-Hamiltonian densities
Λ, Λ^*	Lagrangian and co-Lagrangian densities
$\mathbf{T}_{ik}$	energy-momentum tensor

Fluid symbols

E	internal energy
h	specific enthalpy
p	pressure
Q	source (or sink)
u, v, w	velocity components
ρ	fluid density
ψ	velocity potential

THERMODYNAMICS AND FIELD THEORY AS A BASIS FOR SYSTEMS DYNAMICS

J. Korn
Faculty of Engineering
Science and Mathematics
Middlesex Polytechnic
Enfield
Great Britain

F. J. Evans *and* D. J. Holding
Department of Electrical and Electronic Engineering
Queen Mary College
University of London
Great Britain

SUMMARY

An approach to systems analysis is given that is based on the natural physical processes implicit in the state space formulation. An extension of these concepts is developed in terms of irreversible thermodynamics by considering an interaction between two classes of system components. This interaction is derived from topological transformations based on network theory. The well-known network potential functions of content and co-content are then shown to be forms of the thermokinetic potential function that is the basis of the thermodynamic approach to non-linear dissipative systems. Network theory provides the means of defining systems structure, and the properties of certain topological matrices are related to the transformational tensors of particle dynamics.

A tensor field interpretation is given for dynamic behaviour, and the metric properties of the coordinate spaces are discussed generally, and in particular with reference to network equations. Electro-mechanical and simple electrical network problems are used to illustrate the principles involved.

1. INTRODUCTION

Attempts to extend the fundamental notions of irreversible thermodynamics as propounded by Onsager, Prigogine, de Groot and others [1, 2, 3], have recently been reported. The fields of application include biological [4], chemical [5] and network systems [6]. In this chapter further extensions are suggested, particularly in the fields of network theory and particle dynamics, and from these a new integrated formulation is given in terms of tensor fields to cover a broad class of dissipative non-linear system.

It is possible to view the natural time behaviour of dynamic systems from a number of different conceptual standpoints, and this can result in greatly increased insight into the actual "mechanics" by which such variations in time occur. Newtonian mechanics was of a causal nature in the sense that motion resulted from the imposition of one or a number of forces on the system. Reformulation of mechanics via d'Alembert's principle led to a far greater generality of the theory which finally resulted in a new view that natural processes behaved in some special economic manner in order to minimize some integral over time. The details and consequences of this approach have been discussed by Sewell and by Jones, Holding and Evans in this volume. Here we shall be concerned with the examination of this central problem of *how* we can explain the time evolution of physical systems, not simply in terms of some numeric solution of a set of equations, but in terms of the underlying physical *processes.*

2. BASIC CONCEPTS

Classical variational mechanics stopped short of non-conservative systems, and it was not until thermodynamics – really thermostatics – was able to provide the means of considering equilibrium states in systems that exhibited some form of dissipation, that non-conservative systems could be examined. This, of course was achieved by the formulation of the concept of entropy. The cornerstones of the theory that follows will rest firmly on the two classical laws of thermodynamics.

However, not until the advent, in recent years, of the study of irreversible thermodynamics was it possible to consider not only the equilibrium states themselves but the nature of the trajectories that connect them. It is from these more recent concepts that we shall proceed to develop a theory, ultimately expressible in the geometry of multi-variable tensor fields, that permits us to examine the characteristics of the dynamic motion, without the explicit solution of the system equations, even in the case of a non-linear problem. Before embarking on the central aspects of this we shall discuss some related topics that provide a background to its development, and also serve to illustrate how certain qualitative aspects of modern control theory have been rather neglected, and how they can form the starting point of useful lines of thought.

2.1. State-space equations

The state-space equations describing the autonomous system

$$\dot{\mathbf{x}} = \mathbf{A}(\mathbf{x}) \,.\, \mathbf{x} \tag{1}$$

invoke a superficial geometric interpretation. However, there is usually no attempt to develop this concept in a more rigorous fashion that demands the unambiguous definition of the reference frame. The functional approaches so far proposed, such as those associated with Liapunov methods, make no stringent demands on a specific choice of co-ordinates. We feel that some attention to the *physical* significance of the space selected for the description of a system can be extremely profitable in terms of understanding.

This can be done initially at a relatively simple level. The state-space equations as expressed in eqn (1) represent a simple statement of the vectorial relationship between a system-velocity and a system-displacement vector. In other words the phase relationship between the two vectors is governed by the nature of the matrix $\mathbf{A(x)}$. Consideration of the most elementary system indicates that if the velocity *never* has a component in phase with the displacement the system can *never* be "unstable". It is only when the velocity vector has, for some time period, a component that is in phase with the displacement vector that the system could become "unstable".

This concept can be generalized to high dimensioned spaces.

If $$\cos\theta = \frac{\langle \mathbf{x}, \dot{\mathbf{x}} \rangle}{||\mathbf{x}|| \,.\, ||\dot{\mathbf{x}}||}$$

then if $\theta > 90^\circ$ then $\langle \mathbf{x}, \dot{\mathbf{x}} \rangle < 0$ and $\langle \mathbf{x}, \mathbf{A(x)x} \rangle$ must be < 0.

Clearly this approach is itself related to the Liapunov method for stability assessment. For if one chooses a Liapunov function of the form

$$V = \langle \mathbf{x}, \mathbf{A(x)x} \rangle$$

then it follows that the sign definite properties of the matrix $\mathbf{A(x)}$ will be intimately related to the system stability. The non-uniqueness of system state equations, and likewise the possible arbitrary choice of a Liapunov function for any given system, both raise the question of the existence of a natural method of selecting equations and functions which might have greater physical significance.

2.2. Simple vector field theory

It is interesting to examine this from another point of view. Any matrix can be written as the sum of two components, one symmetric and the other skew-symmetric.

$$\mathbf{A} = \mathbf{A}_1 + \mathbf{A}_2 \tag{2}$$

where $\mathbf{A}_1$ is symmetric and $\mathbf{A}_2$ skew-symmetric.

Then if we consider the simple relationship

$$\mathbf{y} = \mathbf{A}\mathbf{x}$$

we can write

$$\mathbf{y} = \mathbf{A}_1\mathbf{x} + \mathbf{A}_2\mathbf{x} \tag{3}$$

The first of these terms, due to its symmetry, possesses the properties which can allow the existence of a scalar potential function, W and can be derived from

$$\mathbf{A}_1\mathbf{x} = \text{Grad}_{\mathbf{x}} W \tag{4}$$

The second, due to its skew-symmetry, possesses properties allowing the existence of a vector potential, $\mathbf{H}$, and can be derived from

$$\mathbf{A}_2 = \text{Curl } \mathbf{H} \tag{5}$$

where the Curl operation is specified in its complete multi-dimensional form (not to be confused with the axial vector representation applicable only to low dimensional examples).

It follows that the vector on the left-hand side of equation (3) can be considered as being composed of the sum of two components, and since

$$\mathbf{x}^T\mathbf{A}_2\mathbf{x} = 0 \tag{6}$$

is always true (because of the skew-symmetry of $\mathbf{A}_2$), the component $\mathbf{A}_2\mathbf{x}$ can always be considered as being orthogonal to the original vector $\mathbf{x}$.

These simple concepts will be expanded later to form the basis of a more rigorous tensor field interpretation for general systems behaviour. To partition the conventional state-space equation $\dot{\mathbf{x}} = \mathbf{A}(\mathbf{x})\,\mathbf{x}$ in this way clearly re-emphasizes that it is the vectorial relationship between the system-velocity vector and system-displacement vector that is important, as stated initially. Also it can be seen that if the symmetrical component is sign definite negative, as in the case of a simple electrical network with positive resistors, then the system must be stable.

It should be noted that this type of approach, like the more sophisticated one that we shall develop, contains no inherent restrictions to linear systems. There is no doubt that the *solution*, and also the analysis, of non-linear problems must always be, to some degree, more difficult; but any overall concepts that embrace both classes of problem has great advantage. It can easily be shown that the linear eigenvalue/eigenvector analysis is also embedded in the simple notions above, as it is also, to some extent in all that follows.

Further, one can extend the notion of dynamic system behaviour as reflected in the motion of a particle in an n-dimensional space, by asking if it is possible to demonstrate the existence of an n-dimensional field that can be considered to generate the system motion *without* explicit solution of the original equations for the system.

3. THE THERMODYNAMICS OF SYSTEM BEHAVIOUR

Thermodynamics is concerned with systems that have an interaction with an external environment with respect to a passage of heat or an application of work. This is conventionally stated in the first law of thermodynamics. We shall be concerned with similar concepts but the boundaries of the system, and any interaction across these boundaries, will be carefully re-examined. We shall consider systems that are bounded and suffer a change of internal energy due to the performance of work. It is proposed that the system boundary be assumed to embrace two groups of components; the first (set A), are identified with the state, in a thermodynamic sense, and the remaining components, (set B), can be considered as bringing about a *change* in that state, as shown in Fig. 1. It is shown that the coupling between these two classes can be related not only to a certain topological transformation but also to a transformation of state functions.

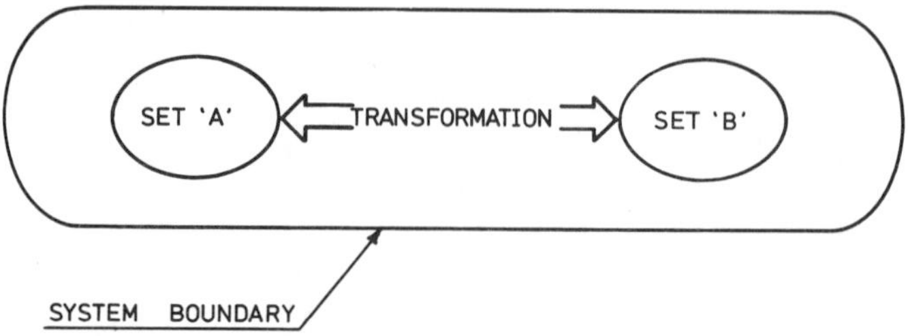

Fig. 1.

It should be emphasized that this partitioning into two subsystems does *not* necessarily coincide with a separation of *all* storage elements from all non-storage (i.e. dissipative) elements. For example, it is possible to conceive that in a simple mechanical system containing only a mass suspended from a spring the partitioning would, by necessity, separate the mass from the spring. It is only in this way that one can formulate any equations of motion. If, on the other hand, this simple system included a frictional damper the subdivision may be accomplished by separating *all* of the storage devices (i.e. mass *and* spring) from the dissipating elements. It is important to state here that the only criterion for the partitioning of a system in this way is that the sub-system, A, contains *at least* sufficient of the system to uniquely define the state. There is therefore usually more than one way in which this can be achieved.

The second law of thermodynamics postulates that the internal entropy S_i, of any system can only vary naturally according to

$$d_iS \geqslant 0 \tag{7}$$

and hence

$$\frac{d_iS}{dt} \geqslant 0 \tag{8}$$

The more recent development of a theory of irreversible thermodynamic processes supplements this basic postulate with relationships between certain sets of physical variables that provide additional conditions for natural behaviour. We feel that it is only with an integrated approach from the points of view of *both* thermodynamics and network theory that the full potential of the underlying concepts can be realized.

In simplest terms the stability of a system is dependent on the relative magnitudes of the rate of entropy production (i.e. dissipation) and the rate at which internal energy varies. Clearly, when viewed as in Fig. 1, these two internal processes are inextricably linked by the topological and constitutive constraints. For a bounded system knowledge of one of these processes will determine the overall system behaviour, and it should be possible to initiate the investigation from either an entropic or energic standpoint. A system cannot be stable if the rate of entropy production is insufficiently large, or if the rate of internal energy storage is continually increasing. This can be stated explicitly as

$$\frac{d}{dt}(\text{stored energy}) \leqslant 0 \tag{9}$$

and the rate of production of internal entropy is governed by the thermodynamic inequality

$$\frac{d_iS}{dt} \geqslant 0 \tag{8}$$

The limiting case, when all changes occurring are of a reversible nature, demands that both be identically zero.

It is necessary to formulate these conditions in terms of measurable variables (thermodynamically these are the *intensive* variables) [7]. However the internal energy of a system results from the storage of *extensities* (e.g. in the case of the electrical network, electrostatic charge and electromagnetic flux). If X_j are the set of extensive variables for a system then the rate of change of these with respect to time, the fluxes, J_j, are one set of intensive variables.

$$J_j = \frac{d}{dt} X_j \tag{10}$$

In irreversible thermodynamics the concepts of fluxes, or flows, resulting from the existence of a "force" or affinity, F, is expressed in the most general form.

In thermodynamics this dependency is postulated in the linear form

$$J_k = L_{kj} F_j \tag{11}$$

in which $L_{kj} = L_{jk}$ as in conventional irreversible thermodynamics, the L_{kj} being the phenomenological coefficients [7]. (Just as in the special case of an electrical resistor we have such a statement in Ohm's Law $v = Ri$.)

Although the fundamental equation

$$U = U(S, X_j) \quad j = 1 \ldots, n \tag{12}$$

expresses the fact that the total internal energy is purely a function of the extensities, including the entropy. For our purposes it is preferable to solve this equation in terms of entropy.

$$S = S(X_j) \quad j = 0 \ldots, n \tag{13}$$

Then the thermodynamic forces are defined by the change in internal entropy, S_i.

$$d_i S = \sum_j \frac{\partial S}{\partial X_j} dX_j = \sum_j F_j \, dX_j \tag{14}$$

and it can be seen that

$$\frac{d_i S}{dt} = \sigma = \sum_j F_j \frac{dX_j}{dt} = \sum_j F_j J_j \tag{15}$$

Equation (11) above, relating the thermodynamic fluxes and forces, is usually restricted in this form to the case of linear systems in which the coefficients are constant. In the much more common situation when this is not so, it is necessary to consider the differential form

$$d\sigma = \sum_j F_j \, dJ_j + \sum_j J_j \, dF_j = d\sigma_J + d\sigma_F \tag{16}$$

In this the first term has no specific sign definite properties, but the second is negative zero. The integral of this term, σ_F, has (when it exists) been termed the thermokinetic potential function. It is not explicitly obtained from the thermodynamics but must be derived by integration [8].

$$d\sigma_F = \sum_j J_j \, dF_j$$

$$\therefore \qquad \sigma_F = \int \mathbf{J}\, d\mathbf{F} \quad \text{and} \quad \mathbf{J} = \text{Grad}_{\mathbf{F}}\, \sigma_F \tag{17}$$

It is the identification of this function, σ_F, that is central to the whole analytic problem. Its integrability is related directly to the more well-known concepts of reciprocity and conservation. In a system in which irreversible processes are taking place the existence of any steady state is dependent on the minimum achievable value of the thermokinetic potential consistent with the constraints imposed by the constituent nature of the system.

So, in conclusion, it could be stated that a study of the transient nature of the system behaviour requires that one firstly identifies an entropic function, defined in terms of a set of measurable intensive coordinates related to the "*B*" set of components. Secondly, that the final description must be in terms of a set of intensive coordinates related to the "*A*" set of components. Implicit in the solution of the problem, therefore, is a transformation from one set of coordinates to another and the ability to define all necessary state functions in terms of either. Network theory provides the structure, and the insight, to make this possible.

4. NETWORK THEORY

The network theory required will initially be presented from an electrical point of view although the existence of well-known analogies will ultimately allow its extension into a wide class of system. Any network is composed of a number of different, well-recognized elements, and it will be assumed at this stage that these fall into the following categories.

Independent Sources: these are independent ideal current and voltage sources.

Dissipators: these dissipate energy and therefore produce internal entropy in the system.

Storage Devices: these have the ability to store the extensities associated with the system, and as a result are "receptacles" for the system's internal energy.

Couplers: these connect the storage and dissipating elements of the system and, in effect, create functional dependencies which arise due to energy conversion and amplification *within* the system.

The constraints that result from the characteristics of the elements listed above are superimposed on the topological description of the network connectivity. This underlying topological structure can be described by a linear graph, to each element of which is allocated two variables – a current and a voltage. This allows Kirchhoff's laws for the network to be stated in the following form

$$\begin{bmatrix} \mathbf{i}_t \\ \mathbf{v}_l \end{bmatrix} = \begin{bmatrix} \mathbf{O} & \mathbf{D} \\ -\mathbf{D}^T & \mathbf{O} \end{bmatrix} \begin{bmatrix} \mathbf{v}_t \\ \mathbf{i}_l \end{bmatrix} = \begin{bmatrix} \mathbf{O} & \mathbf{D} \\ \mathbf{D}^T & \mathbf{O} \end{bmatrix} \begin{bmatrix} -\mathbf{v}_t \\ \mathbf{i}_l \end{bmatrix} \tag{18}$$

where the suffices t and l refer to tree and link branches; the matrix $\mathbf{D}$ is called the dynamical transformation matrix.

As a result of this equation it can be shown that

$$\langle \mathbf{i}, \mathbf{v} \rangle = 0 \quad \text{(over all branches)} \tag{19}$$

for the complete network, and this represents a form of Tellegen's theorem [9].

An expansion of this equation is

$$\langle \mathbf{i}_t, \mathbf{v}_t \rangle + (\mathbf{i}_l, \mathbf{v}_l) = 0 \tag{20}$$

and therefore

$$\langle \mathbf{i}_t, \mathbf{v}_t \rangle = -(\mathbf{i}_l, \mathbf{v}_l) \tag{21}$$

The choice of a tree for any network constitutes, in fact, the choice of a set of independent variables. Since the matrix **D** of eqn (18) is in general singular, the variables $\mathbf{v}_t$ and $\mathbf{i}_l$ must be taken as sets of topologically independent variables. If these variables are considered, at this stage, to be completely independent of each other they can be considered to form a set of independent variables for the system. Any state functions that may eventually be used to describe the system must therefore be expressed as a function of these variables.

It would therefore be expected that eqn (17) which expresses the fluxes, J, as the gradient of an entropic (dissipative), function; (the gradient being taken with respect to the independent variables, F_j), would enable the analogous electrical function to be obtained.

If the tree currents are taken as the thermodynamic fluxes of the electrical network; the tree voltages as providing the corresponding affinities, then the tree co-content $J_t(\mathbf{v}_t)$ [10] enables this to be done as

$$dJ_t(\mathbf{v}_t) = \mathbf{i}_t, d\mathbf{v}_t \tag{22}$$

can be expressed in the form of eqn (17)

$$\mathbf{i}_t = \mathrm{Grad}_{\mathbf{v}_t} J_t(\mathbf{v}_t) \tag{23}$$

Similarly, if the link voltages are taken as the network "fluxes" and the link currents as the corresponding affinities then the link content, $G_l(\mathbf{i}_l)$ can be used to obtain an expression of the form of equation (17)

where
$$dG_l(\mathbf{i}_l) = \mathbf{v}_l, d\mathbf{i}_l \tag{24}$$

and
$$\mathbf{v}_l = \mathrm{Grad}_{\mathbf{i}_l} G_l(\mathbf{i}_l) \tag{25}$$

Both eqns (23) and (25) can be taken as forms of eqn (17) depending on the choice of forces and fluxes.

Using eqn (18)

$$dJ_t(\mathbf{v}_t) = \mathbf{i}_t, d\mathbf{v}_t = \mathbf{D}\mathbf{i}_l, d\mathbf{v}_t = \mathbf{i}_l, d\mathbf{D}^T\mathbf{v}_t = -\mathbf{i}_l, d\mathbf{v}_l = -dJ_l(\mathbf{v}_l) \tag{26}$$

and

$$dG_l(\mathbf{i}_l) = \mathbf{v}_l, d\mathbf{i}_l = -\mathbf{D}^T\mathbf{v}_t, d\mathbf{i}_l = -\mathbf{v}_t, d\mathbf{D}\mathbf{i}_l = -\mathbf{v}_t, d\mathbf{i}_t = -dG_t(\mathbf{i}_t) \tag{27}$$

and it is therefore possible to rewrite eqns (23) and (25)

$$-\mathbf{v}_l = \mathrm{Grad}_{\mathbf{i}_l} G_t(\mathbf{i}_l) \tag{28}$$

$$-\mathbf{i}_t = \mathrm{Grad}_{\mathbf{v}_t} J_l(\mathbf{v}_t) \tag{29}$$

and we can also see that

$$\mathbf{v}_l = \text{Grad}_{\mathbf{i}_l}\, G_l(\mathbf{i}_l) = -\mathbf{D}^T\, \text{Grad}_{\mathbf{i}_t}\, G_t(\mathbf{i}_t) \tag{30}$$

$$\mathbf{i}_t = \text{Grad}_{\mathbf{v}_t}\, J_t(\mathbf{v}_t) = \mathbf{D}\, \text{Grad}_{\mathbf{v}_l}\, J_l(\mathbf{v}_l) \tag{31}$$

A parallel can now be drawn between the generalized forms developed in the previous section and the more familiar variables of conventional electrical networks. In these the storage devices are capacitors and inductors which store the extensive entities of charge, q_c, and electromagnetic flux, ψ_L, respectively. Hence the associated "thermodynamic" fluxes will be the capacitive current, i_c, and the inductive voltage, v_L.

Networks do not in general consist solely of energy storage devices. Under these circumstances the gradient equations derived above (eqns (30) and (31)) become statements of Kirchhoff's mesh and cut-set laws and do not constitute explicit forms of the dynamic equations of the network.

For reasons discussed earlier it is the extensities that determine the energy state of a system, and it is the intensive variables associated with the storage elements that are required to describe the system.

The tree content and link co-content functions can be made completely resistive (i.e. dissipative or entropic), if the resistive branches are selected to form the tree or co-tree respectively. Under these circumstances (provided that the network contains no charge meshes, flux cut-sets, or resistor meshes; all tree resistors being current controlled and link resistors voltage controlled), by proper selection, the system independent coordinates can be made to coincide with the intensive variables of the storage elements (see Appendix).

Then for a network which has a tree consisting of resistors (and independent voltage sources), the links containing all the storage elements, other resistors (and independent current sources), we can write

$$-\mathbf{v}_{sl} = \text{Grad}_{\mathbf{i}_{sl}}\, G_t(\mathbf{i}_l) \tag{32}$$

where $\mathbf{v}_{sl}$ and $\mathbf{i}_{sl}$ are the link storage-element voltages and currents respectively.

Similarly, for a network the links of which contain resistors (and independent current sources) only; the tree containing all the storage elements, other resistors (and independent voltage sources), then we can write

$$-\mathbf{i}_{st} = \text{Grad}_{\mathbf{v}_{st}}\, J_l(\mathbf{v}_t) \tag{33}$$

where $\mathbf{i}_{st}$ and $\mathbf{v}_{st}$ are the tree storage-element currents and voltages respectively. If a further constraint, namely that the network contains no independent current and voltage sources, is added to the restrictions on equations (32) and (33) then these equations can be simplified, and the new, less general, **D** matrix can be utilized to give the following equations.

Resistive tree:

$$\begin{bmatrix} \mathbf{v}_{sl} \\ \mathbf{i}_{Rt} \end{bmatrix} = \begin{bmatrix} \mathbf{O} & \mathbf{D}^T \\ \mathbf{D} & \mathbf{O} \end{bmatrix} \begin{bmatrix} \mathbf{i}_{sl} \\ (-\mathbf{v}_{Rt}) \end{bmatrix} \tag{34}$$

$$\mathbf{v}_{sl} = \text{Grad}_{\mathbf{i}_{sl}} - G_t(\mathbf{i}_{sl}) = \mathbf{D}^T\, \text{Grad}_{\mathbf{i}_{Rt}} - G_t(\mathbf{i}_{Rt}) \tag{35}$$

Resistive co-tree:

$$\begin{bmatrix} \mathbf{i}_{st} \\ \mathbf{v}_{Rl} \end{bmatrix} = \begin{bmatrix} \mathbf{O} & \mathbf{D} \\ \mathbf{D}^T & \mathbf{O} \end{bmatrix} \begin{bmatrix} (-\mathbf{v}_{st}) \\ \mathbf{i}_{Rl} \end{bmatrix} \tag{36}$$

$$\mathbf{i}_{st} = \text{Grad}_{(-\mathbf{v}_{st})} J_l(-\mathbf{v}_{st}) = \mathbf{D}\ \text{Grad}_{\mathbf{v}_{Rl}} J_l(\mathbf{v}_{Rl}) \tag{37}$$

where the suffix R refers to resistive branch variables.

It should be noted that, in general, the dynamical transformation matrices appearing in eqns (34, 35) and (36, 37) do not have the same elements, due to the constraints imposed on the selection on tree and link branches for each formulation.

Although eqns (32), (33) and (35), (37) above have the required gradient form of eqn (17) and are expressed in terms of the intensive variables of the storage elements, the associated fluxes and affinities are not strictly assigned physically unless the storage elements consist solely of inductors in the case of the resistive tree, and capacitors in the case of the resistive co-tree. This is consistent with the existence of only one form of extensity storage in classical thermodynamics.

In order to obtain a correct physical assignment of the "thermodynamic" fluxes for networks with mixed storage elements it is necessary to choose a preferred tree consisting of capacitors, resistors (and independent voltage sources); the links consisting of inductors, other resistors (and independent current sources). In this case it is not possible to express the "fluxes" directly as gradients of the tree and link content and co-content functions; they can however be obtained as derivatives of a topologically extended form of the Brayton–Moser mixed potential function, P_T, [11, 12], where

$$dP_T = -\mathbf{i}_t, d\mathbf{v}_t + \mathbf{v}_l, d\mathbf{i}_l = -dJ_t + dG_l \tag{38}$$

and

$$\mathbf{v}_{sl} = \frac{\partial(P_T)}{\partial \mathbf{i}_{sl}} \tag{39}$$

$$\mathbf{i}_{st} = \frac{\partial(-P_T)}{\partial \mathbf{v}_{st}} \tag{40}$$

These equations, (39) and (40), are not a true gradient formulation and are discussed in detail later. It will then be shown that the "fluxes" are the resultant of the gradient of a dissipative (entropic) function *and* a solenoidal field caused by the existence of two "modes" of extensity storage.

It follows therefore that the simple electrical network can be described by the "thermodynamic" gradient formulation of eqn (17) in both a topological/intensive-variable form, and a storage-element/intensive-variable form. Associated with each formulation is a corresponding assignment of network "fluxes" and "affinities"; the scalar potential functions common to both forms being the tree-content and link co-content. However, when these "fluxes" and "affinities" are assigned on a strictly physical basis the resultant equations are recognizable as derivatives of an extended form of the Brayton–Moser mixed potential function.

The full significance of the transformational aspects of network structure, both in terms of variables and potential functions can only be appreciated by resort to an examination of the properties of transformations: these concepts are embodied in the discussion of classical particle dynamics which follows, Section 5.

4.1. Couplers and functional dependency

It will be recalled that certain constraints were placed on the phenomenological coefficients, L, in eqn (11). These constraints are reflected in the well-known relationships existing in the partitioned matrices that occur in network descriptions. These conditions have implications on the integrability of state functions [13].

The existence of naturally occurring state functions is also intimately related to the concept of passivity in networks. Tellegen's theorem, virtually a statement of the conservation of power in the network, rests on the skew symmetry of the matrices in eqn (18) and this, in turn, arises simply from the topology and the existence of the superimposed sets of completely independent variables. If, however, some additional constraints are now imposed in the form of some functional dependency the skew symmetry in the transformations on independent variables could be destroyed, and hence the ability to postulate power conservation within the system is lost.

In broadest terms a functional dependency can occur in two ways.

(*i*) A branch variable can be dependent on the other variable associated with the *same* branch.

(*ii*) A branch variable can be dependent on a branch variable associated with *another* branch. A dependency of this type need not be between like variables and can also govern *both* variables of a branch.

The first of these is usually generated by a simple resistive element in a branch and has a characteristic governed by Ohm's law relating its voltage and current.

The second type of dependency can be defined by a multi-port, or more frequently by a two-port, element or coupler and can be represented by an equation of the form

$$\begin{bmatrix} v_1 \\ i_1 \end{bmatrix} = \begin{bmatrix} a_{11} & a_{12} \\ a_{21} & a_{22} \end{bmatrix} \begin{bmatrix} v_2 \\ i_2 \end{bmatrix} \tag{41}$$

Depending on the nature of the element, eqn (41) can, for example, take the following form.

(*i*) $a_{12} = 0 = a_{21}$ and $a_{22} \neq 1/a_{11}$.
This is defined as a non-conservative transformer type coupler since summation of power over all the ports is not zero.

(*ii*) $a_{12} = 0 = a_{21}$ and $a_{22} = 1/a_{11}$
This is a conservative transformer type coupler. (The armature controlled "perfect" d.c. motor and "perfect" electrical transformer serve as examples.)

(*iii*) $a_{11} = 0 = a_{22}$ and $a_{21} \neq 1/a_{12}$
This is a non-conservative and non-reciprocal coupler.

(*iv*) $a_{11} = 0 = a_{22}$ and $a_{21} = 1/a_{12}$

This is commonly known as a gyrator type coupler; such couplers are found in electrical, mechanical and fluid power engineering [14]. As will be shown later in the example, the coupler concept is not restricted to a relationship between variables of a like nature and thus can be used to relate mixed-variable sub-systems (electrical/mechanical). Furthermore for the actual construction of equivalent networks it will be shown that the coupler itself appears as a combination of modulated sources distributed throughout a set of disconnected sub-networks of which the complete system is composed.

Clearly if only one element of the matrix in eqn (41) is non-zero the relationship would represent a perfect source, dependent on or modulated by another branch variable, as would occur if a *conventional* feedback term were present.

If conditions exist in the system descriptive equations such that Tellegen's theorem is no longer satisfied then the infringement that it suggests in the basic law of conservation of power implies that there are sources on which the system is drawing power that have not been stated explicitly in the description. In such cases it can be considered that the system "net power" is no longer zero, and it will be shown that this has a most important bearing on the transient stability.

Up to this point all matrix transformations have been considered to be linear (i.e. the elements of the topological **D** matrices are 0 or 1, and elements of eqn (41) constant).

In order to accommodate at a later stage functional dependencies (e.g. feedback) it will be necessary to remove this constraint with respect to eqn (41). Therefore, although a matrix formulation is adequate for the simpler linear forms it is not so for the non-linear cases, and it is necessary to introduce differential forms of the tensor transformation as is used in conventional particle dynamics.

5. PARTICLE DYNAMICS

It is essential in most problems concerning the dynamic behaviour of a mechanical system to have the ability to transform the coordinates of the system in a systematic manner. Conventionally the transformation from a set of generalized coordinates q^γ to any other set of coordinates x^α can be expressed in the following differential form

$$dx^\alpha = \frac{\partial x^\alpha}{\partial q^\gamma} dq^\gamma = C^\alpha_\gamma dq^\gamma \qquad (42)$$

It is not possible to separate from the concept of coordinate transformation the necessity to define those functions that are invariant to such transformations. It is clearly necessary that for a small variation of coordinate that the resulting work done should satisfy the following relationship.

$$(f_Q)_\nu dq^\gamma = (f_x)_\sigma dx^\alpha \qquad (43)$$

In this equation $(f_Q)_\nu$ and $(fx)_\sigma$ are the forces expressed with respect to the x and q frames of reference.

In order that this equality be satisfied, the transformations that must hold for both displacements and forces can be jointly expressed in the matrix equation

$$\begin{bmatrix} (f_Q)_\nu \\ dx^\alpha \end{bmatrix} = \begin{bmatrix} 0 & C_\nu^{\cdot\sigma} \\ C^\alpha_{\cdot\gamma} & 0 \end{bmatrix} \begin{bmatrix} dq^\gamma \\ (fx)_\sigma \end{bmatrix} \qquad (44)$$

in which the transformation tensors must also satisfy

$$C^\alpha_{\cdot\gamma} = C_\nu^{\cdot\sigma} \qquad (45)$$

Note: The position of the indices indicates co-variance or contra-variance in the conventional manner.

It can clearly be seen that eqn (44) is isomorphic to eqn (18) derived in the context of network theory. The formulation of network equations based on resistive trees can be seen as a reduced case of equation (44) in which the co-tree provides the generalized reference frame (eqn (34)). Alternatively the resistive co-tree formulation can be used in which the network tree provides the generalized reference frame (eqn (36)).

The possibility has now been established that a problem in the field of particle dynamics may be mapped into one in the field of network theory. But this only "closes the loop", so to speak, as it has also been demonstrated earlier that the network equations also reflect important thermodynamic characteristics, and lastly, that thermodynamic theory had already adopted simple mechanical concepts of forces, flows and coordinates.

If the following differential forms are postulated.

$$dI = (f_Q)_\nu \, dq^\gamma \tag{46}$$

$$dV = (f_x)_\sigma \, dx^\alpha \tag{47}$$

then it is possible to generate a further relationship of the form

$$(\text{Grad}_{\mathbf{q}}\, I)^\nu = C_\nu^{;\sigma} (\text{Grad}_{\mathbf{x}}\, V)_\sigma \tag{48}$$

and it can be seen that this can be reduced to the form of the earlier equations (30), (31), (35) and (37).

The conditions for the integrability of these differential forms must reside in the "structure" of the transformation tensors $C_\nu^{;\sigma}$ and $C^\alpha_{.\gamma}$. This fact is also reflected in the network interpretation of the problem by the matrix relationships necessary for the satisfying of Tellegen's theorem, as previously stated.

6. TENSOR FIELD THEORY

The transformational properties of equations of the form of eqns (30), (31), (35), (37) and (48) will now be examined in more detail. We shall consider the particular case of eqn (48).

$$(f_Q)_\nu = (\text{Grad}_{\mathbf{q}}\, I(q))_\nu = C_\nu^{;\sigma} (\text{Grad}_{\mathbf{x}}\, V(x))_\sigma \tag{49}$$

and recalling a notion introduced in Section 2 of this chapter, it will be assumed that a partitioning is possible to separate the symmetric and skew-symmetric components, as in eqn (2).

Hence

$$(f_Q)_\nu = (\text{Grad}\, W)_\nu + (\text{Curl}\, \mathbf{H})_{\nu\gamma}\, q^\gamma \tag{50}$$

where W is a scalar potential function arising from the symmetrical part, and **H** a vector potential function due to the skew-symmetric part of eqn (49).

The second part of the equation, $(\text{Curl}\, \mathbf{H})_{\nu\gamma}\, q^\gamma$, represents a solenoidal vector, the integrability condition on **H** being the divergence condition

$$\text{Div Curl}\, \mathbf{H} = 0 \tag{51}$$

6.1. Force tensor derived from content and co-content functions

When the force tensor of eqn (49) is derived from considerations of content and co-content then for passive networks, or for coupled networks with conservative reciprocal couplers,

$$\text{Curl } \mathbf{H} = 0 \tag{52}$$

Equation (49) gives the co-variant components of the force tensor expressed in the appropriate generalized coordinate system. The coordinate system is derived from considerations of the metric tensor which in turn is derivable from the coordinate transformation tensor, $C^{\alpha}_{\cdot\gamma\cdot}$.

In network analysis the coordinate transformation tensor may take the form of $\mathbf{D}$ or $\mathbf{D}^T$ (or, to show the inclusion of elements other than 1 and 0, $\mathbf{\Delta}_1$ or $\mathbf{\Delta}_2$ for coupled networks) depending on whether the node or loop equations are required by the analysis.

With reference to eqn (42) we are now looking for a relationship between the reference and generalized coordinate systems.

Thus rewriting eqn (42) in vector form

$$d\mathbf{x}^{\alpha} = \mathbf{a}\, d\mathbf{q}^{\gamma} \tag{53}$$

in which the matrix $\mathbf{a}$ is composed of $\mathbf{a}^{\gamma}$ column vectors, the elements derived from $C^{\alpha}_{\cdot\gamma\cdot}$.

Defining the metric in the reference coordinate system

$$(dx^{\alpha})^2 = d\mathbf{x}^{\alpha} \,.\, d\mathbf{x}^{\alpha} \tag{54}$$

and in the generalized coordinate system

$$(dx^{\alpha})^2 = d\mathbf{q}^{\gamma}\, \mathbf{a} \,.\, \mathbf{a}\, d\mathbf{q}^{\gamma} \tag{55}$$

There will also be a coordinate system reciprocal to the generalized coordinate system expressed by

$$(dx^{\alpha})^2 = d\mathbf{q}_{\gamma}\, \mathbf{a} \,.\, \mathbf{a}\, d\mathbf{q}_{\gamma} \tag{56}$$

From eqns (55) and (56) the metric and reciprocal metric tensors are defined as

$$\mathbf{g} = \mathbf{a}^T \,.\, \mathbf{a} \tag{57}$$

$$\mathbf{g}' = \mathbf{a}'^T \,.\, \mathbf{a}' \tag{58}$$

Also, the relationship between the coordinate vectors in the generalized and reciprocal coordinate system is

$$\mathbf{a}'^T \,.\, \mathbf{a} = \mathbf{I} \tag{59}$$

$$\mathbf{a}' \,.\, \mathbf{a}^T = \mathbf{I} \tag{60}$$

The coordinate systems are generated by the connection properties of a network which in turn are determined by the functional relationships caused by the couplers. This is demonstrated in the example of Appendix I.

When the elements of the transformation tensor are constant the coordinate system is *oblique* and when they are not the coordinate system is curvi-linear. In general, the generalized coordinate system is *Cartesian* only when the analysis is concerned with physically identified "fluxes", generally associated with the network state equations.

The equations above enable one to construct coordinate systems in which, at least in two dimensions, the force vectors and the fields created by these forces may be sketched.

6.2. Mixed-potential formulation and state-space equations

It was shown in Section 4 that the thermodynamic fluxes and affinities were not correctly assigned (in a physical sense) to networks with more than one form of energy storage medium unless the mixed-potential form of network description was used. Physically identifiable "fluxes" and affinities could also be used in the content and co-content formulations when only one form of energy storage was present. In the previous section it was shown that these network descriptions (the network state equations using sets of independent coordinates) naturally coincide with an orthogonal system of generalized coordinates.

The network state equations and the constitutive relationships of the extensity storage elements enable the state description to be extended directly into the state-space equations of the system, while retaining the physical significance of the variables and potential functions involved [15].

Considering first the mixed potential formulation, in which the fluxes are expressed as derivatives of the extended Brayton–Moser mixed potential function, eqns (39) and (40). These equations can be expressed explicitly in terms of the network state variables.

The general transformational properties of the network, at a topological level, are given by eqn (18)

$$\begin{bmatrix} (i_t)_N \\ (v_l)^M \end{bmatrix} = \begin{bmatrix} 0 & D_N^{\cdot\sigma} \\ -D^M_{\cdot\rho} & 0 \end{bmatrix} \begin{bmatrix} (v_t)^\rho \\ (i_l)_\sigma \end{bmatrix} \tag{61}$$

and the network topological variables, $\mathbf{v}_t$ and $\mathbf{i}_l$ can be expressed in terms of the network state variables, $\mathbf{i}_{sl}$ and $\mathbf{v}_{st}$ and the network independent sources e and j.

$$\begin{bmatrix} (v_t)^N \\ (i_l)_M \end{bmatrix} = \begin{bmatrix} \alpha^N_{\cdot n} & \alpha^{Nm} \\ \alpha_{Mn} & \alpha_M^{\cdot m} \end{bmatrix} \begin{bmatrix} (v_{st})^n \\ (i_{sl})_m \end{bmatrix} + \begin{bmatrix} \beta^N_{\cdot\lambda} & \beta^{N\pi} \\ \beta_{M\lambda} & \beta_M^{\cdot\pi} \end{bmatrix} \begin{bmatrix} e^\lambda \\ j_\pi \end{bmatrix} \tag{62}$$

Considering for stability purposes the transient behaviour of the undriven system

$$\begin{bmatrix} (i_t)_N \\ (v_l)^M \end{bmatrix} = \begin{bmatrix} 0 & D_N^{\cdot\sigma} \\ -D^M_{\cdot\rho} & 0 \end{bmatrix} \begin{bmatrix} \alpha^\rho_{\cdot n} & \alpha^{\rho m} \\ \alpha_{\sigma n} & \alpha_\sigma^{\cdot m} \end{bmatrix} \begin{bmatrix} (v_{st})^n \\ (i_{sl})_m \end{bmatrix} = \begin{bmatrix} \Delta_{Nm} & \Delta_N^{\cdot m} \\ \Delta^M_{\cdot n} & \Delta^{Mm} \end{bmatrix} \begin{bmatrix} (v_{st})^n \\ (i_{sl})_m \end{bmatrix} \tag{63}$$

The network state equation is embedded within eqn (63) and will be examined in detail.

$$\begin{bmatrix} (i_{st})_\nu \\ (v_{sl})^\mu \end{bmatrix} = \begin{bmatrix} A_{\nu n} & A_\nu^{\cdot m} \\ A^\mu_{\cdot n} & A^{\mu m} \end{bmatrix} \begin{bmatrix} (v_{st})^n \\ (i_{sl})_m \end{bmatrix} \tag{64}$$

When the network fluxes are correctly assigned, the left-hand side of eqn (64) is equal to the derivatives of the Brayton–Moser potential function, eqns (39) and (40) are also the time rate of change of the network extensities, thus leading directly to the network state-space equations.

In order to identify the scalar and vector potential functions for the network we shall consider a network consisting of a preferred tree, no resistive loops, and functional dependencies of type (i) in Section 4.1.

Then eqn (64) can be written as

$$\begin{bmatrix} (i_c)_\nu \\ (v_L)^\mu \end{bmatrix} = \begin{bmatrix} A_{\nu n} & A_\nu^{\cdot m} \\ A^\mu_{\cdot n} & A^{\mu m} \end{bmatrix} \begin{bmatrix} (v_c)^n \\ (i_L)_m \end{bmatrix} \tag{65}$$

where

$$A_{\nu n} = -D_\nu^{\cdot \epsilon} G_{\epsilon\epsilon} D^\epsilon_{\cdot n}; \quad A^{\mu m} = -D^\mu_{\cdot \delta} R^{\delta\delta} D_\delta^{\cdot m}$$
$$A_\nu^{\cdot m} = D_\nu^{\cdot m}; \quad A^\mu_{\cdot n} = -D^\mu_{\cdot n}$$

and m, μ are summed over the resistive branches; n, ν over the capacitive branches, ϵ over the link conductances and δ over the tree resistances.

Defining the elements of the $(n + m)$ dimensional state vector, $\mathbf{x}$, in an orthogonal $(n + m)$ dimensional Riemanian space, $V_{(n+m)}$, by

$$x^n = (v_c)^n \quad \text{with metric } g_{nn} = (C)_{nn}$$
$$x_m = (i_L)_m \quad \text{with metric } g^{mm} = (L)^{mm}$$

then

$$x_n = (q_c)_n \quad \text{and} \quad x^m = (\psi_L)^m \tag{66}$$

the fundamental form for V_{n+m} being defined as

$$g_{ik} x^i x^k = (v_c)^n (q_c)_n + (\psi_L)^m (i_L)_m \tag{67}$$

where i, k are summed over $1 \ldots (n + m)$.

These equations, (65) and (66) enable the co-variant, contra-variant, and mixed variant forms of the network state equations to be obtained. These equations, corresponding to the intensive, extensive, and mixed variable forms of the state-space equations all refer to the *same* system and are all characterized by the same scalar and vector potential functions.

Examining the form

$$\begin{bmatrix} \dot{x}_\nu \\ \dot{x}_\mu \end{bmatrix} = \begin{bmatrix} A_{\nu n} & A_{\nu m} \\ A_{\mu n} & A_{\mu m} \end{bmatrix} \begin{bmatrix} x^n \\ x^m \end{bmatrix} \tag{68}$$

this equation can be separated into symmetric and skew-symmetric parts

$$\begin{bmatrix} \dot{x}_\nu \\ \dot{x}_\mu \end{bmatrix} = \begin{bmatrix} \theta_{\nu n} & \theta_{\nu m} \\ \theta_{\mu n} & \theta_{\mu m} \end{bmatrix} \begin{bmatrix} x^n \\ x^m \end{bmatrix} + \begin{bmatrix} \phi_{\nu n} & \phi_{\nu m} \\ \phi_{\mu n} & \phi_{\mu m} \end{bmatrix} \begin{bmatrix} x^n \\ x^m \end{bmatrix} \tag{69}$$

where $\theta_{ij} = \theta_{ji}$ and $\phi_{ij} = -\phi_{ji}$.

Then we can define scalar and vector potential functions W and $\mathbf{H}$ by

$$\theta_{ij}\, x^j = (\text{Grad } W)_i = \frac{\partial W}{\partial x^i} \tag{70}$$

and

$$\phi_{ij} = (\text{Curl } \mathbf{H})_{ij} = \frac{\partial \mathbf{H}_i}{\partial x^j} - \frac{\partial \mathbf{H}_j}{\partial x^i} \tag{71}$$

Provided that the conditions for the existence of W and $\mathbf{H}$, namely

$$\text{Curl Grad } W = 0 \tag{72}$$

$$\text{Div Curl } \mathbf{H} = 0 \tag{73}$$

are satisfied. The scalar potential function is then given by

$$W = \int \theta_{ij}\, x^j\, dx^i \tag{74}$$

Obviously the method above, eqns (68)–(74) can, by analogy, be applied to any state-space formulation satisfying the existence conditions for the potential functions.

Continuing the application to eqn (65), then since

$$\theta_{\nu m} = 0 = \theta_{\mu n}$$

$$W = \int \theta_{\nu n}\, x^n\, dx^\nu + \int \theta_{\mu m}\, x^m\, dx^\mu$$

$$= -\int D_\nu^{\cdot\epsilon}\, G_{\epsilon\epsilon}\, D_{\cdot\eta}^{\epsilon} (v_c)^n\, dv_c^\nu - \int (L^{-1})_{\mu\mu} D_{\cdot\delta}^{\mu}\, R^{\delta\delta}\, D_\delta^{\cdot m} (L^{-1})_{mm}\, \psi_L^m\, d\psi_L^\mu$$

$$= -\int (i_{Rl})_\epsilon\, d(v_{Rl})^\epsilon - \int (v_{Rt})^\delta\, d(i_{Rt})_\delta \tag{75}$$

Hence

$$W = -\sum_{\text{links}} J_R - \sum_{\text{tree}} G_R \tag{76}$$

Hence the scalar potential function, W, can be identified as the sum of the tree content and link co-content. This function can be evaluated at any point in the state-space without explicit solution of the system equations.

The Curl of the associated vector potential function, $\mathbf{H}$, has components

$$\phi_{\nu n} = 0 = \phi_{\mu m}$$

$$\phi_{\nu m} = D_\nu^{\cdot m} (L^{-1})_{mm} = \frac{\partial \mathbf{H}_\nu}{\partial (\psi_L)^m} - \frac{\mathbf{H}_m}{\partial (v_c)^\nu}$$

$$\phi_{\mu n} = -(L^{-1})_{\mu\mu}\, D_{\cdot n}^{\mu} = \frac{\partial \mathbf{H}_\mu}{\partial (v_c)^n} - \frac{\partial \mathbf{H}_n}{\partial (\psi_L)^\mu} \tag{77}$$

This has the solution

$$2\,.\,Hn = Dn^{\cdot m}(i_L)\,m$$

$$2\,.\,H^m = -D^m\,.\,n(v_c)^n \tag{78}$$

which is associated with the extensity storage elements only. Indeed in the case of the energy conservative network, then

$$\text{Grad } W = 0 \tag{79}$$

$$\mathbf{H} = \text{Curl } \mathbf{H}\,.\,\mathbf{x}; \quad \text{and} \quad \mathbf{H}\,.\,\mathbf{x} = 0 \tag{80}$$

i.e. the system velocity vector is always orthogonal to the system state vector, and coincides with the vector potential itself.

The use of the scalar and vector potential functions in constructing the fields and state-space trajectories of a system is shown in Appendix II.

7. CONCLUSIONS

It has been shown that the transient behaviour of systems can be "generated" from a multi-dimensional tensor field, the form of which can be systematically derived without the explicit solution of equations. There is no restriction, in principle, to linearity in this method, nor to the dimensional order of the system. It is hoped that this approach can form the basis of computer aided analysis by the visual presentation of projections of such fields, and the establishment of adequate criteria relating real performance to actual contour shape.

APPENDIX I

Properties of tensor transformations and stability: example

Consider the speed control system shown schematically in Fig. 2. This system consists of an external source; "ideal" amplifier, field-controlled d.c. motor, and tachogenerator; a mechanical load including a frictional element, and a feedback network. The components are identified conventionally.

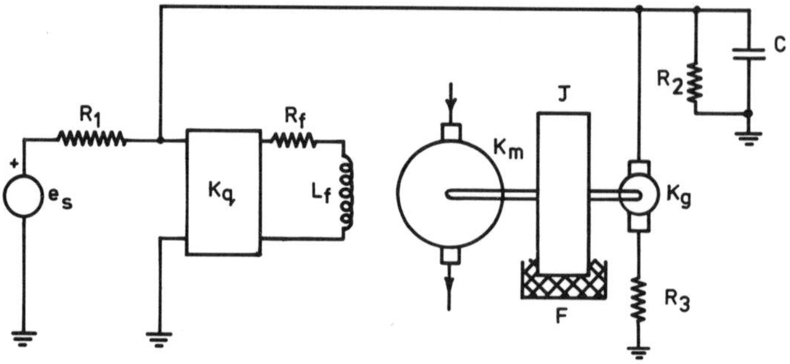

Fig. 2.

The network representation of the system, without external sources, is shown in Fig. 3. In this diagram the role of the "couplers" between the various parts of the network is clearly shown.

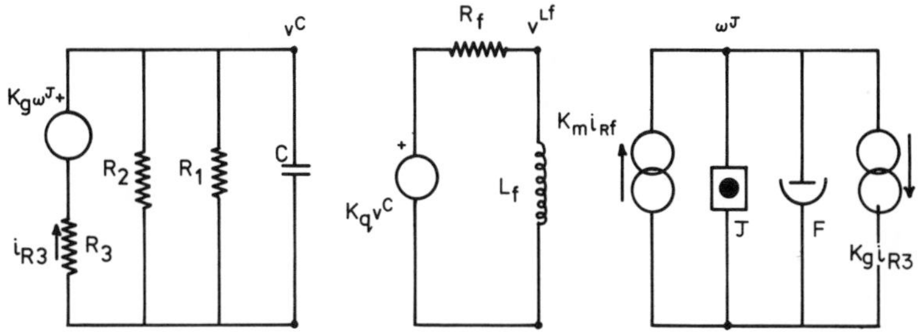

Fig. 3.

In order to derive the network node equations the tree is selected as in Fig. 4, the co-tree branches being indicated by dotted lines.

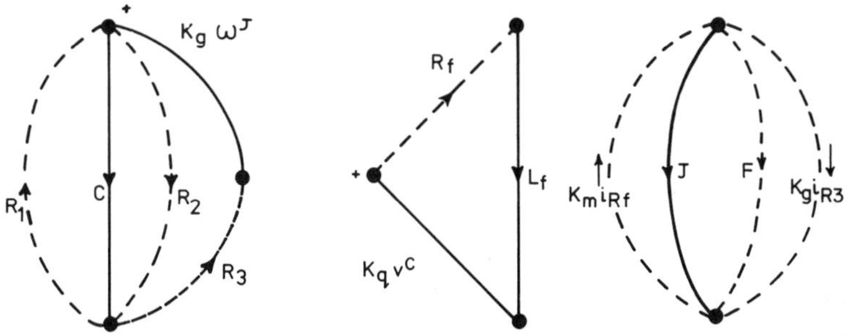

Fig. 4.

From these networks the transformational equations corresponding to eqns (42) and (43) can be obtained.

Equation (43): $(f_Q)_\nu = C_\nu^{\cdot\sigma}(fx)_\sigma$ becomes:

$$\begin{bmatrix} i_C \\ \tau_J \\ i_{Lf} \end{bmatrix} = \begin{bmatrix} 1 & -1 & 1 & 0 & 0 \\ 0 & 0 & -Kg & Km & -1 \\ 0 & 0 & 0 & 1 & 0 \end{bmatrix} \begin{bmatrix} i_{R1} \\ i_{R2} \\ i_{R3} \\ i_{Rf} \\ \tau_F \end{bmatrix} \tag{81}$$

where τ denotes torque

and equation (42); $dx^\alpha = C^\alpha_{.\gamma}\, dq^\gamma$ becomes

$$\begin{bmatrix} v^{R1} \\ v^{R2} \\ v^{R3} \\ v^{Rf} \\ \omega^F \end{bmatrix} = \begin{bmatrix} -1 & 0 & 0 \\ 1 & 0 & 0 \\ -1 & \text{Kg} & 0 \\ \text{Kq} & 0 & -1 \\ 0 & 1 & 0 \end{bmatrix} \begin{bmatrix} v^C \\ \omega^J \\ v^{Lf} \end{bmatrix} \tag{82}$$

If the dissipative elements are assumed to have linear characteristics then the link co-content, J_l, is given by

$$J_l = \frac{1}{2R_1}(v^{R1})^2 + \frac{1}{2R_2}(v^{R2})^2 + \frac{1}{2R_3}(v^{R3})^2 + \frac{1}{2R_f}(v^{Rf})^2 + \frac{F}{2}(\omega^F)^2 = V(x) \tag{83}$$

Defining the tree co-content, dJ_t, by

$$dJ_t = \langle i_t, dv^t \rangle \tag{84}$$

c.f. eqn (46): $dI = (f_Q)_\nu\, dq^\gamma$.

Then since $(f_Q)_\nu = C_\nu^{;\sigma}(f_x)_\sigma$,

$$(f_x)_\sigma = \text{grad}_x\, V(x)_\sigma \text{ and } d(f_x)_\sigma = G_{\sigma\sigma}\, dx^\sigma$$
$$= G_{\sigma\sigma}\, C^\sigma_{.\rho}\, dq^\rho$$

where $G_{\sigma\sigma}$ is the conductance of the dissipative elements, then

$$dI = C_\nu^{;\sigma}\, G_{\sigma\sigma}\, C^\sigma_{.\rho}\, q^\rho, dq^\gamma \tag{85}$$

from which,

$$(f_Q)_\nu = \text{Grad}_{q^\gamma} J_t = C_\nu^{;\sigma}\, G_{\sigma\sigma}\, C^\sigma_{.\rho} q^\rho \tag{86}$$

In this problem this equation becomes

$$i_t = \begin{bmatrix} i_C \\ \tau_J \\ i_{Lf} \end{bmatrix} = \begin{bmatrix} -\left(\frac{1}{R_1} + \frac{1}{R_2} + \frac{1}{R_3}\right) & \frac{Kg}{R_2} & 0 \\ \left(\frac{Kg}{R_2} + \frac{KqKm}{R_f}\right) & -\left(F + \frac{Kg^2}{R_2}\right) & -\left(\frac{Km}{R_f}\right) \\ \left(\frac{Kq}{R_f}\right) & 0 & -\left(\frac{1}{R_f}\right) \end{bmatrix} \begin{bmatrix} v^C \\ \omega^J \\ v^{Lf} \end{bmatrix} \tag{87}$$

This equation is *not* symmetrical about the leading diagonal.

The stability of the system is examined by considering the rate of entropy production, the rate of change of stored energy, and an examination of the system steady state, (if any).

(i) The system dissipation function, the link co-content, is sign definite positive, and therefore the rate of internal entropy production is also positive.

(ii) The sign definite properties of the rate of change of stored energy, or the work done by the force tensor in the generalized coordinate system, are determined from

$$(f_Q)_\gamma q^\gamma = (\text{Grad}_q W)_\gamma q^\gamma$$

where W is the scalar potential function in equation 50.

In this case, if we let $R = R_1 = R_2 = R_3$, we obtain

$$\text{Power} = [v^C \, \omega^J v^{Lf}] \begin{bmatrix} -\dfrac{3}{R} & \left(\dfrac{K_g}{2} + \dfrac{K_q K_m}{2R_f}\right) & \dfrac{K_q}{2R_f} \\[2ex] \left(\dfrac{K_g}{R} + \dfrac{K_q K_m}{2R_f}\right) & -\left(F + \dfrac{K_g{}^2}{R}\right) & -\dfrac{K_m}{2R_f} \\[2ex] \dfrac{K_q}{2R_f} & -\dfrac{K_m}{2R_f} & -\dfrac{1}{R_f} \end{bmatrix} \begin{bmatrix} v^C \\[2ex] \omega^J \\[2ex] v^{Lf} \end{bmatrix} \quad (88)$$

Then, by adopting the following values

$$R = 1\Omega,\, R_f = 2\Omega,\, K_m = 1.5\ \text{Nm/A},\, K_g = 0.5\ \text{Vs},\, F = 0.1\ \text{Nms},$$

if $K_q < 1.81$ the function given by equation 88 will be sign definite negative.

(iii) Increasing the value of the amplifier gain K_q beyond that required for absolute stability results in the 'opening' of the equi-potential lines which, in this three dimensional case, now form a set of hyperboloids in the generalized coordinate system. A value of the amplifier gain is possible at which the force field becomes such that it maintains a steady state. This condition can be determined by equating the fluxes of the energy storage elements in equation 87 to zero.

$$\begin{bmatrix} 0 \\[2ex] 0 \\[2ex] i_{Lf} \end{bmatrix} = \begin{bmatrix} -\dfrac{3}{R} & \dfrac{K_g}{2} & 0 \\[2ex] \left(\dfrac{K_g}{R} + \dfrac{K_q K_m}{R_f}\right) & -\left(F + \dfrac{K_g{}^2}{R}\right) & -\dfrac{K_m}{R_f} \\[2ex] \dfrac{K_q}{R_f} & 0 & -\dfrac{1}{R_f} \end{bmatrix} \begin{bmatrix} v^C \\[2ex] \omega^J \\[2ex] 0 \end{bmatrix} \quad (89)$$

In this all components of the force tensor are zero except one, in the coordinate system in which the network is being viewed, and this one component is given by

$$i_{Lf} = \frac{K_q}{R_f} v^C$$

and we obtain the condition for the steady state to exist

$$\begin{bmatrix} -\dfrac{3}{R} & \dfrac{K_g}{R} \\[2ex] \left(\dfrac{K_g}{R} + \dfrac{K_q K_m}{R_f}\right) & -\left(F + \dfrac{K_g{}^2}{R}\right) \end{bmatrix} \begin{bmatrix} v^C \\[2ex] \omega^J \end{bmatrix} = 0 \quad (90)$$

Using the values given above we therefore obtain

$$K_q = 2{\cdot}14 \text{ V/V}.$$

Increasing the gain of the amplifier beyond this value will result in instability.

The solution to this problem could also be derived from consideration of other scalar potential functions, such as that in equations 39 and 40. Analytic nonlinearities can easily be accommodated. For example by introducing the non-linear torque characteristic

$$\tau = K_F(\omega^F)^2$$

the link co-content becomes

$$J_l = \frac{1}{2R}(v^R)^2 + \frac{1}{2R_f}(v^{Rf})^2 + \frac{K_F}{4}(\omega^F)^4 \tag{91}$$

It is then possible to generate conditions governing stability similar to those obtained before.

APPENDIX II

State-space equations and tensor fields: example

Consider the circuit of Fig. 5 in which the components take the values

$$C = 1F \qquad r_1 = R_1(1 - (i_L/I)^2)$$
$$L = 1H \qquad R_1 = 2\Omega$$
$$r_2 = R_2 = 2\Omega \qquad I = 2A$$

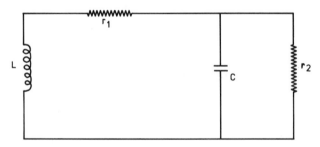

Fig. 5.

The network state equation is given by eqn (65).

$$\begin{bmatrix} i_C \\ v_L \end{bmatrix} = \begin{bmatrix} -G_l & 1 \\ -1 & -R_t \end{bmatrix} \begin{bmatrix} v_C \\ i_L \end{bmatrix} = \begin{bmatrix} -0.5 & 1 \\ -1 & (0.5i_L^2 - 2) \end{bmatrix} \begin{bmatrix} v_C \\ i_L \end{bmatrix} \tag{92}$$

The associated intensive-variable state space equation being

$$\begin{bmatrix} \dot{v}_C \\ i_L \end{bmatrix} = \begin{bmatrix} -C^{-1}G_l & C^{-1} \\ -L^{-1} & -L^{-1}R_t \end{bmatrix} \begin{bmatrix} v_C \\ i_L \end{bmatrix} = \begin{bmatrix} -0.5 & 1 \\ -1 & (0.5i_L^2 - 2) \end{bmatrix} \begin{bmatrix} v_C \\ i_L \end{bmatrix} \tag{93}$$

Defining the elements of the state-vector as in eqn (66) gives

$$x^1 = v_C; \quad g_{11} = C = 1F; \quad x_1 = q_c$$
$$x_2 = i_L; \quad g^{22} = L = 1H; \quad x^2 = \psi_L$$

Then the scalar potential function, W, defined by equation (74) becomes

$$W = -\sum_l J_R - \sum_t G_R = -\int i_{r2} dv_{R2} - \int v_{r1}\, di_{r1}$$

$$= -\int G_{r2} v_c dv_c - \int R_{r1} i_L\, di_L = 0.125\,(i_L^4 - 8i_L^2 - 2v_c^2) \qquad (94)$$

Contours of W are shown plotted in Fig. 6. The gradient of this potential function provides one component of the resultant tensor field, as shown in Fig. 8.

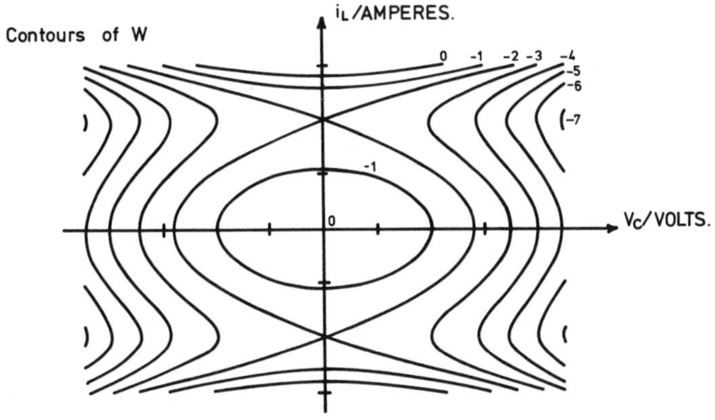

Fig. 6. Contours of W.

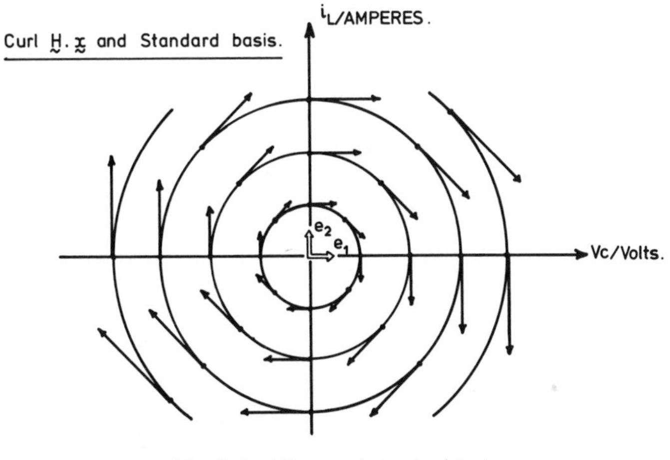

Fig. 7. Curl **H** . **x** and standard basis.

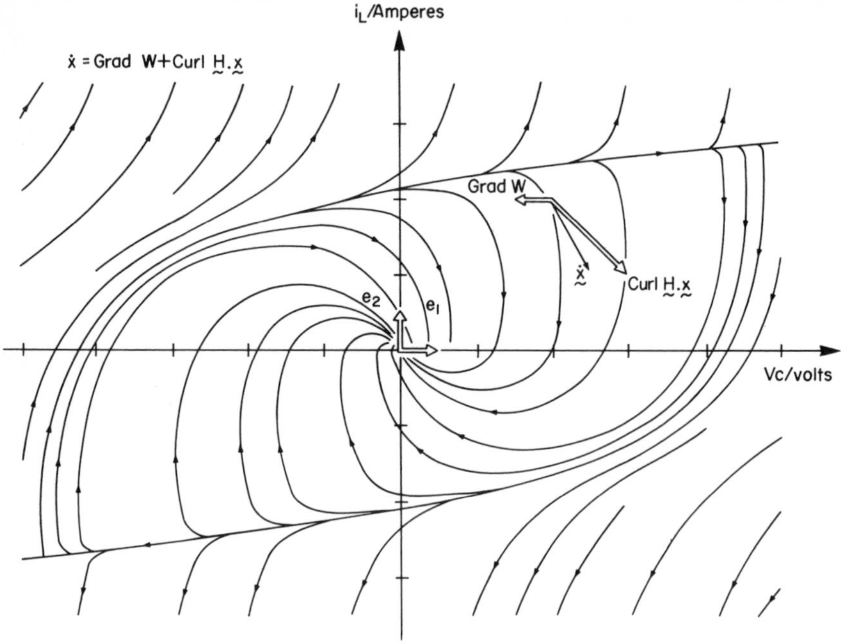

Fig. 8. $\dot{x}$ = Grad W + Curl **H . x.**

The vector potential function, **H**, defined by eqn (78) has components.

$$H_1 = i_L; \quad H^2 = v_c \tag{95}$$

The product of the Curl of the vector potential and the state vector, Curl **H . x**, as shown in Fig. 7, provides the other component of the tensor field for the system.

The two components of the tensor field for the system combine to provide a resultant which defines the motion of the system. This resultant vector, and the system state-space trajectories, are shown in Fig. 8.

REFERENCES

1. Onsager, L. "Reciprocal relations in irreversible processes". *Phys. Rev. 37, 38,* 1931.
2. Prigogine, I. *Introduction to the thermodynamics of irreversible processes.* Charles C. Thomas, 1955.
3. Groot, S. R. de. *Thermodynamics of irreversible processes.* Interscience, New York, 1952.
4. Oster, G. F. and Auslander, D. M. "Topological Representations of Thermodynamic Systems-I and II". *J. Franklin Inst. 292*, 1971, 1–17, 77–92.
5. Wei, J. "Irreversible thermodynamics in engineering". *Industrial & Engineering Chem. 58.*

6. Meixner, J. "Thermodynamics of electric networks and the Onsager–Casimir reciprocal relations". *J. Math. Phys. 4*, 1963.
7. Callen, H. B. *Thermodynamics*, Wiley, New York, 1960.
8. Li, J. C. M. "Caratheodory's principle and the thermokinetic potential in irreversible thermodynamics". *J. Phys. Chem. 66*, 1962.
9. Tellegen, B. D. H. "General network theorem with applications". *Philips Res. Rep. 7*, 1952.
10. Millar, W. "Some general theorems for non-linear systems possessing resistance". *Phil. Mag. 42*, 1951.
11. Jones, D. L. and Evàns, F. J. "Variational analysis of electrical networks". *J. Franklin Inst. 295*, 1, 1973.
12. Brayton, R. K. and Moser, J. K. "A theory of nonlinear networks". *Quart. App. Maths 22*, 1, 1964.
13. Korn, J. *Invariants and field concepts in engineering systems analysis*. Ph.D. Thesis, University of London, 1972.
14. Korn, J. *et al. Hydrostatic transmission systems.* Intertext, London, 1970.
15. Holding, D. J. and Evans, F. J. "Tensor field interpretation of the dynamic behaviour of networks". *Electronic Letters* (I.E.E., London), *9*, 1973.

NOMENCLATURE

$\mathbf{A}$	system matrix
a^γ, a_γ	basis vectors
$A_{\nu}^{\cdot\mu}, A_{\nu\mu}$ etc.	system tensors
$C_{\nu}^{\cdot\sigma}, C_{\cdot\gamma}^{\alpha}$	coordinate transformation tensors
C	capacitor (and suffix)
$D_{N}^{\cdot\sigma}, D_{\cdot\rho}^{\mu}$	topological transformation tensors
$\mathbf{D}$	dynamical transformation matrix
F_j	thermodynamic force or affinity
f_x, f_Q	co-variant forces
G	content
$\mathbf{g}, \mathbf{g}'$	metric tensor and inverse with elements g_{ij} and g^{ij} resp.
G	conductance matrix, content
H	vector potential
I	scalar potential function
i	current
J_j	thermodynamic flux
J	co-content
L	inductor (and suffix)
l	suffix, link
L_{kj}	thermodynamic phenomological coefficient
P_T	topologically extended form of mixed potential function
q^γ	generalized coordinate
q_C	capacitive charge
R	resistor (and suffix)
S	suffix, storage
S_i	internal entropy
t	suffix, tree
U	internal energy
v	voltage
W	scalar potential
V	scalar potential
$\mathbf{x}$	state vector
X_j	thermodynamic extensity
x^α	contravariant coordinate
σ_F	thermokinetic potential function
ψ	inductor flux
ω	angular velocity
τ	torque

CONSEQUENCES OF A DISSIPATION INEQUALITY IN THE THEORY OF DYNAMICAL SYSTEMS

J. C. Willems
Mathematical Institute
University of Groningen
Groningen, The Netherlands

SUMMARY

This paper discusses some system theoretic questions which arise in the context of dissipative dynamical systems, that is, systems for which all motions satisfy a given inequality constraint. Bounds on the internal storage function are derived together with variational expressions for the extremal storage functions. The compatibility of external and internal qualitative properties of such systems as, for example, dissipativeness and symmetry are proven and illustrated.

1. INTRODUCTION

A majority of the contemporary problems in applied mathematics, be they based on physics, economics, system theory, computer science, etc., concern the analysis and synthesis of *dynamical* systems. Thus much research effort has gone into modelling these systems in such a way that their relevant qualitative properties may be deduced effectively and into synthesizing systems which need to meet certain specifications.

In many applications one may *a priori* postulate that the system has certain qualitative properties. Typically these conclusions are justified as implications of some general physical laws which apply to the case under consideration. In this paper we will consider a frequently occurring situation of this kind, namely when it may be postulated that all motions of the system obey a given inequality. We will call such systems "*dissipative*". Typical examples where this occurs are thermodynamic systems where the second law provides such an inequality or mechanical and/or electrical energy transmission systems where the losses (due to, say, heat or radiation) are not explicitly modelled and where the resulting loss of energy leads to the desired inequality.

This paper is written from a "modern" system theoretic point of view. This implies that we will be employing state space methods for modelling systems with memory. It will be seen that this framework is indeed the appropriate one for treating problems of the type discussed here.

The present paper is mainly a re-exposition of the results in reference [1] but the conceptual framework is somewhat more logical and a few of the results have been generalized. We have dispensed with most of the formal proofs. These are either identical or simple extensions of those in [1].

2. DISSIPATIVE DYNAMICAL SYSTEMS: DEFINITIONS

In this section we will put forward a rational and yet very general definition of dissipativeness. This definition involves a dynamical system, a supply rate which expresses the supply flow into the system from its environment, the storage function which represents the accumulation inside the system, and finally the dissipation inequality which postulates that only part of what is supplied from the outside is stored in the system, while the remainder is dissipated.

Formally, let $\Sigma = \{U, \mathcal{U}, Y, \mathcal{Y}, X, \phi, r\}$ be a given (continuous time) *dynamical system.* In here,

U = the set of input values;
$\mathcal{U}$ = the input space (a collection of U-valued functions of time);
Y = the set of output values;
$\mathcal{Y}$ = the output space;
X = the state space;
ϕ = the state transition function. Thus $\phi(t_1, t_0, x_0, u)$, defined for $t_1 \geqslant t_0$, denotes the state at time t_1 obtained from the initial state x_0 at time t_0 by applying input $u \in \mathcal{U}$; and
r = the read out function. Thus $r(x, u)$ denotes the output value resulting from the presence of the state x and the input value u.

We will only be concerned with stationary systems (this restriction has, to some extent, already been used by postulating that the function r does not depend on time explicitly). The system Σ satisfies a standard series axioms expressing the various postulates of stationary dynamical systems in state space form. These may be found with discussions in the literature [2, 3].

Assume now that in addition to the dynamical system Σ we are given a function $S : X \to R$ (the real numbers), and a function $w : U \times Y \to R$ which satisfies

$$\int_{t_0}^{t_1} |w(t)|\, dt < \infty \; (\forall -\infty < t_0 \leqslant t_1 < \infty),$$

where $w(t) = w(u(t), y(t))$ and u and y are arbitrary elements of $\mathcal{U}$ and $\mathcal{Y}$ respectively. The function S will be called the *storage function* and w will be called the *supply rate.*[1]

In terms of these notions a dissipative dynamical system may be defined as follows:

[1] The elements of U and Y are typically power variables each either an *across-variable* (voltage, displacement, pressure difference, etc.) or a *through-variable* (current, force, flow, etc.). The function w then equals $\mathbf{u}^T\mathbf{y}$ and has dimension of power and S is the stored energy and has dimension of energy. The dissipation inequality then expresses that the increase in energy stored cannot be larger than the energy supplied from the outside. Another situation is when one uses *scattering variables* where the elements of U denote the magnitude of the *incident wave* and those of Y the magnitude of the *reflected wave.* The function w then equals $||\mathbf{u}||^2 - ||\mathbf{y}||^2$ i.e., the difference of the power in the incident wave and in the reflected wave, while S still denotes the stored energy.

Definition 1: The triplet $\{\Sigma, w, S\}$ is said to define a *dissipative dynamical system* if

(*i*) $S \geqslant 0$; and

(*ii*) The *dissipation inequality:*

$$S(x_0) + \int_{t_0}^{t_1} w(t)\,dt \geqslant S(x_1) \qquad (DIE)$$

is satisfied ($\forall t_1 \geqslant t_0, x_0 \in X, u \in \mathscr{U}$). In here $w(t) = w(u(t), r(\phi(t, t_0, x_0, u), u(t)))$ and $x_1 = \phi(t_1, t_0, x_0, u)$.

The integral form (*DIE*) of the dissipation inequality is of course equivalent to its more common differential form: $\frac{d}{dt} S(x)_{\substack{\text{along}\\ \text{solutions}\\ \text{of } \Sigma}} \leqslant w(u, r(x, u))\,(Vx \in X, u \in U)$,
where it must be assumed that the identicated derivative exists.

Condition (*i*) of *Definition 1* may, without essential consequences (since the dissipation inequality is unaffected by an additive constant in S), be replaced by:

(*i*)′ S is bounded from below.

In some sense, the condition $S \geqslant 0$ thus simply corresponds to a normalization.

Some useful related notions are those of the *dissipation rate* $d(= w - \dot{S})$, of a *conservative* (or *lossless*) dissipative system in which the dissipation inequality becomes an equality (thus conservative systems may be regarded as limit cases of dissipative systems), and of a *strongly* dissipative system in which equality in (*DIE*) holds exceptionally. Since we will make very little use of these notions, we will not bother with a formal definition of these concepts. Typical questions related to dissipative systems which have important applications, are:

1. *What constraints does dissipativeness impose on the constitutive equations ϕ and r? Which qualitative properties (e.g. stability) result from the dissipation hypothesis?*

As an illustration of such implications, consider the dynamical system governed by the equations:

$$\dot{\mathbf{x}} = \mathbf{u}; \quad \mathbf{y} = \mathbf{g}(\mathbf{x})$$

with $\mathbf{x}, \mathbf{u}, \mathbf{y} \in R^n$ (R^n denotes n-dimensional Euclidean space), and let $w = \mathbf{u}^T\mathbf{y}$ (T denotes transpose). This dynamical system may be regarded as a model for an elastic mechanical system (with $\mathbf{y}$ the force and $\mathbf{x}$ the displacement), or of a (non-linear) capacitive electrical n-port network (with $\mathbf{x}$ the charge and $\mathbf{y}$ the voltage). We are trying to illustrate that dissipativeness may impose some fairly strong restrictions on the constitutive equations, in this case on $\mathbf{g}$.

A simple calculation shows that there exists an S such that $\{\Sigma, w, S\}$ is dissipative *if and only if*

(*i*) $\frac{\partial g_i}{\partial x_j} = \frac{\partial g_j}{\partial x_i}$; and

(*ii*) the path integral $P(\mathbf{x}) \triangleq \int_0^{\mathbf{x}} \mathbf{g}^T(\mathbf{x})\,d\mathbf{x}$ is bounded from below.

Moreover, S is essentially unique (up to a normalization at $x = 0$) and differs from the above path integral only by the additive constant $S(0)$. The constitutive equations may be expressed in terms of S as

$$g(.) = \frac{\partial S}{\partial x}(.)$$

and S thus plays the role of a potential. Note that the symmetry condition (i) follows directly from the dissipation hypothesis.

2. *Given Σ and w, when does there exist a function S such that $\{\Sigma, w, S\}$ defines a dissipative dynamical system?*

This question is relevant in several situations, for example:

(*i*) in thermodynamics[2] where the second law postulates the existence of a storage function (the entropy) without, however, giving an expression for it in terms of the physical state of the system. It will be shown, in fact, that in general the second law does not specify the entropy function uniquely;

(*ii*) in situations in which the dynamical equations (Σ) and the interaction (u, y and w) of a system with its environment are known together with the fact that the system is dissipative (existence of S). This occurs when the system is a "black box" whose dynamic terminal behaviour has been deduced from experimentation, but when in addition it is known from physical considerations that the system is internally dissipative. Since however the internal structure of the system is unknown, this only implies the existence of S. Thus one may want to know what the possible storage functions are.

(*iii*) in synthesis problems in which the input/output behaviour of a system is prescribed and where one wants to realize this behaviour by interconnecting certain standard components. Often these components are constrained to be dissipative and thus the question is to synthesize, if possible, a prescribed input/output behaviour using dissipative components. As will be indicated in the sequel, such synthesis problems often involve, as a first step, the specification of a suitable storage function.

3. *Is dissipativeness compatible with certain other qualitative assumptions?*

Most dissipative physical systems obey at the same time other qualitative properties, for example certain symmetry relations, which may involve the storage function. Thus in such situations the storage function need not only satisfy the dissipation inequality but also certain other conditions and the question then arises whether or not such a storage function exists. We will demonstrate that the Onsager–Casimir relations may be interpreted in this context as laws which postulate the compatibility of dissipativeness and internal symmetry.

3. BOUNDS AND UNIQUENESS OF THE STORAGE FUNCTION

In this section certain general properties of the storage function will be derived. It will be assumed throughout that Σ and w are given but that S is unknown. To

[2] Thermodynamic systems are, strictly speaking, not dissipative in the sense of *Definition 1* since the entropy function need not be bounded from below or above. This will be discussed in more detail in Section 9, (*iii*).

indicate the empirical basis of this assumption, assume that we start with an input/output model of a dynamical system defined by $\Sigma_{I/O} = \{U, \mathscr{U}, Y, \mathscr{Y}, F\}$, where $U, \mathscr{U}, Y, \mathscr{Y}$ are as before and $F: U \to Y$ is the *system function*. Thus $y = Fu$ denotes the output which results from applying input u. This system description is assumed to satisfy in its turn a series of axioms which express, among other things, the non-anticipatory nature of the system function F and which make precise what happens at "$t = -\infty$".

The *state space realization problem* is the representation of $\Sigma_{I/O}$ by a dynamical system Σ and a *ground state* x^* chosen in such a way that the output of Σ with initial state x^* at "$t = -\infty$" will, for every input $u \in \mathscr{U}$, be the same as the output of $\Sigma_{I/O}$. This problem is presently very well understood [2, 3, 4], both on an abstract level and for specific classes of linear systems.

The assumption that Σ and w are given may thus be interpreted as specifying the external behaviour of a system and the question is to find from there what, if any, functions $S: X \to R$ there are which will make $\{\Sigma, w, S\}$ dissipative.

The first theorem gives variational inequalities for the existence of a function S such that $\{\Sigma, w, S\}$ is dissipative, together with *the best possible* lower and upper bounds for S.

Definition 2: Let $S_a: X \to R_e$ (the extended real number system) be defined by:

$$S_a(x_0) \stackrel{\Delta}{=} \sup_{\substack{t_1 \geqslant t_0 \\ x_0 \to}} -\int_{t_0}^{t_1} w(t)\,dt$$

where $\sup\limits_{\substack{t_1 \geqslant t_0 \\ x_0 \to}}$ denotes the supremum over all $u \in \mathscr{U}$ and $t_1 \geqslant t_0$, with motions of Σ starting at $x(t_0) = x_0$ and with w evaluated along these motions.

Definition 3: The state space X of Σ is said to be *reachable* from the point $x^* \in X$ if for any given $x \in X$ there exists a $u \in \mathscr{U}$ and $t_{-1} \leqslant t_0$ such that $x = \phi(t_0, t_{-1}, x^*, u)$. Assume that X is reachable from x^*. Let $S_r^*: X \to R_e$ be defined by:

$$S_r^*(x_0) \stackrel{\Delta}{=} \inf_{\substack{t_{-1} \leqslant t_0 \\ x^* \to x_0}} \int_{t_{-1}}^{t_0} w(t)\,dt$$

where $\inf\limits_{\substack{t_{-1} \leqslant t_0 \\ x^* \to x_0}}$ denotes the infimum over all $u \in \mathscr{U}$ and $t_{-1} \leqslant t_0$ with motions of Σ starting at $x(t_{-1}) = x^*$ and ending at $x(t_0) = x_0$, and with w evaluated along these motions.

The function S_a will be called the *available storage*: it represents the maximum amount of internal storage which may be recovered from a system via its external terminals. The function S_r^* represents the minimum amount of external supply which needs to be delivered to the system in executing the transfer from state x^* to its present state. If $\{\Sigma, w, S\}$ is dissipative and if $0 = S(x^*) = \min\limits_{x \in X} S(x)$ then S_r^* will be called the *required supply* and denoted simply by S_r: it represents the minimum amount of external supply which has to be delivered via the external terminals of a dissipative system while bringing it to a particular state from an initial state of no (i.e. minimum) internal storage.

The following two theorems relate the existence of a storage function to the finiteness of S_a and S_r:

Theorem 1 (*Existence of S and a lower bound for S*): *Let* Σ *and w be given. Then there exists S such that* $\{\Sigma, w, S\}$ *is dissipative* if and only if $S_a < \infty$. *Moreover, if so, then* $\{\Sigma, w, S_a\}$ *is dissipative and* $0 \leqslant S_a \leqslant S$.

Theorem 1 is, in a sense, an input/output test for dissipativeness. Moreover it gives a variational expression for evaluating one possible storage function (S_a) and expresses the intuitive fact that not more than what is stored can be extracted through the external terminals of a dissipative system.

Theorem 2 (*Existence of S and an upper bound for S*): *Let* Σ *and w be given and assume that the state space X of* Σ *is reachable from* x^*. *Then there exists S such that* $\{\Sigma, w, S\}$ *is dissipative* if and only if S_r^* *is bounded from below on X. Moreover, if so, then* $\{\Sigma, w, S_r^* - \inf_{x \in X} S_r^*(x)\}$ *is dissipative and* $S \leqslant S_r^* + S(x^*)$.

Finally, if $0 = S(x^*) = \min_{x \in X} S(x)$, *then* $0 \leqslant S_a \leqslant S \leqslant S_r$.

Theorem 2 gives an alternative variational expression for evaluating one possible storage function (S_r) and it expresses the intuitive fact that more than what is stored has previously been supplied to a dissipative system through its external terminals.

Note that *both* S_a and S_r are themselves possible storage functions for a dissipative dynamical system. This implies that the "black box" Σ could very well be constructed in such a way that all its internal storage is available through its terminals or in such a way that it is possible to generate any set of initial conditions without causing dissipation. As a rule, $S_a \neq S_r$. Since the dissipation inequality is closed under convexification there are thus an infinite number of possible storage functions (functions satisfying $S \geqslant 0$ and the dissipation inequality) whenever $S_a \neq S_r$.

In some physical systems it follows from some fundamental laws that S exists without, however, specifying it. Theorems 1 and 2 then give possible S-functions and an upper and lower bound. Moreover, they indicate that S is *only exceptionally* uniquely defined by the dissipation inequality.

It follows from the above theorem that a suitable storage function always exists as long as a system behaves dissipatively when viewed from its terminals and that one may define such a storage function (e.g., S_a) abstractly. This fact has been recognized before. For example, in [5] an axiomatic framework for the study of materials with memory is given based on the concept of available storage. One should however be cautious in using the notion of available storage (or required supply): in most cases it will *not* correspond to the actual storage and other qualitative properties may result in S_a and/or S_r becoming unsuitable as storage functions.

Another interesting property of dissipative systems which may be formalized (see [1] for details) is the fact that the storage function is an *extensive* variable. Thus if $\{\Sigma_\alpha, w_\alpha, S_\alpha\}$, $\alpha \in A$ (= some index set), denotes a family of dissipative dynamical systems which are interconnected by a "neutral" interconnection N (i.e. an interconnection which itself is lossless) then $\{\prod_{\alpha \in A} \Sigma_\alpha | N, w, \sum_{\alpha \in A} S_\alpha\}$ also defines a dissipative dynamical system. Here $\prod_{\alpha \in A} \Sigma_\alpha | N$ denotes the interconnected system obtained by interconnecting the systems Σ_α by means of the interconnection law N.

4. FINITE DIMENSIONAL LINEAR SYSTEMS

A class of dynamical systems which may be studied in great detail are the stationary finite dimensional linear systems described by:

$$\Sigma\colon \dot{\mathbf{x}} = \mathbf{A}\mathbf{x} + \mathbf{B}\mathbf{u};\ \ \mathbf{y} = \mathbf{C}\mathbf{x} + \mathbf{D}\mathbf{u}, \qquad (FDLS)$$

with $\mathbf{x} \in R^n$, $\mathbf{u} \in R^m$, $\mathbf{y} \in R^p$, and $\{\mathbf{A}, \mathbf{B}, \mathbf{C}, \mathbf{D}\}$ matrices of appropriate dimensions. In here $\mathscr{U}$ ($\mathscr{Y}$) is the space of those $R^m(R^p)$-valued functions on $(-\infty, +\infty)$ which are locally square integrable and U, Y, X, ϕ and r may be formally defined in a way which is obvious from the equation (*FDLS*).

It is well-known [3, 4] that the state space of Σ is reachable if and only if Rank $[\mathbf{B}|\mathbf{AB}| \ldots |\mathbf{A}^{n-1}\mathbf{B}] = n$ and that it is observable (i.e. $\mathbf{x}(t_0)$ may be deduced by observing $\mathbf{u}(t)$ and $\mathbf{y}(t)$ for $t \geqslant t_0$) if and only if Rank $[\mathbf{C}^T|\mathbf{A}^T\mathbf{C}^T| \ldots |(\mathbf{A}^T)^{n-1}\mathbf{C}^T] = n$. *It will be assumed throughout that these conditions are satisfied.* The *transfer function* of Σ is given by $\mathbf{G}(s) = \mathbf{D} + \mathbf{C}(\mathbf{I}s - \mathbf{A})^{-1}\mathbf{B}$. However, Σ is not the only system with this transfer function. In fact all the models (*FDLS*) which have the given transfer function $\mathbf{G}(s)$ and which are reachable and observable may be obtained from $\{\mathbf{A}, \mathbf{B}, \mathbf{C}, \mathbf{D}\}$ as elements on its orbit under the transformation group

$$\{\mathbf{A}, \mathbf{B}, \mathbf{C}, \mathbf{D}\} \xrightarrow[\det \mathbf{S} \neq 0]{\mathbf{S}} \{\mathbf{S}\mathbf{A}\mathbf{S}^{-1}, \mathbf{S}\mathbf{B}, \mathbf{C}\mathbf{S}^{-1}, \mathbf{D}\}$$

with $\mathbf{S}$ any element of the general linear group $GL(n)$ (i.e. any invertible $(n \times n)$ matrix). The assumption of reachability and observability is equivalent to the so-called *minimality* property, i.e. the state space has minimal dimensions n (subject, of course, to the form (*FDLS*) and the fact that the transfer function $\mathbf{G}(s)$ is prescribed).

The *impulse response* of the system (*FDLS*) is given by $\mathbf{D}\delta(t) + \mathbf{C}e^{\mathbf{A}t}\mathbf{B}$ $(t \geqslant 0)$ (δ denotes the delta-function). Let $\mathbf{W}(t) = \mathbf{C}e^{\mathbf{A}t}\mathbf{B}$ denote the continuous part of this impulse response. A convenient characterization of this part of the system (*FDLS*) is by its so-called *Hankel matrix* [2, 4] defined by

$$\mathbf{H}(t) \stackrel{\Delta}{=} \left[\begin{array}{c|c|c|c|c|c}
\mathbf{W}(t) & \mathbf{W}^{(1)}(t) & \mathbf{W}^{(2)}(t) & \cdots & \mathbf{W}^{(n)}(t) & \cdots \\ \hline
\mathbf{W}^{(1)}(t) & \mathbf{W}^{(2)}(t) & \cdot & \cdots & \cdot & \cdots \\ \hline
\mathbf{W}^{(2)}(t) & \cdot & \cdot & \cdots & \cdot & \cdots \\ \hline
\vdots & \vdots & \vdots & \vdots\vdots\vdots & \vdots & \vdots\vdots\vdots \\ \hline
\mathbf{W}^{(n)}(t) & \cdot & \cdot & \cdots & \mathbf{W}^{(2n)}(t) & \cdots \\ \hline
\vdots & \vdots & \vdots & \vdots\vdots\vdots & \vdots & \vdots\vdots\vdots
\end{array}\right]$$

where the superscripts denote derivatives.

We will now consider this system with the supply rate $w = \mathbf{u}^T\mathbf{y}$. This supply rate is chosen as a representative example of a quadratic functional on $U \times Y$. It corresponds to the usual case considered in mechanics, in viscoelasticity, and in electrical network theory when one uses an impedance, admittance, or hybrid description for the network.

The first question which arises is whether or not there exists a function S such that $\{\Sigma, w, S\}$ is dissipative. It is readily shown that this is the case if and only if $\int_0^T \mathbf{u}^T\mathbf{y}\, dt \geq 0$ for all $T \geq 0$ whenever $\mathbf{x}(0) = \mathbf{0}$. The translation of this inequality into a condition on the transfer function $\mathbf{G}(s)$ is a classical problem in the theory of Laplace transforms. It is well-known [6] that this requires $\mathbf{G}(s)$ to be *positive real* i.e. $\mathbf{G}(\sigma + j\omega) + \mathbf{G}^T(\sigma - j\omega) \geq \mathbf{O}$ for all $\sigma \geq 0$ and all real ω such that $\sigma + j\omega \neq \lambda(\mathbf{A})$ ($\lambda(\mathbf{A})$ denotes an arbitrary eigenvalue of $\mathbf{A}$). This is formalized in the following theorem:

Theorem 3 *Consider the system (FDLS) and* $w = \mathbf{u}^T\mathbf{y}$. *Then there exists a function S such that* $\{\Sigma, w, S\}$ *is dissipative* if and only if $\mathbf{G}(s) = \mathbf{D} + \mathbf{C}(\mathbf{I}s - \mathbf{A})^{-1}\mathbf{B}$ *is a positive real function of s.*

The next theorem gives a time domain condition for dissipativeness of (*FDLS*). It may be shown that any storage function satisfies the inequality $S(\mathbf{x}) \geq S(\mathbf{0})$ with equality holding only if $\mathbf{x} = \mathbf{0}$. It is then easily seen from optimal control that $S_a(\mathbf{x})$ and $S_r(\mathbf{x})$, if they are finite, are quadratic functions of $\mathbf{x}$. Thus if any storage function exists, a quadratic one exists. Moreover, it follows immediately from the differential form of the dissipation inequality that $\frac{1}{2}\mathbf{x}^T\mathbf{Q}\mathbf{x}$ defines a possible storage function if and only if $\mathbf{x}^T\mathbf{Q}(\mathbf{A}\mathbf{x} + \mathbf{B}\mathbf{u}) \leq \mathbf{u}(\mathbf{C}\mathbf{x} + \mathbf{D}\mathbf{u})$ ($\forall \mathbf{x} \in R^n, \mathbf{u} \in R^m$). This leads to the following generalization of the Kalman-Yacubovich-Popov lemma [7, 8]:

Theorem 4 *Consider the system (FDLS) and* $w = \mathbf{u}^T\mathbf{y}$. *Then there exists a function S such that* $\{\Sigma, w, S\}$ *is dissipative* if and only if *there exists a solution* $\mathbf{Q} = \mathbf{Q}^T \geq \mathbf{O}$ *to the matrix inequality:*

$$\left[\begin{array}{c|c} \mathbf{A}^T\mathbf{Q} + \mathbf{Q}\mathbf{A} & \mathbf{Q}\mathbf{B} - \mathbf{C}^T \\ \hline \mathbf{B}^T\mathbf{Q} - \mathbf{C} & -\mathbf{D} - \mathbf{D}^T \end{array}\right] \leq \mathbf{O} \qquad (MIE)$$

Every such solution defines a dissipative dynamical system $\{\Sigma, w, \frac{1}{2}\mathbf{x}^T\mathbf{Q}\mathbf{x}\}$ *and all possible quadratic storage functions are thus recovered. Moreover, the set of solutions to (MIE) is convex, compact and attains its minimum,* $\mathbf{Q}^-$, *and its maximum,* $\mathbf{Q}^+$. *In fact, if there exists any solution* $\mathbf{Q} = \mathbf{Q}^T \geq \mathbf{O}$ *to (MIE) then* $\mathbf{Q}^- > \mathbf{O}$, $S_a = \frac{1}{2}\mathbf{x}^T\mathbf{Q}^-\mathbf{x}$, $S_r = \frac{1}{2}\mathbf{x}^T\mathbf{Q}^+\mathbf{x}$ *and every possible storage function satisfies* $\frac{1}{2}\mathbf{x}^T\mathbf{Q}^-\mathbf{x} \leq S(\mathbf{x}) - S(\mathbf{0}) \leq \frac{1}{2}\mathbf{x}^T\mathbf{Q}^+\mathbf{x}$.

The maximum and minimum in the above theorem are to be taken in the positive definite sense, i.e., in the partial ordering induced by the positive semi-definite cone on the space of symmetric matrices.

Two important special cases are:

(*i*) If $\mathbf{D} = \mathbf{O}$, then (*MIE*) reduces to $\mathbf{A}^T\mathbf{Q} + \mathbf{Q}\mathbf{A} \leq \mathbf{O}$ and $\mathbf{Q}\mathbf{B} = \mathbf{C}^T$. This yields $\mathbf{B}^T\mathbf{C}^T = \mathbf{C}\mathbf{B} \geq \mathbf{O}$ as a well-known necessary condition for dissipativeness in this case;

(*ii*) If $\mathbf{D} + \mathbf{D}^T > 0$, then (*MIE*) is equivalent to the quadratic matrix inequality $\mathbf{A}^T\mathbf{Q} + \mathbf{Q}\mathbf{A} - (\mathbf{Q}\mathbf{B} - \mathbf{C}^T)(\mathbf{D} + \mathbf{D}^T)^{-1}(\mathbf{B}^T\mathbf{Q} - \mathbf{C}) \leqslant \mathbf{O}$. The equality solution to this matrix inequality thus satisfy an algebraic matrix Riccati equation and correspond to those solutions of (*MIE*) which make its left-hand side of minimum rank. These equality solutions are of particular interest and $\mathbf{Q}^-$ and $\mathbf{Q}^+$, among others, satisfy this inequality with equality. This follows rather easily from the theory of least squares stationary optimal control, as a standard result there shows how to obtain $\mathbf{Q}^-$ and $\mathbf{Q}^+$ from this algebraic Riccati equation. In network synthesis applications, these equality solutions lead to realizations using a minimum number of resistors.

5. SYMMETRIC SYSTEMS

Consider a system of the form (*FDLS*), and let $\mathbf{G}(s) = \mathbf{D} + \mathbf{C}(\mathbf{I}s - \mathbf{A})^{-1}\mathbf{B}$ be its transfer function. Let $\mathbf{\Sigma}_e$ be a signature matrix (a diagonal matrix with diagonal entries either +1 or −1). This matrix will be called the *external signature matrix* of (*FDLS*) and, since it depends on the external behaviour, we consider it as given. We now have the following formal definition:

Definition 4: The system (*FDLS*) is said to be *externally reciprocal* with external signature matrix $\mathbf{\Sigma}_e$ if $\mathbf{G}(s) = \mathbf{\Sigma}_e\mathbf{G}^T(s)\mathbf{\Sigma}_e$.

The input (output) variables corresponding to the +1 entries in $\mathbf{\Sigma}_e$ are called inputs (outputs) of *even parity* and those corresponding to the −1 entries are said to be of *odd parity.*

The concept of reciprocity is a rather evasive one. It says something about the bilateral character of a system: the *j*th output due to the *i*th input equals (or equals the negative) of what would be the *i*th output if this input were to be taken as the *j*th input. An interesting way of interpreting *Definition 4* is in terms of time-reversal. Thus if we let R denote time-reversal (i.e. the operator which maps $f(t)$ into $f(-t)$) then it may be shown [9] that in the case $D = 0$, reciprocity is equivalent to commutativity of the diagram

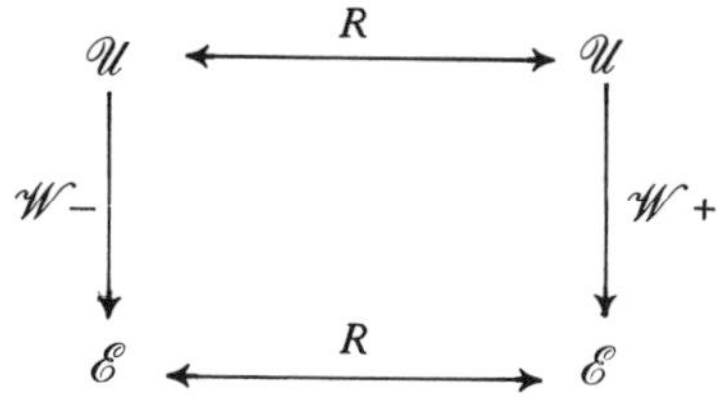

where $(\mathscr{W}_- u)(t) = \int_{-\infty}^{t} \mathbf{u}^T(\tau)\,\mathbf{y}(\tau)\,d\tau$ is the absorbed supply and $(\mathscr{W}_+ u)(t) =$

$\int_{t}^{\infty} \mathbf{u}^T(\tau)\,\mathbf{y}(\tau)\,d\tau$ is the "to be absorbed" supply.

Since reciprocity implies certain equality relations in a system it is clear that it imposes stringent conditions on the constitutive equations of a system but that it may also yield certain useful qualitative and quantitative laws on the dynamic behaviour of the system. Also, since one can see this property to be satisfied in most electrical,[3] most mechanical, and in chemical systems, for example, it may very well have an analogue and yield as yet unexplored implications in other areas (e.g. economics, biology). One difficulty however, is that it remains very difficult to generalize this concept to non-linear or time-varying systems, although one would expect such a general property to admit a non-linear generalization. Reciprocity seems tied up with ideas of time reversal, duality, topological constraints (Tellegen's theorem) and dissipativeness (see e.g. the examples in Section 2). Some interesting work on reciprocity for non-linear systems may be found in [10, 11, 12, 13].

The implications of reciprocity on $\{\mathbf{A}, \mathbf{B}, \mathbf{C}, \mathbf{D}\}$ are stated in the following theorem:

Theorem 5 *The system (FDLS) is externally reciprocal with external signature matrix Σ_e if and only if there exists a (unique) nonsingular matrix $\mathbf{T} = \mathbf{T}^T$ such that $\mathbf{A} = \mathbf{T}^{-1}\mathbf{A}^T\mathbf{T}$ and $\mathbf{B} = T^{-1}\mathbf{C}^T\mathbf{\Sigma}_e$.*

6. THE COMPATIBILITY OF DISSIPATIVENESS AND SYMMETRY

In this section we continue the study of the minimal system (*FDLS*) with the external reciprocity assumption $\mathbf{G}(s) = \mathbf{\Sigma}_e\mathbf{G}^T(s)\mathbf{\Sigma}_e$. Let $\mathbf{Q}$ be such that $\frac{1}{2}\,\mathbf{x}^T\mathbf{Q}\mathbf{x}$ represents a quadratic storage function and let $\mathbf{T} = \mathbf{T}^T$ be the (unique) matrix which satisfies the relations $\mathbf{A} = \mathbf{T}^{-1}\mathbf{A}^T\mathbf{T}$ and $\mathbf{B} = T^{-1}\mathbf{C}^T\mathbf{\Sigma}_e$ (see Theorem 5).

In this section we will investigate whether or not there exists such a $\mathbf{Q}$ which also satisfies $\mathbf{T}\mathbf{Q}^{-1}\mathbf{T} = \mathbf{Q}$.

Before solving the problem just posed we show how this question arises. We consider two related motivations.

(i) Reactance extraction

Let $\mathbf{G}(s)$ be the transfer function of a "physical" system. Assume that this system may be considered as the interconnection of lossless reciprocal storage elements and non-dynamic reciprocal dissipative elements. It may help the reader in following the manipulations below by considering electrical *RLC* networks in which we extract the capacitors and the inductors (assumed n in number) and then write the equations of the remaining (resistive) part of the network with the connections leading to the capacitors and inductors considered as external inputs. The interconnection constraint then simply states how these capacitors and inductors fit into the network.

We assume that the storage elements are described by the n-dimensional characteristic

$$\mathbf{y}_1 = \mathbf{M}\mathbf{z}$$

with $\mathbf{z}$ the storage variables (e.g. the charge on capacitors, the flux in inductors, the strain in a viscoelastic element, the displacement of a spring, etc.) and $\mathbf{y}_1$ the supply variables (e.g. the voltage across the capacitors, the current through the inductors,

[3] The external signature matrix of a electrical n-port will have a +1 if the ith input is (say) a current and a −1 if it is a voltage.

the stress of a viscoelastic element, the force exercised by the spring, etc.). The work rate into the storage elements is assumed to be given by

$$w_1 = \langle \dot{\mathbf{z}}, \mathbf{y}_1 \rangle$$

It is easy to see that the storage elements then define a dissipative system if and only if $\mathbf{M} = \mathbf{M}^T \geqslant \mathbf{O}$, in which case it is conservative system, and the storage function is uniquely given by $\frac{1}{2}\mathbf{z}^T\mathbf{M}\mathbf{z}$.

We further postulate that the storage elements are reciprocal with signature matrix $\mathbf{\Sigma}_i$ (which, for reasons which will become apparent later, we will call the *internal signature matrix*). Thus we have $\mathbf{M} = \mathbf{\Sigma}_i\mathbf{M}^T\mathbf{\Sigma}_i$, i.e. if the vector $\mathbf{z}$ is arranged in such a way that $\mathbf{\Sigma}_i = \left[\begin{array}{c|c} \mathbf{I}_{n_+} & \mathbf{O} \\ \hline \mathbf{O} & -\mathbf{I}_{n_-} \end{array}\right]$ then $\mathbf{M}$ is conformably block-diagonal, i.e. $\mathbf{M} = \left[\begin{array}{c|c} \mathbf{M}_+ & \mathbf{O} \\ \hline \mathbf{O} & \mathbf{M}_- \end{array}\right]$ with $\mathbf{M}_+ = \mathbf{M}_+^T \geqslant \mathbf{O}$ and $\mathbf{M}_- = \mathbf{M}_-^T \geqslant \mathbf{O}$.

We assume that the dissipative elements are described by the $(n + m)$-dimensional characteristic

$$\left[\begin{array}{c} \mathbf{u}_2 \\ \hline \mathbf{y} \end{array}\right] = \left[\begin{array}{c|c} \tilde{\mathbf{A}} & \tilde{\mathbf{B}} \\ \hline \tilde{\mathbf{C}} & \tilde{\mathbf{D}} \end{array}\right] \left[\begin{array}{c} \mathbf{y}_2 \\ \hline \mathbf{u} \end{array}\right].$$

The supply rate into the storage elements is assumed to be given by $w = \mathbf{u}_2^T\mathbf{y}_2 + \mathbf{u}^T\mathbf{y}$. Thus dissipativeness requires that $\left[\begin{array}{c|c} \tilde{\mathbf{A}} & \tilde{\mathbf{B}} \\ \hline \tilde{\mathbf{C}} & \tilde{\mathbf{D}} \end{array}\right] + \left[\begin{array}{c|c} \tilde{\mathbf{A}} & \tilde{\mathbf{B}} \\ \hline \tilde{\mathbf{C}} & \tilde{\mathbf{D}} \end{array}\right]^T \geqslant \mathbf{O}$. We also postulate that the dissipative elements are reciprocal with signature matrix[4] $\left[\begin{array}{c|c} -\mathbf{\Sigma}_i & \mathbf{O} \\ \hline \mathbf{O} & \mathbf{\Sigma}_e \end{array}\right]$. Thus $\tilde{\mathbf{A}} = \mathbf{\Sigma}_i\tilde{\mathbf{A}}^T\mathbf{\Sigma}_i$, $\tilde{\mathbf{B}} = -\mathbf{\Sigma}_i\tilde{\mathbf{C}}^T\mathbf{\Sigma}_e$, and $\tilde{\mathbf{D}} = \mathbf{\Sigma}_e\tilde{\mathbf{D}}^T\mathbf{\Sigma}_e$.

We finally assume that when the storage elements are interconnected with the dissipative elements via the neutral interconnection constraint $\mathbf{u}_2 = \mathbf{\Sigma}_i\mathbf{u}_1$; $\mathbf{y}_2 = -\mathbf{\Sigma}_i\mathbf{y}_1$, then we obtain a representation of the original system. Thus it is assumed that

$$\begin{aligned} \dot{\mathbf{z}} &= -\mathbf{\Sigma}_i\tilde{\mathbf{A}}\mathbf{\Sigma}_i\mathbf{M}\mathbf{z} \quad -\mathbf{\Sigma}_i\tilde{\mathbf{B}}\mathbf{u} \\ \mathbf{y} &= \tilde{\mathbf{C}}\mathbf{\Sigma}_i\mathbf{M}\mathbf{z} + \tilde{\mathbf{D}}\mathbf{u} \end{aligned}$$

is a realization of the trans function $\mathbf{G}(s)$. Hence by the state space isomorphism theorem there exists a non-singular matrix $\mathbf{S}$ such that $-\mathbf{\Sigma}_i\tilde{\mathbf{A}}\mathbf{\Sigma}_i\mathbf{M} = \mathbf{S}\mathbf{A}\mathbf{S}^{-1}$, $-\mathbf{\Sigma}_i\tilde{\mathbf{B}} = \mathbf{S}\mathbf{B}$, $\tilde{\mathbf{C}}\mathbf{\Sigma}_i\mathbf{M} = \mathbf{C}\mathbf{S}^{-1}$, and $\tilde{\mathbf{D}} = \mathbf{D}$. Note also that the storage function $\frac{1}{2}\mathbf{z}^T\mathbf{M}\mathbf{z}$ is, in terms of $\mathbf{x} = \mathbf{S}^{-1}\mathbf{z}$, given by $\frac{1}{2}\mathbf{x}^T\mathbf{S}^T\mathbf{M}\mathbf{S}\mathbf{x}$.

The question of the possibility of the reactance extraction and the assumed

[4] These sign conventions differ from those in [1]. Here we have chosen the even and odd variables in such a way that if we consider electrical networks and extract the resistive elements then the voltage (current) controlled ports have all the same parity.

reciprocity of the storage elements and the dissipative elements thus reduces to finding:

(*i*) a storage matrix **Q** satisfying the conditions put forward in Theorem 4;
(*ii*) an internal signature matrix $\mathbf{\Sigma}_i$; and
(*iii*) a nonsingular matrix **S**,

such that

$$\mathrm{QAS}^{-1}\mathbf{\Sigma}_i\mathrm{S} = \mathrm{S}^T\mathbf{\Sigma}_i(\mathrm{S}^T)^{-1}\mathrm{A}^T\mathrm{Q},$$
$$\mathrm{QB} = \mathrm{S}^T\mathbf{\Sigma}_i(\mathrm{S}^T)^{-1}\mathrm{C}^T\mathbf{\Sigma}_e,$$
$$\mathrm{QS}^{-1}\mathbf{\Sigma}_i\mathrm{S} = \mathrm{S}^T\mathbf{\Sigma}_i(\mathrm{S}^T)^{-1}\mathrm{Q}.$$

(ii) The Onsager–Casimir relations

The Onsager–Casimir relations which play an important role in several areas of physics may roughly be stated as follows:

In a dissipative dynamical system there exists a set of variables, **j**, called the *force,* and a corresponding set of variables, **f**, called the *flux* such that $\langle \mathbf{f}, \mathbf{j}\rangle$ = the rate of dissipation and such that the flux is given in terms of the force by $\mathbf{f} = \mathbf{Lj}$ where **L** is a matrix which is signature symmetric for some appropriately choosen signature matrix $\mathbf{\Sigma}$, i.e. $\mathbf{L} = \mathbf{\Sigma}\mathrm{L}^T\mathbf{\Sigma}$.

As has been clearly demonstrated by Truesdell [14, Lecture 7, see especially p. 129] the above statement becomes essentially void of content if no additional constraints are placed on the choice of the force and the flux. We will present here one method of modifying the above law in a rational way so that it becomes equivalent to the question posed in the beginning of this section which, as we shall see later, imposes some meaningful constraints on the system and on the storage function.

Our interpretation of the Onsager–Casimir relations is the following:

Consider a "physical" system described by the dynamical equations $\Sigma = (FDLS)$ with supply rate $w = \mathbf{u}^T\mathbf{y}$ and storage function $S = \frac{1}{2}\,\mathbf{x}^T\mathbf{Q}\mathbf{x}$. Thus $\{\Sigma, w, S\}$ defines a dissipative dynamical system. Assume also that the system is externally reciprocal with external signature matrix $\mathbf{\Sigma}_e$. Then there exists a set of variables, **j**, called the force, and a corresponding set of variables, **f**, called the flux, such that:

(*i*)[5] $\mathbf{j} = \begin{bmatrix}\mathrm{F}\mathbf{x}\\ \hline \mathbf{u}\end{bmatrix}$ and $\mathbf{f} = \begin{bmatrix}\mathrm{G}\dot{\mathbf{x}}\\ \hline \mathbf{y}\end{bmatrix}$ for some nonsingular matrices **F** and **G**. Let **L** be such that $\mathbf{f} = \mathbf{Lj}$;

[5] The statement of the Onsager–Casimir relations given here has a much more general appearance than the one given in [1]. It is to be interpreted as follows: for the external quantities (**u**, **y**) one knows what are the fluxes and forces, with their corresponding signature, and for the internal variables we postulate that the force is "like" the state and that the flux is a combination of the time-derivatives of the force. We remark that if one were to let the internal force ($\mathbf{j}_i$) and flux ($\mathbf{f}_i$) be related to **x** and $\dot{\mathbf{x}}$ by the more general law $\begin{bmatrix}\mathbf{j}_i\\ \hline \mathbf{f}_i\end{bmatrix} = \mathrm{H}\begin{bmatrix}\mathbf{x}\\ \hline \dot{\mathbf{x}}\end{bmatrix}$ with H a nonsingular ($2n \times 2n$) matrix then, as shown by Truesdell [14, p. 129], the symmetry requirement becomes essentially meaningless.

(*ii*) $\langle \mathbf{f}, \mathbf{j} \rangle = \mathbf{f}^T\mathbf{j}$ = the dissipation rate

$$= \mathbf{u}^T\mathbf{y} - \frac{d}{dt}\, \mathbf{x}^T\mathbf{Q}\mathbf{x}.$$

Thus $\mathbf{L} + \mathbf{L}^T \geqslant \mathbf{O}$;

(*iii*) the $(n+m) \times (n+m)$ matrix $\mathbf{L}$ is signature symmetric, i.e. $\mathbf{L} = \mathbf{\Sigma}\mathbf{L}^T\mathbf{\Sigma}$, with $= \left[\begin{array}{c|c} -\mathbf{\Sigma}_i & \mathbf{O} \\ \hline \mathbf{O} & \mathbf{\Sigma}_e \end{array}\right]$ where $\mathbf{\Sigma}_i$ is called the *internal signature matrix*. The fluxes (forces) corresponding to a +1 in $\mathbf{\Sigma}_i$ or $\mathbf{\Sigma}_e$ are said to be of *even parity* while those corresponding to a −1 in $\mathbf{\Sigma}_i$ or $\mathbf{\Sigma}_e$ are said to be of *odd parity*;

(*iv*) (see, e.g. [15, p. 85]: "*the internal energy is even against time reversal*") the storage function $\frac{1}{2}\,\mathbf{x}^T\mathbf{Q}\mathbf{x}$ is invariant under the transformation $\mathbf{f} \to \mathbf{\Sigma}\mathbf{f}$ (i.e. $\mathbf{F}\mathbf{x} \to -\mathbf{\Sigma}_i\mathbf{F}\mathbf{x}$).

After some elementary manipulations it may be seen that the Onsager–Casimir relations thus stated claim the existence of a storage function $\frac{1}{2}\,\mathbf{x}^T\mathbf{Q}\mathbf{x}$ satisfying the conditions put forward in Theorem 4, an internal signature matrix $\mathbf{\Sigma}_i$, and a nonsingular matrix $\mathbf{S}$, such that

$$\mathbf{QAS}^{-1}\mathbf{\Sigma}_i\mathbf{S} = \mathbf{S}^T\mathbf{\Sigma}_i(\mathbf{S}^T)^{-1}\mathbf{A}^T\mathbf{Q}$$
$$\mathbf{QB} = \mathbf{S}^T\mathbf{\Sigma}_i(\mathbf{S}^T)^{-1}\mathbf{C}^T\mathbf{\Sigma}_e$$
$$\mathbf{QS}^{-1}\mathbf{\Sigma}_i\mathbf{S} = \mathbf{S}^T\mathbf{\Sigma}_i(\mathbf{S}^T)^{-1}\mathbf{Q}$$

The first two equations originate from condition (*iii*) since

$\mathbf{L} = \left[\begin{array}{c|c} \mathbf{G} & \mathbf{O} \\ \hline \mathbf{O} & \mathbf{I} \end{array}\right] \left[\begin{array}{c|c} \mathbf{A} & \mathbf{B} \\ \hline \mathbf{C} & \mathbf{D} \end{array}\right] \left[\begin{array}{c|c} \mathbf{F}^{-1} & \mathbf{O} \\ \hline \mathbf{O} & \mathbf{I} \end{array}\right]$, while the third equation is simply condition (*iv*). The matrix $\mathbf{F} = \mathbf{S}$ and $\mathbf{G}$ is related to it by the equation $\mathbf{G}^T\mathbf{F} = -\mathbf{Q}$, which follows from condition (*ii*).

Thus the Onsager–Casimir relations impose identical conditions as those obtained by postulating the reactance extraction. The following lemma reduces these equations to a condition on $\mathbf{Q}$ alone:

Lemma 1: A necessary condition for there to exist a signature matrix $\mathbf{\Sigma}_i$ *and a nonsingular matrix* $\mathbf{S}$ *such that*

$$\mathbf{QAS}^{-1}\mathbf{\Sigma}_i\mathbf{S} = \mathbf{S}^T\mathbf{\Sigma}_i(\mathbf{S}^T)^{-1}\mathbf{A}^T\mathbf{Q}$$
$$\mathbf{QB} = \mathbf{S}^T\mathbf{\Sigma}_i(\mathbf{S}^T)^{-1}\mathbf{C}^T\mathbf{\Sigma}_e$$
$$\mathbf{QS}^{-1}\mathbf{\Sigma}_i\mathbf{S} = \mathbf{S}^T\mathbf{\Sigma}_i(\mathbf{S}^T)^{-1}\mathbf{Q}$$

is that (FDLS) is reciprocal with external signature matrix $\mathbf{\Sigma}_e$. *Assuming this to be the case, then the solvability of these equations is equivalent to the existence of a matrix* $\mathbf{Q} = \mathbf{Q}^T$ *which satisfies (MIE) and* $\mathbf{TQ}^{-1}\mathbf{T} = \mathbf{Q}$, *where* $\mathbf{T}$ *is the (unique, symmetric) matrix defined in Theorem 5.*

Lemma 1 is easily proven if one realizes, as is shown in [1, p. 372], that $\mathbf{TQ}^{-1}\mathbf{T} = \mathbf{Q}$ implies that there exists a matrix $\tilde{\mathbf{S}}$ such that $\tilde{\mathbf{S}}^T\tilde{\mathbf{S}} = \mathbf{Q}$ and $\tilde{\mathbf{S}}^T\mathbf{\Sigma}\tilde{\mathbf{S}} = \mathbf{T}$. Since one wants to show in Lemma 1 that $\mathbf{QS}^{-1}\mathbf{\Sigma}_i\mathbf{S} = \mathbf{T}$, the result follows from there by taking $\mathbf{S} = \tilde{\mathbf{S}}$.

Several other related results of interest are:

(*i*) the matrix $\mathbf{\Sigma}_i$ is an invariant (modulo permutation of its diagonal entries). Its signature equals the signature of the (nonsingular, symmetric) matrix $\mathbf{T}$. It may be shown that also non-minimal reciprocal realizations have internal signature matrices with at least as many +1's and as many −1's as $\mathbf{\Sigma}_i$.

(*ii*) the quantity $\frac{1}{2}\mathbf{x}^T\mathbf{T}\mathbf{x}$, called the *Lagrangian* or the *coenergy*, equals, in the reactance extraction interpretation, $\frac{1}{2}\mathbf{z}_+^T\mathbf{M}_+\mathbf{z}_+ - \frac{1}{2}\mathbf{z}_-^T\mathbf{M}_-\mathbf{z}_-$ thus *showing the invariance*[6] *of the electrical* minus *the magnetic energy or of the kinetic* minus *the potential energy.*

The following theorem states the solvability of the equation $\mathbf{T}\mathbf{Q}^{-1}\mathbf{T} = \mathbf{Q}$. Using the relations of Theorem 5 it is easily seen that the solutions to (*MIE*) are closed under the map $\mathbf{Q} \to \mathbf{T}\mathbf{Q}^{-1}\mathbf{T}$. The question thus is to show that this map has a fixed point. Since the set of solutions to (*MIE*) is convex compact this fixed point property is an immediate consequence (and a most interesting application) of Brouwer's fixed point theorem.

Theorem 6 *Assume that the system (FDLS) is externally reciprocal with external signature matrix* $\mathbf{\Sigma}_e$, *let* $\mathbf{T}$ *be as defined in Theorem 5, and assume furthermore that* $\mathbf{G}(s)$ *is positive real. Then there exists a matrix* $\mathbf{Q}$ *which simultaneously satisfies (MIE) and* $\mathbf{T}\mathbf{Q}^{-1}\mathbf{T} = \mathbf{Q}$.

The proof of Theorem 6 as indicated above has the disadvantage of not being constructive. One may however compute a solution to (*MIE*) and $\mathbf{T}\mathbf{Q}^{-1}\mathbf{T} = \mathbf{Q}$ in terms of an arbitrary solution $\mathbf{Q}_1$ of (*MIE*). In fact, the geometric mean of $\mathbf{Q}_1$ and $\mathbf{T}\mathbf{Q}_1^{-1}\mathbf{T}$, defined by $\sqrt{\mathbf{Q}_1}\sqrt{\sqrt{\mathbf{Q}_1}^{-1}\mathbf{T}\mathbf{Q}_1^{-1}\mathbf{T}\sqrt{\mathbf{Q}_1}^{-1}}\sqrt{\mathbf{Q}_1}$ will satisfy both $\mathbf{T}\mathbf{Q}^{-1}\mathbf{T} = \mathbf{Q}$ and (*MIE*). This then leads to a less elegant but more useful proof of the theorem.

7. REPRESENTATIONS OF DISSIPATIVE AND SYMMETRIC SYSTEMS

Recall that for a system with transfer function $\mathbf{G}(s)$ there are many ways of representing it in the form (*FDLS*). If we only consider minimal realizations (*FDLS*) then all such realizations are elements on the orbit of $\{\mathbf{A}, \mathbf{B}, \mathbf{C}, \mathbf{D}\}$ under the transformation group $\{\mathbf{A}, \mathbf{B}, \mathbf{C}, \mathbf{D}\} \xrightarrow[\det \mathbf{S} \neq 0]{\mathbf{S}} \{\mathbf{S}\mathbf{A}\mathbf{S}^{-1}, \mathbf{S}\mathbf{B}, \mathbf{C}\mathbf{S}^{-1}, \mathbf{D}\}$. We will now exploit this freedom in representing a particular system such that its dissipativeness and/or reciprocity become obvious. These "canonical" representations are especially useful in network synthesis [16].

Definition 5: The system (*FDLS*) is said to define an *internally passive* dynamical system if

$$\left[\begin{array}{c|c} -\mathbf{A} & -\mathbf{B} \\ \hline \mathbf{C} & \mathbf{D} \end{array}\right] + \left[\begin{array}{c|c} -\mathbf{A} & -\mathbf{B} \\ \hline \mathbf{C} & \mathbf{D} \end{array}\right]^T \geqslant 0$$

$\left(\text{i.e. } \frac{d}{dt}\frac{1}{2}\mathbf{x}^T\mathbf{x} \leqslant \mathbf{u}^T\mathbf{y}\right)$.

[6] In other words, for reciprocal electrical *N*-port networks which consist also internally of reciprocal elements it follows that the *difference* of the electrical and the magnetic energy depends only on the input/output behaviour of the *N*-port, but not on the specific realization of the *N*-port. On the other hand the total energy, i.e. the sum of the electrical and the magnetic energy, does depend on the specific internal structure of the *N*-port.

The following theorem is only concerned with minimal realizations and follows readily from the results of Section 5.

Theorem 7 *Let* G(*s*) *be given. Then* G(*s*) *admits a passive realization if and only if* G(*s*) *is positive real. All passive realizations may be obtained by transforming a given realization* $\{A, B, C, D\}$ *into* $\{SAS^{-1}, SB, CS^{-1}, D\}$ *with* S^TS *a solution of (MIE). All passive realizations which correspond to the* same *storage function may be obtained from one by operating with an orthogonal transformation and thus form a group which is isomorphic to the orthogonal group.*

Thus the class of similarity transformations which transform $\{A, B, C, D\} \overset{S}{\rightarrow} \{SAS^{-1}, SB, CS^{-1}, D\}$ into a passive dynamical system forms a subset (in general not a group) of the general linear group. The reason for this is that only those similarity transformations which are such that S^TS satisfies (*MIE*) are allowed. Note that Theorem 7 shows that fixing the storage function is like choosing the norm on the state space.

The concept of internal reciprocity may be defined as follows:

Definition 6: The system (*FDLS*) is said to define an *internally reciprocal* dynamical system if there exists a signature matrix Σ such that $\Sigma \left[\begin{array}{c|c} -A & -B \\ \hline C & D \end{array}\right]$ is symmetric. The matrix Σ may be written as $\Sigma = \left[\begin{array}{c|c} -\Sigma_i & 0 \\ \hline 0 & \Sigma_e \end{array}\right]$ with Σ_i an $(n \times n)$ signature matrix which is called the *internal signature matrix* and Σ_e an $(m \times m)$ signature matrix which is called the *external signature matrix.*

The following theorem is an immediate consequence of Theorem 5 and shows the compatibility of internal and external passive reciprocal realizations.

Theorem 8 *Let* G(*s*) *and* Σ_e *be given. Then* G(*s*) *admits a reciprocal realization if and only if* $G(s) = \Sigma_e G^T(s) \Sigma_e$. *Moreover,* Σ_i *is invariant up to permutation of its diagonal elements and is equal to the signature of the (unique, symmetric, non-singular) solution* T *to* $A = T^{-1}A^TT$, $B = T^{-1}C^T\Sigma_e$. *All reciprocal realizations may be obtained by transforming a given realization* $\{A, B, C, D\}$ *into* $\{SAS^{-1}, SB, CS^{-1}, D\}$ *with* S *a solution of* $S^T\Sigma_iS = T$. *All such realizations may be obtained from one by operating by the similarity transformation* S_1 *with* $S_1^T\Sigma_iS_1 = \Sigma_i$ *and thus form a group which is isomorphic to the* Lorentz group $LO(n_+, n_- = n - n_+)$ *where* n_+ = *the number of positive eigenvalues of* T. *Finally the Lagrangian or coenergy* $\frac{1}{2}x^T\Sigma_ix$ *is an invariant (i.e. it depends on* u *but not on the actual representation) of the reciprocal realizations of* G(*s*). *In terms of a (not necessarily reciprocal) realization* $\{A, B, C, D\}$ *it equals* $\frac{1}{2}x^TTx$.

Assume that G(*s*) is positive real and that $G(s) = \Sigma_e G^T(s) \Sigma_e$. It then follows that there exists at least one realization which is internally passive and at least one realization which is internally reciprocal. The problem considered next concerns the compatibility of these two properties, i.e. *does there exist a realization which is simultaneously internally passive and internally reciprocal?* In view of what has been established before it is clearly necessary for such a realization to exist that G(*s*) be positive real and that $\Sigma_e G(s)$ be symmetric. To show that these conditions are also sufficient it suffices to prove the existence of a matrix S such that S^TS satisfies (*MIE*) *and* such that $S^T\Sigma_iS = T$. The matrix S may be eliminated from these equations and exists *if and only if* there exists a solution $Q = Q^T > 0$ to (*MIE*) and

$\mathbf{TQ}^{-1}\mathbf{T} = \mathbf{Q}$. The solvability of this equation is in fact shown in Theorem 6. This leads to:

Theorem 9 *Let* $\boldsymbol{\Sigma}_e$ and G(s) *be given. Then there exists a minimal realization of* G(s) *which is simultaneously internally reciprocal and internally passive* if and only if $\boldsymbol{\Sigma}_e$G(s) *is symmetric and* G(s) *is positive real.*

It follows from Theorem 9 that if a system is externally dissipative and reciprocal, then it admits a representation with:

$$\mathbf{A} = \left[\begin{array}{c|c} \mathbf{A}_{++} & \mathbf{A}_{+-} \\ \hline -\mathbf{A}_{+-}^T & \mathbf{A}_{--} \end{array}\right], \quad \mathbf{B} = \left[\begin{array}{c|c} \mathbf{B}_{++} & \mathbf{B}_{+-} \\ \hline \mathbf{B}_{-+} & \mathbf{B}_{--} \end{array}\right], \quad \mathbf{C} = \left[\begin{array}{c|c} \mathbf{B}_{++}^T & -\mathbf{B}_{-+}^T \\ \hline -\mathbf{B}_{+-}^T & \mathbf{B}_{--}^T \end{array}\right], \quad \mathbf{D} = \left[\begin{array}{c|c} \mathbf{D}_{++} & \mathbf{D}_{+-} \\ \hline -\mathbf{D}_{+-}^T & \mathbf{D}_{--} \end{array}\right],$$

where $\mathbf{A}_{++}$, $\mathbf{A}_{--}$, $\mathbf{D}_{++}$, and $\mathbf{D}_{--}$ symmetric matrices and with dissipation matrix:

$$\boldsymbol{\Delta} = \left[\begin{array}{c|c|c|c} -\mathbf{A}_{++} & \mathbf{0} & \mathbf{0} & -\mathbf{B}_{+-} \\ \hline \mathbf{0} & -\mathbf{A}_{--} & -\mathbf{B}_{-+} & \mathbf{0} \\ \hline \mathbf{0} & -\mathbf{B}_{-+}^T & \mathbf{D}_{++} & \mathbf{0} \\ \hline -\mathbf{B}_{+-}^T & \mathbf{0} & \mathbf{0} & \mathbf{D}_{--} \end{array}\right] \geqslant \mathbf{0}.$$

Thus the instantaneous coupling between states of the same parity is responsible for the dissipation. The instantaneous coupling between states of opposite parity introduces an exchange of storage between elements of opposite parity and yields the oscillatory part of the response function.

In terms of the Onsager–Casimir relations, Theorem 9 states that one may choose the state in such a way that $\mathbf{j} = \left[\begin{array}{c} \mathbf{x} \\ \hline \mathbf{u} \end{array}\right]$ and $\mathbf{f} = \left[\begin{array}{c} -\dot{\mathbf{x}} \\ \hline \mathbf{y} \end{array}\right]$. There are in general many ways to obtain such a representation and the even (and odd) variables in one such representation will in general be linear combinations of the even *and* odd variables in another representation. However for those representations which leave invariant the internal storage $\frac{1}{2}\mathbf{x}^T\mathbf{x}$ or, equivalently, the dissipation rate $\langle f, j\rangle$ (in addition to the Lagrangian $\frac{1}{2}\mathbf{x}^T\boldsymbol{\Sigma}_i\mathbf{x}$) one can show that the even (or odd) variables on one representation are recombinations only of the even (and odd) variables in another representation.

8. FURTHER QUALITATIVE ASSUMPTIONS

There are several other qualitative assumptions which may be imposed and which lead to canonical representations of various interesting classes of dynamical systems

(*i*) *Lossless systems* which are systems for which the dissipation inequality holds with equality. For lossless systems the equality $S_a = S_r$ holds and thus the storage function is uniquely determined by Σ and w. Linear lossless systems admit a minimal realization with $\mathbf{A} + \mathbf{A}^T = \mathbf{0}$, $\mathbf{B} = \mathbf{C}^T$, and $\mathbf{D} + \mathbf{D}^T = \mathbf{0}$.

(*ii*) *Reversible systems* which are systems of the type (*FDLS*) which have the additional property that when the zero initial condition response to $\mathbf{u}(t)$ is $\mathbf{y}(t)$ then the response to $\boldsymbol{\Sigma}_e\mathbf{u}(-t)$ will be $-\boldsymbol{\Sigma}_e\mathbf{y}(-t)$. It may be shown that dissipativeness and reversibility $\Leftrightarrow$ losslessness *and* reciprocity with external signature $\boldsymbol{\Sigma}_e$. They admit the representation

$$\mathbf{A} = \left[\begin{array}{c|c} \mathbf{O} & \mathbf{A}_{+-} \\ \hline -\mathbf{A}_{+-}^T & \mathbf{O} \end{array}\right], \quad \mathbf{B} = \mathbf{C}^T = \left[\begin{array}{c|c} \mathbf{B}_{++} & \mathbf{O} \\ \hline \mathbf{O} & \mathbf{B}_{--} \end{array}\right], \quad \mathbf{D} = \left[\begin{array}{c|c} \mathbf{O} & \mathbf{D}_{+-} \\ \hline -\mathbf{D}_{+-}^T & \mathbf{O} \end{array}\right]$$

Thus the states of the same parity are instantaneously uncoupled in a reversible system. The coupling between the states of opposite parity leads to the oscillations which are characteristic for such systems. Moreover, every representation of $\mathbf{G}(s)$ having the above form is such that the energy $\frac{1}{2}\mathbf{x}^T\mathbf{x}$ and the coenergy $\frac{1}{2}\mathbf{x}^T\mathbf{\Sigma}_i\mathbf{x}$ are invariants. They may thus all be obtained from one by the similarity transformation $\mathbf{S} = \left[\begin{array}{c|c} \mathbf{S}_+ & \mathbf{O} \\ \hline \mathbf{O} & \mathbf{S}_- \end{array}\right]$: the space of the even and odd variables are thus invariants in the sense that in any representation the even (odd) variables are recombinations of the even (odd) variables of another representation. The dissipation matrix of a reversible system is zero.

A reversible system also satisfies an internal reversibility property which states that there exists a representation such that if $\mathbf{u}(t)$ and $\mathbf{x}(0)$ lead to $\mathbf{x}(t)$ and $\mathbf{y}(t)$ then $\mathbf{\Sigma}_e\mathbf{u}(-t)$ and $-\mathbf{\Sigma}_i\mathbf{x}(0)$ lead to $-\mathbf{\Sigma}_i\mathbf{x}(-t)$ and $-\mathbf{\Sigma}_e\mathbf{y}(-t)$. Thus external and internal reversibility are compatible properties.

Strictly speaking the notion of a reversible system calls for a more elaborate set of axioms than the axioms of dynamical systems which we have been using so far, since forward and backward transitions need be defined, i.e. ϕ needs to have the group property (rather than only the semi-group property).

(*iii*) *Relaxation systems* which are systems of the type (*FDLS*) which have an impulse response $\mathbf{D}\delta(t) + \mathbf{C}e^{\mathbf{A}t}\mathbf{B}$ which is a completely monotonic function on $[0, \infty)$ (i.e., $\mathbf{W}(t) = \mathbf{W}^T(t)$ and $(-1)^n \dfrac{d^n\mathbf{W}(t)}{dt^n} \geqslant \mathbf{O}$ for $t \geqslant 0$ and $n = 0, 1, 2, \ldots$). It may be shown that this is the case if and only if the Hankel matrix $\mathbf{H}(t)$ of (*FDLS*) and its shift[7] $-\sigma\mathbf{H}(t)$ are symmetric nonnegative definite ($\forall t \geqslant 0$). Examples of relaxation systems are chemical reactions, models of viscoelastic materials in which inertia is neglected, and *RL* or *RC* networks. A relaxation system admits a minimal realization with $\mathbf{A} = \mathbf{A}^T \leqslant \mathbf{O}$, $\mathbf{B} = \mathbf{C}^T$, and $\mathbf{D} = \mathbf{D}^T \geqslant \mathbf{O}$. Thus $\mathbf{\Sigma}_e = \mathbf{I}$ and $\mathbf{\Sigma}_i = \mathbf{I}$.

Relaxation systems obey Onsager's law[8] (i.e., the matrix $\mathbf{L}$ introduced in Section 7 satisfies $\mathbf{L} = \mathbf{L}^T \geqslant \mathbf{O}$). As may be seen from Theorems 5 and 6 we have that for a relaxation system $\mathbf{T} = \mathbf{T}^T > \mathbf{O}$ and the only matrix $\mathbf{Q}$ which satisfies $\mathbf{T}\mathbf{Q}^{-1}\mathbf{T} = \mathbf{Q}$ is $\mathbf{Q} = \mathbf{T}$. Thus the storage function is an invariant of the internally reciprocal representations of a relaxation system. This interesting property in fact does not depend on the minimality assumption. Thus if we make the assumption that a system which behaves externally as a relaxation system has the above property as a result of internal symmetry (which in most applications would seem to be a very reasonable hypothesis) then we may prove the following formula:

The internal storage at $t = 0$ due to the input $\mathbf{u}(t)$ on $(-\infty, 0]$ $= \frac{1}{2}\int_0^\infty \mathbf{y}^T(t)\mathbf{u}(-t)\,dt$

[7] The matrix $-\sigma\mathbf{H}(t)$ is the Hankel matrix of $-W^{(1)}(t)$.

[8] Onsager's law thus holds for systems which have only one kind of energy storage elements as, for example, chemical reactions. Casimir thus extended the validity of these symmetry relations to arbitrary reciprocal systems, thus making these laws much more globally applicable.

where $\mathbf{y}(t)$ is the output which results from this input with $\mathbf{u}(t) = \mathbf{O}$ for $t > 0$,

i.e. $\mathbf{y}(t) = \int_{-\infty}^{0} \mathbf{W}(t - \tau)\mathbf{u}(\tau)\, d\tau$.

This formula also holds for non-minimal systems.

9. APPLICATIONS

In this section we will briefly describe several areas of application of the ideas of this paper.

(i) Electrical network synthesis [16]

Assume that $\mathbf{G}(s)$ is given as the hybrid port description of an electrical m-port network which is to be synthesized using passive R, L, C, T and G's. Assume that the m_1 first ports are current controlled and that the m_2 $(= m - m_1)$ last ports are voltage controlled. Let $\mathbf{\Sigma}_e = \left[\begin{array}{c|c} I_{m_1} & 0 \\ \hline 0 & -I_{m_2} \end{array}\right]$.

In terms of the framework introduced in this paper we may hence assume that we are given a system $(FDLS)$ (where $\mathbf{x}$ does not have any physical significance other than the fact that it is a memory function). The power (= the supply rate) into the network is given by $w = \mathbf{u}^T\mathbf{y}$. If the system is realizable by interconnecting passive elements then there must exist a function S such that $\{\Sigma, w, S\}$ is dissipative. Thus, as shown in Theorem 3, positive realness of $\mathbf{G}(s)$ is a necessary condition for realizability.

Theorem 5 enables one to eliminate the dynamic element out of this synthesis problem by bringing the question back to the synthesis of a memoryless network with hybrid port description $\left[\begin{array}{c|c} -\mathbf{A} & -\mathbf{B} \\ \hline \mathbf{C} & \mathbf{D} \end{array}\right]$ which, when terminated with (say) unit capacitors, yields a synthesis. The synthesis of this network using passive R, T and G's is quite easily carried out, provided, as is implied by Theorem 7, the symmetric part of this hybrid port matrix is nonnegative definite. Note that we have also shown that there always exists at least one synthesis which has all its internal storage available as its terminals $(S = S_a)$ and at least one synthesis which can be brought from rest into any state without dissipation $(S = S_r)$.

If we are concerned with gyratorless synthesis then it follows from Lemma 1 that $\mathbf{G}(s) = \mathbf{\Sigma}_e\mathbf{G}^T(s)\mathbf{\Sigma}_e$ is a necessary condition for such a synthesis. Theorem 8 shows that this condition is also sufficient since the hybrid matrix $\left[\begin{array}{c|c} -\mathbf{A} & -\mathbf{B} \\ \hline \mathbf{C} & \mathbf{D} \end{array}\right]$ may indeed be realized without gyrators and will lead to a synthesis if the ports corresponding to +1 in $\mathbf{\Sigma}_i$ are terminated by unit capacitors and those corresponding to −1 in $\mathbf{\Sigma}_i$ by unit inductors. Note that for gyratorless realizations using the minimal number of reactive elements we have that the number of capacitors (n_+) and inductors (n_-) are both invariants and that for such realizations using a non-minimal number of reactive elements we have that the number of capacitors is at least n_+

and that the number of inductors is at least n_-. Another interesting fact which we have shown here is that the energy stored in the capacitors *minus* the energy stored in the inductors depends on the transfer function $\mathbf{G}(s)$ and on the past of u only.

(ii) Mechanics

Assume that we are given a linear *mechanical system* which we consider as a mapping from $\mathbf{f}$ (the *force*) to $\mathbf{v}$ (the *velocity*) described by the minimal finite dimensional system

$$\Sigma\colon \dot{\mathbf{x}} = \mathbf{A}\mathbf{x} + \mathbf{B}\mathbf{f};\ \ \mathbf{v} = \mathbf{C}\mathbf{x}$$

We assume that the system has the following properties:

(*i*) it is externally reciprocal with $\mathbf{\Sigma}_e = \mathbf{I}$. Thus $\mathbf{G}(s) = \mathbf{C}(\mathbf{I}s - \mathbf{A})^{-1}\mathbf{B}$ is symmetric;

(*ii*) there exists S such that $\{\Sigma, w, S\}$ is dissipative, where $w = \mathbf{f}'\mathbf{v}$. Thus $\mathbf{G}(s)$ is positive real; and

(*iii*) the dissipation vanishes whenever $\mathbf{v} = \mathbf{0}$; i.e., $\dfrac{d}{dt} S(\mathbf{x}(t)) = 0$ whenever $\mathbf{v}(t) = \mathbf{0}$.

These axioms imply, as is shown in Theorem 9, that the system Σ admits a representation (*FDLS*) with

$$\mathbf{A} = \left[\begin{array}{c|c} \mathbf{A}_{++} & \mathbf{A}_{+-} \\ \hline -\mathbf{A}_{+-} & \mathbf{A}_{--} \end{array}\right], \quad \mathbf{B} = \left[\begin{array}{c} \mathbf{B}_{+} \\ \hline \mathbf{B}_{-} \end{array}\right], \quad \mathbf{C} = [\mathbf{B}_+^T \ \vdots\ -\mathbf{B}_-^T],$$

and $\mathbf{A}_{++} = \mathbf{A}_{++}^T$, $\mathbf{A}_{--} = \mathbf{A}_{--}^T$, and such that the dissipation matrix

$$\left[\begin{array}{c|c|c} -\mathbf{A}_{++} & \mathbf{O} & \mathbf{O} \\ \hline \mathbf{O} & -\mathbf{A}_{--} & -\mathbf{B}_- \\ \hline \mathbf{O} & -\mathbf{B}_-^T & \mathbf{O} \end{array}\right]$$

is non-negative definite. This condition reduces to $-\mathbf{A}_{++} \geqslant \mathbf{O}$, $-\mathbf{A}_{--} \geqslant \mathbf{O}$, and $\mathbf{B}_- = \mathbf{O}$. Moreover, from (*iii*) it follows that $\mathbf{B}_+^T\mathbf{x}_+ = \mathbf{0}$ must imply that $\mathbf{x}_+^T\mathbf{A}_{++}\mathbf{x}_+ = \mathbf{x}_-^T\mathbf{A}_{--}\mathbf{x}_- = 0$. This shows that $\mathbf{A}_{--} = \mathbf{0}$. Thus the system admits the minimal representation.

$$\begin{aligned} \dot{\mathbf{x}}_+ &= \mathbf{A}_{++}\mathbf{x}_+ + \mathbf{A}_{+-}\mathbf{x}_- + \mathbf{B}_+\mathbf{f} \\ \dot{\mathbf{x}}_- &= -\mathbf{A}_{+-}^T\mathbf{x}_+ \\ \mathbf{v} &= \mathbf{B}_+^T\mathbf{x}_+ \end{aligned}$$

From this it is easily seen that the system admits the canonical form

$$\ddot{\mathbf{z}} + \mathbf{P}\dot{\mathbf{z}} + \mathbf{Q}\mathbf{z} = \mathbf{M}\mathbf{f};\ \ \mathbf{d} = \mathbf{M}^T\mathbf{z}$$

where $\mathbf{P} = \mathbf{P}^T \geqslant \mathbf{O}$ and $\mathbf{Q} = \mathbf{Q}^T \geqslant \mathbf{O}$ and where we have put

$$\mathbf{d} \triangleq \int \mathbf{v}\, dt \ \text{(}\mathbf{d}\text{ is the } \textit{displacement}\text{)}.$$

The stored energy is given by $S = \frac{1}{2}\dot{\mathbf{z}}^T\dot{\mathbf{z}} + \frac{1}{2}\mathbf{z}^T\mathbf{Q}\mathbf{z}$, the Lagrangian (or coenergy) is given by $L = \frac{1}{2}\dot{\mathbf{z}}^T\dot{\mathbf{z}} - \frac{1}{2}\mathbf{z}^T\mathbf{Q}\mathbf{z}$, and the dissipation rate is given by $d = \dot{\mathbf{z}}'\mathbf{P}\dot{\mathbf{z}}$. The Lagrangian is independent of the actual representation and depends on the past of f only, but S and d depend in general on the representation. If, in addition, the system is lossless then it is called a *Hamiltonian* system and it becomes a reversible system which admits the representation.

$$\ddot{\mathbf{z}} + \mathbf{Q}\mathbf{z} = \mathbf{M}\mathbf{f}; \quad \mathbf{d} = \mathbf{M}^T\mathbf{z} \quad \mathbf{Q} = \mathbf{Q}^T \geqslant \mathbf{O}$$

Both the energy and the coenergy are now invariants. Thus the *kinetic energy* $\frac{1}{2}\dot{\mathbf{z}}'\dot{\mathbf{z}}$ and the *potential energy* $\frac{1}{2}\mathbf{z}'\mathbf{Q}\mathbf{z}$ are input/output quantities for Hamiltonian systems. In fact, all the representations of the above form may in this case be obtained from each other by the transformation $\tilde{\mathbf{z}} = \mathbf{S}\mathbf{z}$ with $\mathbf{S}$ any orthogonal matrix.

(iii) Thermodynamics

Consider the thermodynamic behaviour of a material particle of unit mass. Let $\boldsymbol{\epsilon}(.)$,$\boldsymbol{\tau}(.)$, and $T(.)$, denote respectively the (symmetric) strain tensor, the (symmetric) stress tensor, and the temperature. In the theory of simple materials with memory one postulates that the strain is a function of the stress and the temperature and of their pasts. Here we will assume that this dynamic dependence takes the form

$$\dot{\mathbf{x}} = \mathbf{f}(\mathbf{x}, T, \boldsymbol{\epsilon}), \quad \boldsymbol{\tau} = \mathbf{g}(\mathbf{x}, T, \boldsymbol{\epsilon})$$

which essentially only assumes finite dimensionality of the memory function. In *thermostatics* one assumes this dependence to be memoryless, i.e.

$$\boldsymbol{\tau} = \mathbf{g}(T, \boldsymbol{\epsilon})$$

Such materials are said to be *elastic* (or *thermoelastic*). The equation of the variation of the temperature is assumed to take the form

$$\dot{T} = h(T, \boldsymbol{\epsilon}, q, \dot{\boldsymbol{\epsilon}})$$

where $q(t)$, $-\infty < t < \infty$, denotes the rate of heat absorbed by the system. The mechanical power performed on the system is given by

$$w = \langle \boldsymbol{\tau}, \dot{\boldsymbol{\epsilon}} \rangle \overset{\Delta}{=} \mathrm{Tr}\{\boldsymbol{\tau}^T \boldsymbol{\epsilon}\}$$

where Tr denotes Trace.

The response function of a thermodynamic particle is constrained by two physical laws, i.e. the law of conservation of energy and the second law (Clausius' law). Thus we may postulate the existence of two functions E, the *energy*, and S, the *entropy*, such that

(*i*) $\dot{E} = q + w$; and

(*ii*) $\dot{S} \geqslant \dfrac{q}{T}$

The first statement may be interpreted in terms of dissipative systems but the interpretation of the second law leads to some difficulties because the entropy needs in general not be bounded from below or above (consider, e.g., an ideal gas).

Let us now briefly describe how the ideas of this paper may be refined to take care of this difficulty. The next few paragraphs are therefore *not* specifically concerned with the application treated in this section.

Definition 1b: The triplet $\{\Sigma, w, S\}$ is said to define a *cyclo-dissipative dynamical system* if the dissipation inequality (*DIE*) is satisfied.

The only difference between dissipativeness and cyclo-dissipativeness is that in the latter case there is no need for the storage function to be nonnegative (or, essentially equivalently, bounded from below). An example of a cyclo-dissipative system is an electrical network with positive resistors and capacitors or inductors of either sign. The following properties of cyclo-dissipative systems generalize the results of Section 3.

1. Assume that Σ and w are given and that there exists a point $x^* \in X$ such that X is reachable from x^* and controllable to x^*. Then there exists S such that $\{\Sigma, w, S\}$ is cyclo-dissipative *if and only if* $\oint w(t)\,dt \geqslant 0$ (hence the terminology "cyclo-dissipative"). The notation $\oint$ means that the integrand is evaluated along an arbitrary closed path in state space, i.e., $\oint \stackrel{\Delta}{=} \int_{t_0}^{t_1}$ with $t_1 \geqslant t_0$ and $u \in \mathscr{U}$ such that $\phi(t_1, t_0, x_0, u) = x_0$.

2. Assume again reachability from x^* and controllability to x^*. Let $S_a^*, S_r^*: X \to R^e$ be defined by

$$S_a^*(x_0) \stackrel{\Delta}{=} \sup_{\substack{t_1 \geqslant t_0 \\ x_0 \to x^*}} -\int_{t_0}^{t_1} w(t)\,dt; \text{ and}$$

$$S_r^*(x_0) \stackrel{\Delta}{=} \inf_{\substack{t_{-1} \leqslant t_0 \\ x^* \to x_0}} \int_{t_{-1}}^{t_0} w(t)\,dt$$

Let Σ and w be given. Then there exists S such that $\{\Sigma, w, S\}$ is cyclo-dissipative *if and only if* $S_r^* > -\infty$ or $S_a^* < \infty$. Moreover, if so, then $\{\Sigma, w, S_a^*\}$ and $\{\Sigma, w, S_r^*\}$ are cyclo-dissipative and $S_a^*(x) \leqslant S(x) - S(x^*) \leqslant S_r^*(x)$.

Let us now return to our thermodynamic system. Since $w = \langle \dot{\boldsymbol{\epsilon}}, \mathbf{w} \rangle$ it is necessary to consider $\dot{\boldsymbol{\epsilon}}$ as an input variable since if $\boldsymbol{\epsilon}$ would be considered as the input variable then it would not be possible to describe part of the instantaneous interaction (w) of the system with its environment in terms of the instantaneous input and output. Thus we arrive at the following formalization of the thermodynamics of an elementary particle of a simple material with memory: the input is[9] $(q, \dot{\boldsymbol{\epsilon}})$, the output is $(T, \boldsymbol{\tau})$, and the state is $(\mathbf{x}, T, \dot{\boldsymbol{\epsilon}})$. The constitutive equations are:

$$\Sigma:\ \dot{\mathbf{x}} = \mathbf{f}(\mathbf{x}, T, \boldsymbol{\epsilon})$$
$$\dot{T} = h(T, \boldsymbol{\epsilon}, q, \mathbf{v})$$
$$\dot{\boldsymbol{\epsilon}} = \mathbf{v}$$
$$\boldsymbol{\tau} = \mathbf{g}(T, \boldsymbol{\epsilon})$$

Note that these equations have a somewhat special structure. We may in

[9] The spaces U and Y satisfy certain assumptions which are not spelled out explicitly. The only essential restriction which we have to impose is that the first component of the output (i.e. T) is assumed to be always > 0.

addition postulate the existence of two functions E and S such that

(*i*) $\{\Sigma, \langle \boldsymbol{\tau}, \mathrm{v}\rangle, E\}$ is conservative; and

(*ii*) $\{\Sigma, -\frac{q}{T}, -S\}$ is cyclo-dissipative.

Assuming smoothness of E and S it follows from these equations that:

(*i*) $\frac{\partial E}{\partial T} \neq 0$, $\frac{\partial S}{\partial T} \neq 0$, and $\frac{\partial E}{\partial T} = T\frac{\partial S}{\partial T}$

(*ii*) $\mathrm{g} = \frac{\partial E}{\partial \boldsymbol{\epsilon}} - T\frac{\partial S}{\partial \boldsymbol{\epsilon}}$

(*iii*) $\left\langle \frac{\partial S}{\partial \mathrm{x}}, \mathrm{f} \geqslant 0 \right\rangle$

The first of these equations implies that one may trade the independent variable T for either E or S. Most treatments of thermodynamics take $(\mathrm{x}, E, \boldsymbol{\epsilon})$ (or $(\mathrm{x}, S, \boldsymbol{\epsilon})$) as the state which is, because of this, at least in principle justified. The reason one likes to do this however is that the equations of motion may then be written more simply with $(\dot{E}, \dot{\boldsymbol{\epsilon}})$ as input, $(T, \boldsymbol{\tau})$ as output and $(\mathrm{x}, E, \boldsymbol{\epsilon})$ as state, in the form:[10]

$$\Sigma:\ \dot{\mathrm{x}} = \mathrm{f}(\mathrm{x}, E, \boldsymbol{\epsilon})$$
$$\dot{E} = k, \dot{\boldsymbol{\epsilon}} = \mathrm{v}$$
$$\boldsymbol{\tau} = \mathrm{g}(\mathrm{x}, E, \boldsymbol{\epsilon}),\ T = \theta(\mathrm{x}, E, \boldsymbol{\epsilon})$$

The second law then postulates that there exists S such that $\left\{\Sigma, \frac{\langle \boldsymbol{\tau}, \mathrm{v}\rangle - k}{T}, -S\right\}$ is cyclo-dissipative. This implies that

(*i*)′ $\theta = \frac{1}{\partial S/\partial E}$

(*ii*)′ $\mathrm{g} = -\theta\frac{\partial S}{\partial \boldsymbol{\epsilon}}$

(*iii*)′ $\left\langle \frac{\partial S}{\partial \mathrm{x}}, \mathrm{f} \geqslant 0 \right\rangle$

The case of thermo-elasticity is of special interest. Thus (*i*)′ and (*ii*)′ are the well-known equations of thermostatics and we see that $\left\{\Sigma, \frac{\langle \boldsymbol{\tau}, \mathrm{v}\rangle - k}{T}, -S\right\}$ is then necessarily conservative. The entropy and the energy are then thus uniquely determined by the constitutive equations and play the role of potential functions. This is however not the case for more general materials.

An important class of materials are those which have what is commonly called *fading memory*. This means that for every given $(E, \boldsymbol{\epsilon})$ there is assumed to exist a unique point $\mathrm{x}^*(E, \boldsymbol{\epsilon})$ such that $\mathrm{f}(\mathrm{x}^*, E, \boldsymbol{\epsilon}) = \mathbf{0}$ and such that every solution of

[10] Note that the symbols f and g are used with two different meanings.

$\dot{\mathbf{x}} = f(\mathbf{x}, E, \boldsymbol{\epsilon})$ satisfies $\lim_{t\to\infty} \mathbf{x}(t) = \mathbf{x}^*$. Thus it is possible to define the *elastic response* of Σ as

$$\Sigma^*: \dot{E} = k, \dot{\boldsymbol{\epsilon}} = \mathbf{v}$$
$$\boldsymbol{\tau} = \mathbf{g}^*(E,\boldsymbol{\epsilon}), \ T = \theta^*(E,\boldsymbol{\epsilon})$$

where $\mathbf{g}^*(E, \boldsymbol{\epsilon}) \triangleq g(\mathbf{x}^*(E, \boldsymbol{\epsilon}), E, \boldsymbol{\epsilon})$ and $\theta^*(E, \boldsymbol{\epsilon}) \triangleq \theta(\mathbf{x}^*(E, \boldsymbol{\epsilon}), E, \boldsymbol{\epsilon})$. The system Σ^* describes the response of Σ to inputs $\dot{E}, \dot{\boldsymbol{\epsilon}}$ which vary sufficiently slowly (which is what is usually understood by *equilibrium thermodynamics*). By considering slow motions it may be shown that this elastic system itself satisfies the first and second law and thus that there exists S^* such that

$(i)^*$ $\theta^* = \dfrac{1}{\partial S^*/\partial E}$

$(ii)^*$ $\mathbf{g}^* = -\theta^* \dfrac{\partial S}{\partial \boldsymbol{\epsilon}}$

Assuming that $\dfrac{\partial \mathbf{f}(\mathbf{x}, E, \boldsymbol{\epsilon})}{\partial(\mathbf{x}, E, \boldsymbol{\epsilon})}$ is of full rank everywhere (which is very likely to be the case since we have assumed existence and uniqueness of $\mathbf{x}^*(E, \boldsymbol{\epsilon})$) then one may show (*using* $(iii)'$!) that in fact $S^*(E, \boldsymbol{\epsilon}) = S(x^*(E, \boldsymbol{\epsilon}), E, \boldsymbol{\epsilon})$. Analogous relations hold for E and T if the other definitions for the state are used. Note that $(i)^*$ and $(ii)^*$ define S^* uniquely (up to an additive constant) in terms of a path integral in $(E, \boldsymbol{\epsilon})$ space.

It is easy to see from the above that for all $(\mathbf{x}, E, \boldsymbol{\epsilon})$, we have that $S(x^*(E, \boldsymbol{\epsilon})E, \epsilon)^* = S^*(E, \boldsymbol{\epsilon}) = \max_{\mathbf{x}} S(\mathbf{x}, E, \boldsymbol{\epsilon})$ since, in fact $S(\mathbf{x}(t), E, \boldsymbol{\epsilon}) \underset{t\to\infty}{\nearrow} S^*(E, \boldsymbol{\epsilon})$ if we taken $\dot{E} = 0$ and $\dot{\boldsymbol{\epsilon}} = \mathbf{0}$. This expresses the well-known fact that of all paths having given values of E and $\boldsymbol{\epsilon}$ the constant path has the maximum entropy at that time.

In order to determine the non-elastic part of the entropy function, assume that the state space of Σ is reachable from $\mathbf{x}^*(E, \boldsymbol{\epsilon})$ and controllable to $\mathbf{x}^*(E, \boldsymbol{\epsilon})$ (usually it will be a consequence of the fading memory assumption that this controllability holds and that these properties hold for all $(E, \boldsymbol{\epsilon})$ if they hold for any).

Consider now the functions defined by

$$\tilde{S}_a(\mathbf{x}, E, \boldsymbol{\epsilon}) \triangleq -\sup \int_0^\infty \frac{q(t)}{T(t)} dt$$

$$\tilde{S}_r(\mathbf{x}, E, \epsilon) \triangleq \sup \int_{-\infty}^0 \frac{q(t)}{T(t)} dt$$

where these suprema are to be taken over all paths which start at $(\mathbf{x}, E, \boldsymbol{\epsilon})$ and end at $(\mathbf{x}^*(E, \boldsymbol{\epsilon}), E, \boldsymbol{\epsilon})$ in the first case and vice versa in the second. Note that $\tilde{S}_r \leqslant \tilde{S}_a \leqslant 0$. From the results in the beginning of this section it may be seen that $S^*(E, \boldsymbol{\epsilon}) + \tilde{S}_r(\mathbf{x}, E, \boldsymbol{\epsilon}) \leqslant S(\mathbf{x}, E, \boldsymbol{\epsilon}) \leqslant S^*(E, \boldsymbol{\epsilon}) + \tilde{S}_a(\mathbf{x}, E, \boldsymbol{\epsilon})$ and these extremals are possible entropy functions. *As a rule,* $\tilde{S}_a \neq \tilde{S}_r$ and thus there are an infinite number of possible entropy functions for non-equilibrium irreversible thermodynamic systems. Note that it also follows (independent of the fading memory assumption) that Clausius' inequality is equivalent to the requirement $\oint \frac{q(t)}{T(t)}\, dt \leqslant 0$. With fading memory we only need to consider closed paths from and to an equilibrium point $(\mathbf{x}^*(E, \boldsymbol{\epsilon}), E, \boldsymbol{\epsilon})$.

In systems with no external inputs (E and $\boldsymbol{\epsilon}$ are thus constant) $\mathbf{x}(t)$ will converge to $\mathbf{x}^*(E, \boldsymbol{\epsilon})$ and $\tilde{S}(\mathbf{x}, E, \boldsymbol{\epsilon}) \triangleq \tilde{S}(\mathbf{x}, E, \boldsymbol{\epsilon}) - S^*(E, \boldsymbol{\epsilon})$ is thus a non-positive Lyapunov function with a nonnegative derivative. Moreover, $\left\{\Sigma, \dot{E}\left(\frac{1}{T^*} - \frac{1}{T}\right) - \left\langle \dot{\boldsymbol{\epsilon}}, \frac{\boldsymbol{\tau}^*}{T^*} - \frac{\boldsymbol{\tau}}{T} \right\rangle, -\tilde{S}\right\}$ defines a dissipative system (rather than only a cyclo-dissipative system). It should be emphasized however that this dissipativeness and the above asymptotic stability properties follow not so much from the thermodynamic laws but rather from the additional fading memory assumption.

The reader may like to compare the results and the approach of this section with those of [5, Chapter 3] and [14, Lecture 3].

10. CONCLUSION

In this paper we have presented a precise and very general approach to the theory of dissipative dynamical systems. The problems discussed and the questions raised are in principle analytical in nature (although some of the results have, as shown, also implications to synthesis problems). Probably the most interesting general result obtained here is an unambiguous proof of the non-uniqueness of the storage function.

An interesting class of applications of the ideas presented here is to the stability analysis and the construction of Lyapunov functions for feedback systems. Among the extensions which are presently being developed are the treatment of non-minimal systems and the generalization of these ideas to stochastic systems. This last extension appears to be a particularly interesting one since it brings into play the role of observations which, in addition to the input and the output, are needed to describe the interaction of such systems with their environment. A desirable refinement of the theory as it presently stands is the elimination of the interpretation of the input and output as "*cause*" and "*effect*". In typical applications they are functions which occur "simultaneously" in pairs and there is no such clear cut implication of cause and effect.

Finally, we hope to have demonstrated the fruitfulness of a system-theoretic approach to the questions of dissipativeness and of internal symmetry. The main advantage of this approach is one's ability to put the various questions and assumptions in a very sharp focus.

As a concluding comment let us mention that the generalization to nonlinear processes of the results on the compatibility of external and internal dissipativeness and symmetry seems to be very subtle, both from a conceptual and a mathematical point of view. For some results in this direction, see the paper by Varaiya [12]. We believe that this problem is a very important one and that it has implications, among other things, to variational principles and to potential theory.

REFERENCES

1. Willems, J. C. "Dissipative Dynamical Systems, Part I: General Theory, Part II: Linear Systems with Quadratic Supply Rates". *Arch. Rational Mechanics and Analysis, 45,* 1972, 321–351 and 352–392.
2. Kalman, R. E., Falb, P. L., and Arbib, M. A. *Topics in Mathematical System Theory.* McGraw-Hill, New York, 1969.
3. Desoer, C. A. *Notes for a Second Course on Linear Systems.* Van Nostrand Reinhold, New York, 1970.
4. Brockett, R. W. *Finite Dimensional Linear Systems.* Wiley, New York, 1970.
5. Day, W. A. *The Thermodynamics of Simple Materials with Fading Memory*, Springer–Verlag, New York, 1972.
6. Newcomb, R. W. *Linear Multiport Synthesis.* McGraw-Hill, New York, 1966.
7. Kalman, R. E. "Lyapunov Functions for the Problem of Lur'e in Automatic Control". *Proc. Nat. Acad. Sci. U.S.A. 49*, 1963, 201–205.
8. Popov, V. M. "Hyperstability and Optimality of Automatic Systems with Several Control Functions". *Rev. Roumaine Sci. Tech. Electrotechn. et Energ. 9*, 1964, 629–690.
9. Day, W. A. "Time-Reversal and the Symmetry of the Relaxation Function of a Linear Viscoelastic Material". *Archive for Rational Mechanics and Analysis, 40,* 1971, 155–159.
10. Desoer, C. A. and Oster, G. F. "Globally Reciprocal Stationary Systems". *Int. J. Engng. Sci. 11,* 1973, 141–155.
11. Oster, G. F. and Desoer, C. A. "Tellegen's Theorem and Thermodynamic Inequalities". *J. Theor. Biol. 32,* 1971, 219–241.
12. Varaiya, P. "Equivalent Non-linear Networks", pp. 141–147 of *Mathematical Aspects of Electrical Network Analysis*, Vol. III of SIAM-AMS Proc., American Math. Soc., Providence, 1971.
13. Brayton, R. K. "Non-linear Reciprocal Networks", pp. 1–15 of *Mathematical Aspects of Electrical Network Analysis*, Vol. III of SIAM-AMS Proc., American Math. Soc., Providence, 1971.
14. Truesdell, C. *Rational Thermodynamics*. McGraw-Hill, New York, 1969.
15. Meixner, J. "Thermodynamic Theory of Relaxation Phenomena", ch. 5 in R. J. Donelly, R. Herman and I. Prigogine (eds.) *Non-Equilibrium Thermodynamics, Variational Techniques and Stability.* University of Chicago Press, 1965.
16. Anderson, B. D. O. and Vongpanitlerd, S. *Network Analysis and Synthesis.* Prentice Hall, Englewood Cliffs, 1973.

NOMENCLATURE

Σ	dynamical system in state space form	U	space of input values
		$\mathscr{U}$	input space
$\Sigma_{I/0}$	dynamical system in input/ output form	Y	space of output values
		$\mathscr{Y}$	output space

X	state space
ϕ	state transition function
F	system function
r	read-out function
w	supply rate
d	dissipation rate
S	storage function
S_a	available storage
S_r	required supply
(*FDLS*)	finite dimensional linear system
A, B, C, D	state space parameters of (*FDLS*)
G(s)	transfer function of (*FDLS*)
W(t)	impulse response of (*FDLS*)
H(t)	Hankel matrix of (*FDLS*)
Q	storage matrix
$\mathbf{\Delta}$	dissipation matrix
T	Lagrangian matrix
$\mathbf{\Sigma}_e$	external signature matrix
$\mathbf{\Sigma}_i$	internal signature matrix
j	force
f	flux
(*DIE*)	dissipation inequality
(*MIE*)	matrix inequality
$LO(n_+, n_-)$	Lorentz group $\triangleq \left\{ \mathbf{S} \mid \mathbf{S}^T \left[\begin{array}{c\|c} \mathbf{I}_{n_+} & \mathbf{0} \\ \hline \mathbf{0} & -\mathbf{I}_{n_-} \end{array}\right] \mathbf{S} = \left[\begin{array}{c\|c} \mathbf{I}_{n_+} & \mathbf{0} \\ \hline \mathbf{0} & -\mathbf{I}_{n_-} \end{array}\right] \right\}$
R^n	n-dimensional Euclidean space
$\frac{d^n}{dt^n}$	n-th derivative
T	transposition
Tr	trace
I	identity matrix
δ	Dirac delta-function

ON APPLICATIONS OF SADDLE-SHAPED AND CONVEX GENERATING FUNCTIONALS

M. J. Sewell
Department of Mathematics
The University of Reading
Berkshire, England

SUMMARY

A closed chain of saddle-shaped and convex functionals linked by Legendre-type transformations is described. A consequent symmetric analytical structure encompassing many theories in applied mathematics is indicated. An illustration in thermostatics is given. The method is then applied in detail to develop a new approach to plastic constitutive equations. This complements existing treatments and carries a well-defined scope for generalization.

1. INTRODUCTION

A theory in applied mathematics endeavours to progress on two legs. One leg is the available experimental evidence. The other is the purely mathematical structure which seems appropriate. In the formative stage the development of these two legs may be at uneven rates. Progress with the theory may then take place haltingly. There may be insufficient experimental data, or there may be incomplete awareness of the type of general mathematical structure which might encompass the particular theory in question. Satisfactory progress requires a judicious balance to be established between the degrees of freedom available to either side. This is best achieved if a sharp and conscious distinction is kept between the extent of the options open to either side.

The object of this paper is to explore with illustrations a particular group of mathematical ideas whose inter-relation leads to an attractive and highly symmetric analytical structure whose full power has never been exploited before. These ideas centre round a closed chain of Legendre-type transformations connecting four generating functionals, each defined on the product of two different inner product spaces, and each being convex and/or concave with respect to the two spaces. Also crucial are certain characteristic governing equations or inequalities in the two spaces. The paper is therefore a study of the flexibility which may be possessed by the mathematical structure referred to above.

In Section 2 we give an account of the mathematical ideas in question. This account abstracts a number of results from a paper of Noble and Sewell [14], which describes some of them for the first time and which gives a very diverse collection of examples. We also take the opportunity to summarize certain general dual extremum principles which are valid in the problems considered, although we shall not take much space to illustrate these principles here. They are emphasized in the paper just mentioned.

In Section 3 we take a widely familiar context in thermostatics and show how the four "thermodynamic potentials" not only are related by the closed chain of Legendre transformations in that very special case (which has long been known), but also that they have the convex and concave or saddle properties possessed by our general analytical structure (which is much less well recognized). Brief remarks about thermal equilibrium of simple materials are made from this viewpoint.

In Section 4 we develop an account of plastic constitutive rate equations. This is intended as a detailed illustration of how the symmetric analytical structure described in Section 2 may be exploited. The approach is new in that context, and so are the modes of expression of the results. It is explained how the theory is complementary to other investigations of plastic constitutive equations, such as that of Hill [8]. The scope for generalization in certain directions is made evident. A full discussion of dual extremum principles in associated boundary value problems of (elasticity and) plasticity has been given from this viewpoint by Sewell [19].

2. BASIC THEORETICAL FRAMEWORK

Consider two inner product spaces E and F. Denote the inner product in E by $(,)$, and that in F by $\langle , \rangle$. Let $x, y, \ldots$ be typical elements of E, and $u, v, \ldots$ be typical elements of F. Consider given functionals $X[x, u]$, $Y[y, u]$, $W[y, v]$, $Z[x, v]$ defined on (subsets of) the Cartesian product space $E \times F$.

The arrows in Fig. 1 indicate the active variables in a closed sequence of Legendre-type transformations between the functionals displayed.

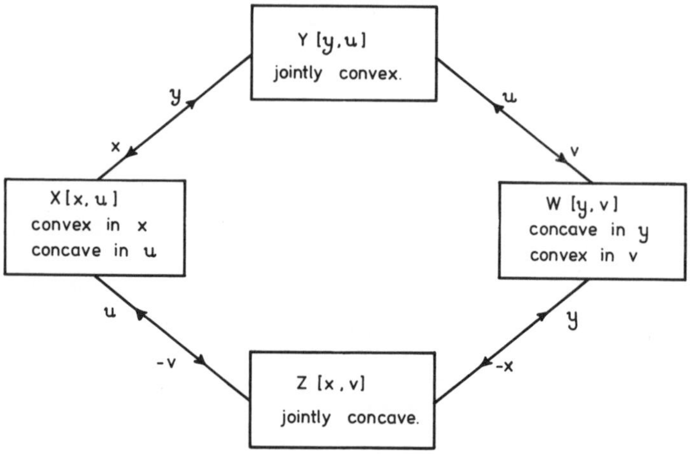

Fig. 1. Legendre transformation links between convex and saddle functionals.

By this we mean that equations of the following type hold. In the transformation between X and Y, for example,

$$X + Y = (x, y), \tag{1}$$

$$x = \frac{\partial Y}{\partial y}, \quad y = \frac{\partial X}{\partial x}, \quad \frac{\partial X}{\partial u} + \frac{\partial Y}{\partial u} = 0. \tag{2}$$

Or again, in the transformation between Y and W,

$$Y + W = \langle u, v \rangle, \tag{3}$$

$$u = \frac{\partial W}{\partial v}, \quad v = \frac{\partial Y}{\partial u}, \quad \frac{\partial Y}{\partial y} + \frac{\partial W}{\partial y} = 0. \tag{4}$$

It follows from such equations that the values of the four functionals must satisfy

$$X + Y + W + Z = 0. \tag{5}$$

The partial derivative symbols indicate gradients of the functionals. These are obtained by expressing a first variation as an inner product of the gradient with the increment in the varied argument; for example, as in

$$W[y, v + \delta v] - W[y, v] \equiv \left(\frac{\partial W}{\partial v}, \delta v\right) + o(\|\delta v\|), \tag{6}$$

where $\|\delta v\| = \sqrt{\langle \delta v, \delta v \rangle}$. Such gradients become simply matrices of the usual partial derivatives in the case of functions defined on finite-dimensional spaces.

The relations between the convexity and concavity properties indicated in Fig. 1 were demonstrated by Noble and Sewell [14, Section 6]. Such properties are not synonymous with the mere existence of the Legendre-type transformation. However, when they do apply, they transform in the way shown. This is an immediate consequence of the following key equation. In the transformation from (say) X to Y, any pair of "points" (elements) x_+, u_+ and x_-, u_- will map into the points y_+, u_+ and y_-, u_- respectively, and vice versa, such that

$$\Delta X - \left(\left.\frac{\partial X}{\partial x}\right|_-, \Delta x\right) - \left\langle \left.\frac{\partial X}{\partial u}\right|_+, \Delta u \right\rangle = -\Delta Y + \left(\left.\frac{\partial Y}{\partial y}\right|_+, \Delta y\right) + \left\langle \left.\frac{\partial Y}{\partial u}\right|_+, \Delta u \right\rangle \tag{7}$$

The prefix Δ will be used as a finite-differencing operator with the effect that

$$\Delta \Rightarrow \text{value at "plus point" less value at "minus point"}.$$

Then (1) and (2) immediately imply (7).

When the right-side of (7) is >0 (or $\geqslant 0$) for all pairs of y, u points the functional $Y[y, u]$ is strictly (or weakly) jointly convex. This extends the version of convexity used by Hill [5] in solid mechanics. When the left-side of (7) is >0 (or $\geqslant 0$) for all pairs of x, u points the functional $X[x, u]$ is a strict (or weak) saddle functional, being convex in x and concave in u. This is the saddle inequality used by Sewell [16, Section 2 (vii)]. These two facts and the equality of the two sides of (7), are the basis of the arguments of Noble and Sewell [14] relating the convexities and concavities displayed in Fig. 1, and in the schematic view shown in Fig. 2.

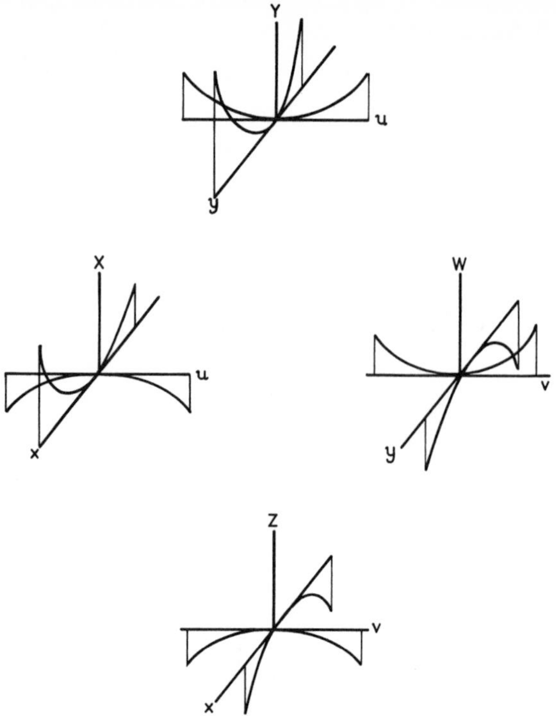

Fig. 2. Relations between convexities and concavities.

If any one of these convexities or concavities is weak, the associated Legendre-type transformation may very well break down, and there will be a break in the otherwise closed sequence of mappings. Not all of the four functionals need then exist. Nevertheless certain useful general results can be obtained, associated with any individual corner of the mapping diagram where gradients can be defined.

For example, suppose that either side of (7) is $\geqslant 0$ for every pair of points. Then by reversing the + and − labels, adding the two inequalities, and using (2) and (4), we get

$$(\Delta x, \Delta y) + \langle \Delta v, \Delta u \rangle \geqslant 0. \tag{8}$$

From the symmetry this is clearly obtainable from any one corner of Fig. 1 by postulating weak convexity and/or concavity there.

Alternatively, we might assert (8) *a priori* as a statement of monotonicity in a relation between two "supervectors" $[x \;\; v]$ and $[y \;\; u]$, both belonging to the space $E \times F$ whose inner product is $(,) + \langle , \rangle$. From this viewpoint there is no reference to the existence of *any* functionals X, Y, W or Z. We shall use this viewpoint in subsections 4(iii) and 4(v) below.

Let us now regard x, y, u and v as representing quantities whose meaning in an applied context is independently given. Associated with any one corner of Fig. 1 are two equations relating these four variables, expressing the two possible choices of active variable as gradients of the generating functional there. For example, the functional $Y[y, u]$ generates the two equations

$$x = \frac{\partial Y}{\partial y}, \quad v = \frac{\partial Y}{\partial u}. \tag{9}$$

We ask what kind of additional conditions will make such a problem determinate. The remarks of Noble and Sewell [14] may be reordered to show that there are four pairs of (sets of) additional conditions which have particular interest for this purpose. According to these, whatever generating functional is being used to give the original two equations, we adjoin either

$$T^*u = y \qquad (\alpha) \tag{10}$$

or

$$T^*u \leqslant y \quad (\alpha), \qquad x \geqslant 0 \quad (\beta), \qquad (x, T^*u - y) = 0, \tag{11}$$

and either

$$Tx = -v \quad (\beta) \tag{12}$$

or

$$Tx \geqslant -v \quad (\beta), \qquad u \geqslant 0 \quad (\alpha), \qquad \langle u, Tx + v \rangle = 0. \tag{13}$$

Here T and T^* are linear operators adjoint in the sense that

$$(x, T^*u) = \langle u, Tx \rangle \tag{14}$$

for all x and u in their respective domains, which in general will be subspaces of E and F. Conditions (11) and (13) each define the boundary of a "quadrant", in $E \times E$ and $F \times F$ respectively. We could also reverse both inequalities in (11), and/or both in (13). The labels (α) and (β) will be explained shortly.

The differences of quantities corresponding to any two distinct solutions of any of the four *sub*-problems (10) with (12) or (13), or (11) with (12) or (13), makes the left-side of (8) non-positive. But the previous interpretation of (8) shows that strict positivity applies there whenever any of the convexities and concavities displayed in Fig. 1 is strict – e.g. by inserting (9) instead of (10)–(13) in (8), and using (7). This contradiction shows that any full problem, completed as described in the last two paragraphs, (e.g. by inserting (9) into (11) and (12)), has at most one solution for whichever of x, y, u, v associates *strict* convexity or concavity with the considered generating functional as displayed in Fig. 1.

Suppose again that the convexity and/or concavity properties of some particular generating functional are weak, and not necessarily strict. Then the associated uniqueness theorem fails. Nevertheless, Noble and Sewell show in this case that dual extremum principles may be constructed to characterize any one of the above full problems associated with that functional. This is done by dividing the governing conditions of the problem into two under-determined subproblems, labelled (α) and (β) (illustrated in (9)–(13)), neither of which need include the orthogonality conditions. It is then shown, from the requirement that expressions like (7) be $\geqslant 0$, that the functionals listed in Table 1 are extremized, among solutions of the indicated α- or β-subproblem, by any solution of the full problem. The minimum and maximum

TABLE 1

Functionals extremized

Generating functional	Functional $J(\alpha)$ minimized	Functional $K(\beta)$ maximized
$Y[y, u]$	Y	$Y - \left(y, \dfrac{\partial Y}{\partial y}\right) - \left\langle u, \dfrac{\partial Y}{\partial u}\right\rangle$
$X[x, u]$	$\left(x, \dfrac{\partial X}{\partial x}\right) - X$	$\left\langle u, \dfrac{\partial X}{\partial u}\right\rangle - X$
$W[y, v]$	$\left\langle v, \dfrac{\partial W}{\partial v}\right\rangle - W$	$\left(y, \dfrac{\partial W}{\partial y}\right) - W$
$Z[x, v]$	$Z - \left(x, \dfrac{\partial Z}{\partial x}\right) - \left\langle v, \dfrac{\partial Z}{\partial v}\right\rangle$	Z

coincide in a common solution value. At any one corner of Fig. 1, i.e. for any one generating functional, the statements $J(\alpha)$ and $K(\beta)$ of the functionals which are extremized are the same for every full problem which may be defined there. In particular $J(\alpha)$ and $K(\beta)$ do not depend on whether or not inequalities are present, but if they are the extrema will not in general be stationary. Notice also that operators such as T and T^* do not appear in $J(\alpha)$ and $K(\beta)$.

Problem (10) with (12) generated by $X[x, u]$ has been considered by Noble [13] and applied by Arthurs [1]. This problem, and also (10) with (13) generated by $X[x, u]$ and by $Y[y, u]$, was considered independently by Sewell [16]. The fully symmetric theory, and many examples, were given by Noble and Sewell [14]. A detailed application to elasticity and plasticity, together with some extensions of the basic theory, including circumstances when more than two inner product spaces are present, has been worked out by Sewell [19].

A particular feature of this general theory is worth drawing out explicitly. Suppose that the under-determined problem implied by any corner of Fig. 1 (e.g. by eqns (9)) is completed by eqns (10) and (12) (instead of by using the inequality conditions (11) or (13)). Then (10) and (12) can be built in to give the determinate version of Fig. 1 shown in Fig. 3.

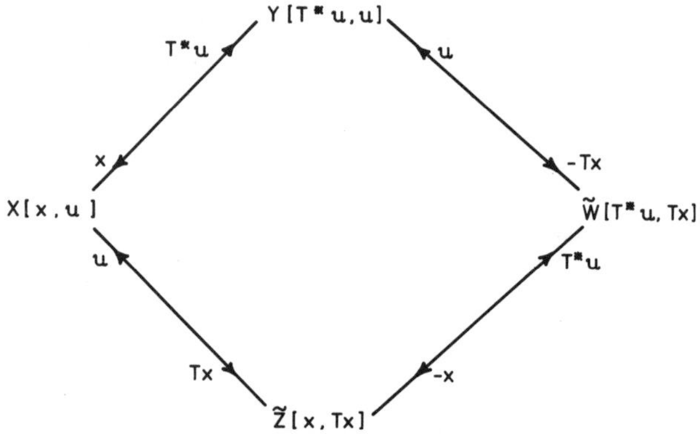

Fig. 3. Chain of determinate problems linked by Legendre transformations.

Here the functionals $\bar{Z}$ and $\bar{W}$ of the indicated arguments are defined by $\bar{Z}[x, Tx] = Z[x, -Tx]$, $\bar{W}[T^*u, Tx] = W[T^*u, -Tx]$. The change of variable induced by the minus sign is merely a reflection, which does not alter the concavity with respect to Tx.

The point of Fig. 3 is that it summarizes a canonical structure – that is, we can immediately read off from it the conclusion that the functionals $Y[T^*u, u]$ and $\bar{Z}[x, Tx]$ both represent "Lagrangian" or "action" functionals in variational principles. This is because the two equations generated by Y (for example) have a generalized Euler–Lagrange structure after elimination of x. Likewise the two equations generated by $\bar{Z}$ also have a generalized Euler–Lagrange structure after elimination of u. The functional $X[x, u]$ plays a role of a generalized Hamiltonian, since it generates generalized Hamiltonian equations. One way of searching for variational principles is therefore to seek to express the governing equations of a given problem in one or more of the alternative forms implied by Fig. 3.

Classical generalized dynamics gives a rather specialized example of Fig. 3, in which the two spaces E and F are essentially the same, and the variables of position $\mathbf{q}$ and momenta $\mathbf{p}$ separate in the generating functionals. The latter are obtained by time-integration of combinations of the kinetic energy $K(\mathbf{p})$, potential energy $V(\mathbf{q})$, kinetic "co-energy" $\bar{K}(\dot{\mathbf{q}})$) and potential "co-energy" $\bar{V}(\dot{\mathbf{p}})$. According to the conventions we have associated with Fig. 1, these functions and their gradients are related by Fig. 4. For example, Lagrange's equations are generated as if by (9) from the Lagrangian $\bar{K}(\dot{\mathbf{q}}) - V(\mathbf{q})$, and Hamilton's equations are generated from the Hamiltonian $K(\mathbf{p}) + V(\mathbf{q})$. Actually, Fig. 4 contains only the "interior" and not the

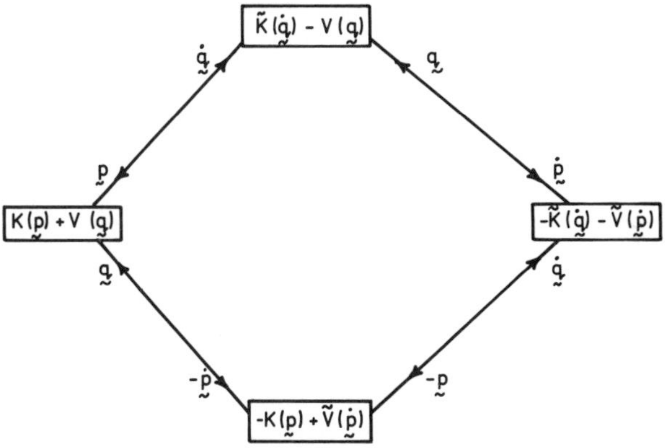

Fig. 4. Dynamical equations generated in four linked ways.

boundary parts of the operators and functionals. That is, it contains the *formally* adjoint differential operators $d/dt \sim T^*$ and $-d/dt \sim T$, and functions which will be integrands of time-integrals – the latter make up the functionals in the general theory.

In certain types of electrical RLC circuits, it is straightforward to show that the relations between inductor and capacitor voltages and currents can be summed up in a diagram like Fig. 1, in terms of generating functions which are the Millar content and co-content, and the Brayton–Moser "mixed potential" function and its dual.

This can be seen, for example, by re-interpreting some equations given by Flower and Evans [4]. Each corner of the diagram generates a pair of governing equations, and the other corners generate alternative statements of this pair. The relations between inductor flux and current, and capacitor charge and voltage, can also be represented by *another* such diagram in terms of the energy and co-energy functions of the capacitors and inductors. Two more differential equations serve to complete a set of six equations connecting the six variables mentioned. One way of seeking variational principles for such a circuit would be to ask whether those six equations can be reduced, by elimination, to two equations having the structure generated by one or other corner of Fig. 3.

We conclude this Section by giving a list of some of the contexts to which the general ideas described here are applicable. The topics mentioned are described more fully in the references already given. No comprehensive search has been made in compiling the list, which is therefore certainly not meant to be complete. It is merely a note of some of the topics which this author has studied from the present viewpoint, with varying degrees of depth. Nevertheless, even baldly stated as here, it serves to illustrate the unifying and rewarding role of the ideas mentioned above.

TABLE 2

Some contexts amenable to the present general theory

General subject	Particular topic
Elasticity	incremental theory, pre-bifurcation
Plasticity	incremental theory, pre-bifurcation limit analysis constitutive "rate" equations
Fluid mechanics	slow viscous flow subsonic compressible inviscid flow cavitation of strictly incompressible fluid
Mathematical programming	linear non-linear
Network theory	electrical circuits pin-jointed frame structures transportation networks
Theoretical physics	diffusion and neutron transport plasma theory
Quantum mechanics	perturbation theory scattering theory
Operator equations	$T^*Tx + f(x) = 0$ differential equations integral equations
Optimization	control theory
Miscellaneous	Steiner

3. THERMOSTATICS

When an application of the foregoing general theory to some specific context is being sought, it may often be that the usual form of the equations governing that context is not immediately recognizable as being appropriate. Some twist or reinterpretation, not necessarily obvious, may be required to cast the equations in an identifiable form, from which the uniqueness and dual extremum principles may then simply be read off, or the sequence of Legendre-type transformations constructed. The paper of Noble and Sewell [14] contains some of these adaptations.

One context not treated there is the problem defined by the equations of local thermal equilibrium of simple materials. Since the four thermodynamic potentials of specific internal energy ϵ, free energy ψ, enthalpy χ, and free enthalpy ζ are probably one of the oldest examples of four functions related by a closed sequence of Legendre transformations, it is natural to ask what is their precise relation with Fig. 1 and the ensuing general theory.

We wish to be brief here, and so will establish this relation in the easiest but not untypical case. This is the case of the simple fluid having specific volume $v(= 1/\text{density})$, entropy η, temperature θ and pressure p. These are all scalars, so the two inner product spaces E and F are both just the real line, and both scalar products mean ordinary multiplication. Associated with the considered particle are given functions $\epsilon[\eta, v]$, $-\psi[\theta, v]$, $\zeta[\theta, p]$ and $-\chi[\eta, p]$. Their partial derivatives will be taken to be equal to the active variables displayed in the long-known

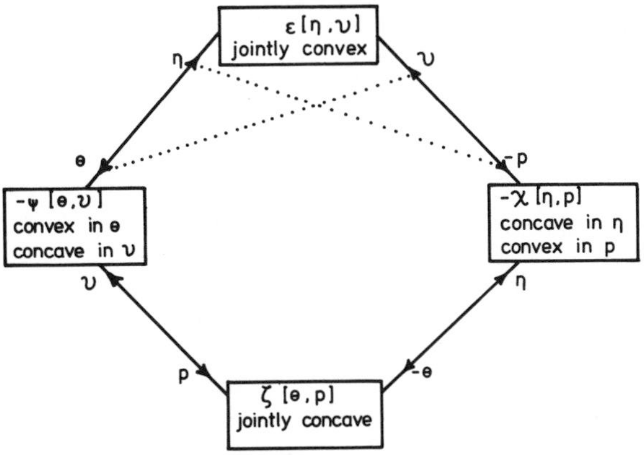

Fig. 5. Relations between thermodynamic potentials.

Legendre transformation links represented by the arrows in Fig. 5, according to the same scheme as that prescribed for Fig. 1. For example, Fig. 5 implies

$$\theta = \frac{\partial \epsilon}{\partial \eta}, \qquad -p = \frac{\partial \epsilon}{\partial v}. \tag{15}$$

When all the links hold, and (for example) $\epsilon[\eta, v]$ is jointly convex, it follows by arguments like that based on (7) that $\zeta[\theta, p]$ is jointly concave, and $-\psi[\theta, v]$ and

$-\chi[\eta, p]$ are the saddle functions indicated. This may be well known, but I have not seen it expressed in that way nor so clearly displayed. Figure 5 may be used as a mnemonic not only for Legendre transformations as already explained, but also for the four Maxwell relations. For example, the dotted lines shown remind us that

$$\left.\frac{\partial\theta}{\partial v}\right|_{\eta\ \text{const}} = \frac{\partial^2\epsilon}{\partial v\partial\eta} = \frac{\partial^2\epsilon}{\partial\eta\partial v} = -\left.\frac{\partial p}{\partial\eta}\right|_{v\,=\,\text{const}} \tag{16}$$

Thus Fig. 5 is different from, and more explicit than, the "thermodynamic square" mnemonic which Callen [2] attributes to Born in 1929. In addition, the convexity and concavity properties are written into it.

The function $\epsilon[\eta, v]$ will be strictly (or weakly) jointly convex if the Hessian matrix of its second derivatives is positive (or non-negative) definite. The usual manipulations show that the positive-definiteness of this Hessian may be concisely expressed in terms of the two "specific heats" c_p and c_v, and the "sound velocity" c, under the following assumptions:

$$\infty > \frac{\partial^2\epsilon}{\partial\eta^2} = \frac{\partial\theta}{\partial\eta} \equiv \frac{\theta}{c_v} > 0, \tag{17}$$

$$\infty > \frac{\partial^2\epsilon}{\partial v^2} = -\frac{\partial p}{\partial v} \equiv \frac{c^2}{v^2} > 0, \tag{18}$$

$$\infty > \frac{c_p}{c_v} \equiv 1 \bigg/ \left[1 - \frac{(\partial^2\epsilon/\partial\eta\partial v)^2}{(\partial^2\epsilon/\partial\eta^2)(\partial^2\epsilon/\partial v^2)}\right] > 1. \tag{19}$$

The two Legendre transformations emanating from $\epsilon[\eta, v]$ in Fig. 5 exist because of (17) and (18) respectively, and the saddle properties of $-\psi[\theta, v]$ and $-\chi[\eta, p]$ therefore follow immediately from (17)–(19) by the previous arguments based on (7).

Having established an analogue of Fig. 1, we have available the two equations (15) relating the four variables v, η, θ and p. The general line of argument developed in Section 2 now suggests certain ways of augmenting (15) to achieve a determinate problem, which may also have dual extremum principles associated with it which extremize the functionals listed in Table 1. Or in other words, by comparing Fig. 5 and Table 1 we can apparently guess the functionals to be extremized before we have augmented (15), provided this augmentation can be made consistent with (10)–(14).

Now in fact the usual governing equations of thermal equilibrium are obtained simply by assigning values θ_0 and p_0 (say) to θ and p respectively in (15). We reconcile these two extra conditions

$$\theta = \theta_0, \qquad p = p_0 \tag{20}$$

with (10) and (12) in the following way. Identify (15) with (9) by writing

$$\theta - \theta_0 = \frac{\partial}{\partial\eta}(\epsilon - \theta_0\eta + p_0 v), \qquad -p + p_0 = \frac{\partial}{\partial v}(\epsilon - \theta_0\eta + p_0 v). \tag{21}$$

Then (10) and (12) may be written

$$T^* v = \eta \quad (\alpha), \qquad T(\theta - \theta_0) = p - p_0 \cdot \quad (\beta) \tag{22}$$

But since both inner product spaces here are the real line, we may satisfy (14) with $T = T^* =$ *any* scalar multiplier. Exploiting this arbitrariness allows (22β) to be equivalent to *both* of (20), while (22α) is no restriction at all on v and η, but simply an unneeded change of variable. (If we impose $v > 0$ at the outset, we can get (22β) from (13) in place of (12).)

The "thermomechanical potential" function

$$\epsilon[\eta, v] - \theta_0 \eta + p_0 v \tag{23}$$

which appears in (21) is therefore to be identified with the $Y[y, u]$ of the general theory. It has the same convexity properties as $\epsilon[\eta, v]$, because the extra linear terms $\theta_0 \eta - p_0 v$ do not affect these. To obtain a *precise* analogue of Fig. 1, fully appropriate to the thermostatic equations in the form (21)–(22), we need to reconstruct Fig. 5 by starting with (23) in place of ϵ, and adding $-p_0 v$, $\theta_0 \eta$ and 0 respectively to $-\psi$, $-\chi$ and ζ. We read off from the first line of Table 1 the extremum principle that (23) is minimized with respect to arbitrary variations of η and v by the solution to the thermostatic equations. These variations are arbitrary because the α-subset (22α) is in effect void. The dual maximum principle read off from Table 1 is that the free enthalpy

$$K(\beta) = \epsilon - \theta \eta + p v \tag{24}$$

is maximized among solutions of the β-subset which is (21) with (22β); but these conditions coincide with those defining the thermal equilibrium state itself, so nothing new is gained. The features of these two principles, and the fact that thermal equilibrium is defined by setting the gradient of (23) to zero, are fully consistent with familiar properties of conservative systems whose potential is a function of a finite number of independent generalized coordinates.

The thermal equilibrium of more general simple materials, such as the hyperelastic solid, has been discussed by Coleman and Noll [3]. Their results which are pertinent to the present discussion are summarized by Truesdell and Noll [22, Sections 81, 87], and in an introductory manner by Truesdell [21]. The internal energy is a given function $\epsilon[\eta, \mathbf{F}]$ of η and of the deformation gradient tensor $\mathbf{F}$. The two inner product spaces are no longer both the real line; one is, but the other is the space of second order tensors with inner product defined as the trace of the tensor product. According to the Coleman–Noll inequality, it is clear that if the approach of Section 2 here is adopted, only a restricted convexity of ϵ with respect to $\mathbf{F}$ should be postulated. In fact their approach is different from the present one. They motivate the introduction of the thermomechanical potential by reference to a homogeneous deformation process, and they *define* thermal equilibrium by a minimum principle (not merely a stationary principle). We arrived at (23) by asserting governing *equations* of thermal equilibrium, and investigating how they could be embedded in the general theoretical framework of Section 2, with the minimum principle emerging as a consequence.

4. CONSTITUTIVE RATE-EQUATIONS IN PLASTICITY

We now wish to use the theoretical structure described in Section 2 to give a more symmetrical statement than is customary of the constitutive rate equations of time-independent plasticity. This statement also contains a desirable kind of flexibility which has not been emphasized before.

(i) Definition of spaces

Let E be the space of second order symmetric tensors $\mathbf{a}, \mathbf{b}, \ldots$ with inner product

$$(\mathbf{a}, \mathbf{b}) \equiv \mathbf{ab} \equiv a_{ij} b_{ji}. \tag{25}$$

Here a common "background frame" of Cartesian coordinates is understood to be available in Euclidean 3-space, and a_{ij} and b_{ji} are the 3×3 matrices of tensor components with respect to those Cartesian coordinates. The inner product in (25) is then defined as the trace of the matrix product of these two matrices (the summation convention on both Latin and Greek repeated suffixes will be used). The matrices of the components of $\mathbf{a}$ and $\mathbf{b}$ with respect to *other* coordinates are expressible via the tensor rule in the usual way, and hence so also is (25) by substitution of the inverses of such expressions. We shall use the simple juxtaposition $\mathbf{ab}$ to denote the scalar product $(\mathbf{a}, \mathbf{b})$, to conform with a common usage in the plasticity context.

The flexibility mentioned above is introduced by leaving the choice of the second inner product space F open at the outset. In particular there is no compelling reason why it should be finite-dimensional, even though this is the most familiar and currently viable case. We denote typical elements of F by μ and γ, and its inner product by

$$\langle \mu, \gamma \rangle \equiv \mu\gamma. \tag{26}$$

Thus we shall use the simple juxtaposition $\mu\gamma$ in place of $\langle \mu, \gamma \rangle$ to put this notation on a similar footing to $\mathbf{ab}$. The context will make it clear whether or not $\mu\gamma$ is just the ordinary product of two scalars. The elements of F do not necessarily have any tensorial property associated with them.

Before proceeding with the theory for general F, we illustrate three possible specific choices of F:

F_1 is the space of real numbers with $\mu\gamma$ as the ordinary product;

F_m is the space of m-tuples typically

$$\begin{aligned} \mu &\equiv \{\mu_\alpha : \alpha = 1, \ldots, m\} \\ &\equiv \{\mu_1, \ldots, \mu_m\} \end{aligned} \tag{27}$$

with real number components, and inner product

$$\begin{aligned} \mu\gamma &\equiv \mu_\alpha \gamma_\alpha \\ &\equiv \mu_1\gamma_1 + \mu_2\gamma_2 + \ldots + \mu_m\gamma_m; \end{aligned} \tag{28}$$

F_∞ is the space of real integrable functions $\mu(\alpha)$ defined on the interval $0 \leqslant \alpha \leqslant 1$ of the real line, and having inner product

$$\mu\gamma = \int_0^1 \mu(\alpha)\gamma(\alpha)\, d\alpha. \tag{29}$$

In succeeding subsections we shall obtain some results which are valid whatever

F may be. The present approach of uncovering and exploiting the *mathematical structure* of plastic constitutive equations is intended to be complementary to other recent investigations. These include the "systemic" approach of Hill [8, 10, 11] which seeks the essence of continuum macro-laws implied by constituent micro-laws, and the thermomechanical approach of Rice [15] who emphasizes the idea of "internal variables". It may be possible to regard the elements of F here as internal variables in the rheological sense, but we shall not need to do so. Instead we regard them simply as useful *intermediate* parameters, which are *ultimately* eliminated by expressing them in terms of the elements of E. In the specific plasticity context we shall ultimately associate F_1, F_m and F_∞ with smooth, pyramidal and conical yield surfaces respectively, but we need not do so at this stage.

(ii) Definition of stress rate and strain rate

Consider a body able to deform continuously in the above Euclidean 3-space. Fix attention on two configurations, the "current" configuration and some (in general distinct) previous "reference" configuration. There will be a triad of material fibres emanating from any considered particle which will be mutually orthogonal in each of the two configurations, but not generally so between them. The local finite displacement will consist of a rotation of this triad, together with length-changes along the fibres of amounts $\lambda_1, \lambda_2, \lambda_3$ (say, λ = "stretch" = current length/reference length). There are many different tensor measures of finite strain. A family of them may be constructed by Hill's [9] procedure. According to this a finite strain tensor is defined as that symmetric tensor whose principal axes are the reference position of the triad, and whose principal values are

$$e_1 = f(\lambda_1), \quad e_2 = f(\lambda_2), \quad e_3 = f(\lambda_3) \tag{30}$$

where $f(\lambda)$ is an arbitrary smooth monotonic function in $\lambda > 0$ which has the properties

$$f(1) = 0, \quad f'(1) = 1. \tag{31}$$

We denote any particular such strain tensor by $\mathbf{e}$. It will belong to the space E. Examples are obtained via the one-parameter family of functions

$$f(\lambda) = \begin{cases} \{\lambda^{2k} - 1\}/2k, & k \neq 0 \\ \ln \lambda, & k = 0. \end{cases} \tag{32}$$

This generates the Green, logarithmic and Almansi strain tensors when $k = 1, 0$ and -1 respectively.

This procedure may be carried out, with the same or different f, to define a strain tensor at *each* of a sequence of current configurations in a virtual displacement path. If the latter is generated by continuous variation of a time-like scalar ϵ, and the same f is used throughout, there results a strain tensor function $\mathbf{e}(\epsilon)$, whose derivative is strain-rate which we write as $\dot{\mathbf{e}}(\epsilon) \equiv \dot{\mathbf{e}}$.

The current virtual work-rate of the stresses, per unit reference volume, will be an inner product

$$\boldsymbol{\tau}\dot{\mathbf{e}} \tag{33}$$

of $\dot{\mathbf{e}}$ with another member $\boldsymbol{\tau}$ of E. Hill describes this tensor of coefficients $\boldsymbol{\tau}$ as an objective stress measure *conjugate* to $\mathbf{e}$. It depends via $\mathbf{e}$ on the choice of f, since (33) itself will not. Also it will be a function $\boldsymbol{\tau}(\epsilon)$ of ϵ on any considered virtual path, and so generates a definition of stress-rate $\dot{\boldsymbol{\tau}}(\epsilon) \equiv \dot{\boldsymbol{\tau}}$ acceptable for use in constitutive rate equations. Hill [10, 11] discusses, for plastic solids in particular, the effects of changes in this definition of $\dot{\boldsymbol{\tau}}$ generated by the different possible choices of $f(\lambda)$.

We suppose that we have a known current mechanical state, so that in particular the current and reference configurations are known. Also *some* acceptable choice of $f(\lambda)$ has been made, leading to known values of $\mathbf{e}$ and $\boldsymbol{\tau}$ and consequent acceptable definitions of $\dot{\mathbf{e}}$ and $\dot{\boldsymbol{\tau}}$. We wish to discuss certain possible relationships between the values of $\dot{\mathbf{e}}$ and $\dot{\boldsymbol{\tau}}$. This discussion will be the same whatever particular choice of $f(\lambda)$ and consequent definition of $\dot{\mathbf{e}}$ and $\dot{\boldsymbol{\tau}}$ employed.

(iii) A monotonicity postulate

The current stress-rate $\dot{\boldsymbol{\tau}}$ and strain-rate $\dot{\mathbf{e}}$ at the considered material particle both belong to the space E. We now incorporate the second and arbitrary space F in the following way. We suppose that elements μ and γ of F can be independently associated with the particle, in a way which suggests that it may be helpful to think in terms of "supervectors" $[\dot{\boldsymbol{\tau}} \ \ -\mu]$ and $[\dot{\mathbf{e}} \ \ \gamma]$. Both supervectors belong to the product space $E \times F$. Eventually, in time-independent plasticity, we shall interpret μ and γ so that they have dimensions as if they are ϵ-derivatives, but for most of what follows we shall not need to suppose that these variables are the derivatives of any function of ϵ. For example, not until equation (41) does γ really take on the character of a strain-rate, although the names attached to (36) and (37) carry this implication to some extent. Thus no physical meaning is yet attached to μ and γ, which are merely anonymous parameters at this stage.

We now introduce a hypothesis of type (8), namely that

$$\Delta\dot{\boldsymbol{\tau}}\Delta\dot{\mathbf{e}} \geqslant \Delta\mu\Delta\gamma. \tag{34}$$

We first interpret this in the spirit of the paragraph *following* (8). That is, we regard it as an *assumption of monotonicity* in an envisaged constitutive relation between the supervectors $[\dot{\boldsymbol{\tau}} \ \ -\mu]$ and $[\dot{\mathbf{e}} \ \ \gamma]$. This shows why it is convenient to think of μ and γ as "stress-like" and "strain-like" parameters respectively. Inequality (34) is

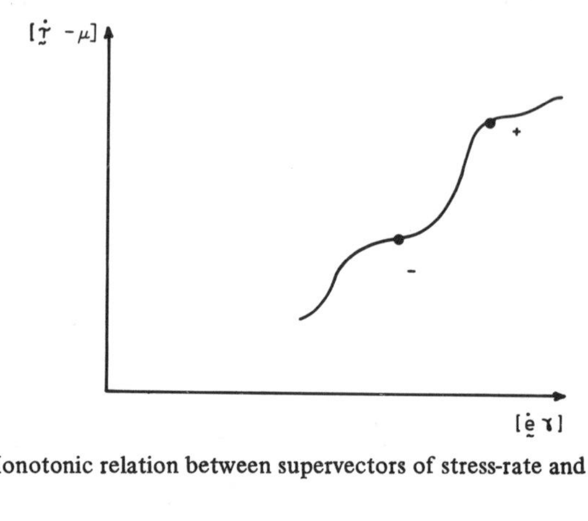

Fig. 6. Monotonic relation between supervectors of stress-rate and strain-rate.

stated in terms of the differences between quantities associated with any two points, labelled + and −, on the monotonic constitutive relation illustrated schematically in Fig. 6. That curve need not even be smooth at this stage; to the extent that it contains horizontal or vertical segments, the inequality in (34) will be weak, and otherwise strict.

(iv) Examples of monotonicity via convex/concave generating functionals

We pursue further the analogy with Section 2 by identifying

$$x, y, u, v \quad \text{with} \quad \dot{\boldsymbol{\tau}}, \dot{\text{e}}, \gamma, -\mu$$

respectively. For example, particular groups of circumstances leading to (34) may be inferred simply by transliterating Fig. 1. This becomes Fig. 7, any one corner of

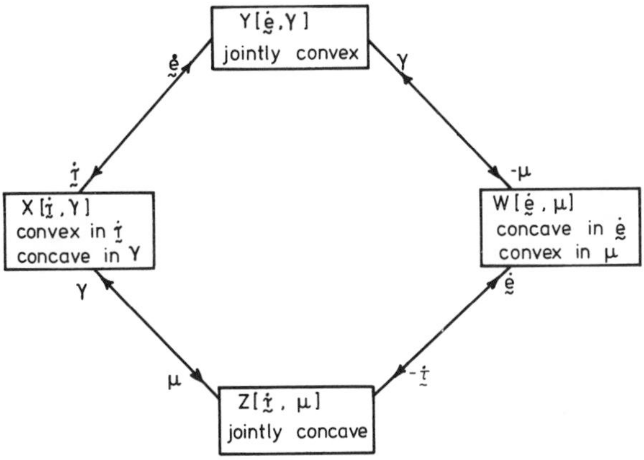

Fig. 7. Linked generating functionals underlying constitutive equations.

which generates two equations between the four variables, as exemplified in (35) or (40) below; and the convexity and/or concavity properties of the generating functional at that corner may then be stated in the form (34), via (7) as explained. For example, if the given functional $X[\dot{\boldsymbol{\tau}}, \gamma]$ exists and is differentiable then

$$\dot{\text{e}} = \frac{\partial X}{\partial \dot{\boldsymbol{\tau}}}, \qquad \mu = \frac{\partial X}{\partial \gamma}, \tag{35}$$

and if $X[\dot{\boldsymbol{\tau}}, \gamma]$ is weakly saddle-shaped this may be expressed via (7) and (35) in the form (34). In such a case Fig. 6 sums up both (35) *and* the saddle property. We can argue similarly from $Y[\dot{\text{e}}, \gamma]$, $W[\dot{\text{e}}, \mu]$ or $Z[\dot{\boldsymbol{\tau}}, \mu]$ if any one of these functionals exists as well or instead. If all the Legendre-type transformations exist, and the convexity and concavity properties are all strict, then the relationships are as displayed in Fig. 7, and (34) becomes a strict inequality. We shall give a concrete illustration of Fig. 7 via the generating functions (55)–(65) listed below.

(v) Uniqueness of constitutive expressions

After these illustrative remarks we return to Fig. 6 and to (34) itself, regarded as a monotonicity statement holding whether or not Fig. 7 exists. It is an under-determined statement about the relations between $\dot{\boldsymbol{\tau}}$, $\dot{\mathbf{e}}$, μ and γ, just as (for example) the two equations (35) are underdetermined and would have to be augmented in ways such as those illustrated in (10)–(14). We have indicated that we wish to regard (34) as a *constitutive* inequality (and by the same token we shall want to regard relations such as (35) as constitutive equations). Such an interpretation will not have content until either we assign a physical meaning to μ and γ directly (as we already have for $\dot{\boldsymbol{\tau}}$ and $\dot{\mathbf{e}}$), or we adjoin further conditions to (34) which allow μ and γ to be expressed in terms of the already meaningful $\dot{\boldsymbol{\tau}}$ and $\dot{\mathbf{e}}$.

Here we take the latter viewpoint. As indicated after (29), we simply regard these elements μ and γ of F as intermediate parameters which are convenient to use for analytical reasons, because they play the role of $-v$ and u respectively in the basic theoretical framework developed in Section 2. If they turn out to have a direct physical meaning (and with the advantage of hindsight we shall see that they may), then so much the better. But our approach does not require this, being abstracted from the analytical framework of Section 2, and so stemming in part from quite other areas of applied mathematics.

We suppose, then, that $\dot{\boldsymbol{\tau}}$, $\dot{\mathbf{e}}$, μ and γ are already connected by a relationship like that of Fig. 6, having the property (34); that either $\dot{\boldsymbol{\tau}}$ or $\dot{\mathbf{e}}$ may be regarded as assigned; and we seek to express μ and γ uniquely in terms of $\dot{\boldsymbol{\tau}}$ or $\dot{\mathbf{e}}$ by adjoining one or other of the following sets of conditions, which are suggested by (13). The names attached to these conditions are associated with the "stress-like" and "strain-like" adjectives already used for μ and γ, and are informed by existing terminology in continuum mechanics. We adjoin either the

Flow Conditions:

$$\mu \leqslant 0, \quad \gamma \geqslant 0, \quad \mu\gamma = 0; \tag{36}$$

or the

Locking Conditions:

$$\mu \geqslant 0, \quad \gamma \leqslant 0, \quad \mu\gamma = 0. \tag{37}$$

These conditions are illustrated schematically in Fig. 8.

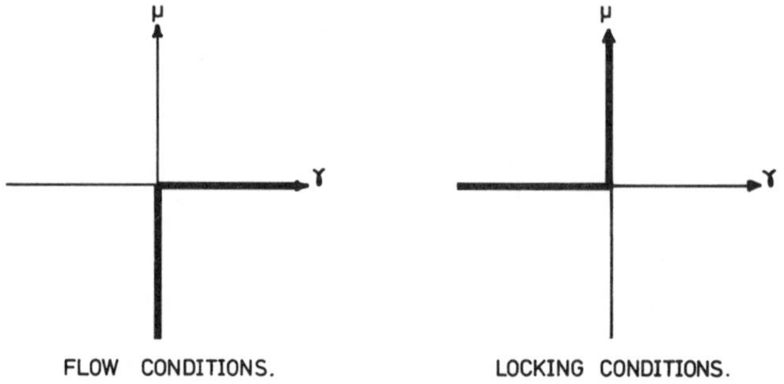

Fig. 8

Evidently a precise formal identification of (36) or (37) with (13) requires the choice $T = 0$ there, and then that (10) should correspond to assigning $\dot{\boldsymbol{\tau}}$ or $\dot{\mathbf{e}}$. The way in which this may be done (and the consequent derivation of extremum principles) is illustrated by Noble and Sewell [14, Section 12(iii)] in a special case – corresponding to a quadratic X on the space $E \times F_m$, as in (55) below. For example, assigned $\dot{\mathbf{e}}$ in (35) requires X to be replaced by $X - \dot{\boldsymbol{\tau}}\dot{\mathbf{e}}$, with $T^* = 0$ in (10); and assigned $\dot{\boldsymbol{\tau}}$ may be treated as if $\dot{\boldsymbol{\tau}}$ were formally absent. Here we shall not need to establish such identifications, but instead we treat the conditions (36) and (37), and those of assigned $\dot{\boldsymbol{\tau}}$ or $\dot{\mathbf{e}}$, directly on their merits.

Consider any μ_+, γ_+ which together happen to satisfy (36). Let μ_-, γ_- be any other solution of (36). Their (generally finite) difference will satisfy

$$\Delta\mu\Delta\gamma \equiv (\mu_+ - \mu_-)(\gamma_+ - \gamma_-)$$
$$= -\mu_+\gamma_- - \mu_-\gamma_+ \geqslant 0. \tag{38}$$

The same applies to the differences of solutions of (37).

Now consider the uniqueness question in the case when the relationship of Fig. 6 actually requires (34) to be strict, so that

$$\Delta\dot{\boldsymbol{\tau}}\Delta\dot{\mathbf{e}} > \Delta\mu\Delta\gamma, \tag{39}$$

in the following different circumstances.

(ai) Suppose (39) holds whenever $\Delta\gamma \neq 0$. Then if there exists a solution $\gamma(\dot{\boldsymbol{\tau}})$ at all when $\dot{\boldsymbol{\tau}}$ is assigned and (36) or (37) are adjoined to the relationship of Fig. 6, this solution must be unique – because $\Delta\dot{\boldsymbol{\tau}} = 0$ and $\Delta\gamma \neq 0$ imply a contradiction between (38) and (39).

(aii) If $\dot{\mathbf{e}}$ is assigned instead of $\dot{\boldsymbol{\tau}}$, the same argument shows that any solution $\gamma(\dot{\mathbf{e}})$ will be unique.

(bi) Suppose that (39) holds whenever $\Delta\mu \neq 0$. The argument of (a) shows that when $\dot{\boldsymbol{\tau}}$ is assigned $\mu(\dot{\boldsymbol{\tau}})$ will be unique.

(bii) Likewise $\mu(\dot{\mathbf{e}})$ is unique, if $\dot{\mathbf{e}}$ is assigned.

(c) Suppose that (39) holds whenever $\Delta\dot{\mathbf{e}} \neq 0$. Then when $\dot{\boldsymbol{\tau}}$ is assigned, any solution $\dot{\mathbf{e}}(\dot{\boldsymbol{\tau}})$ will be unique – because $\Delta\dot{\boldsymbol{\tau}} = 0$ and $\Delta\dot{\mathbf{e}} \neq 0$ imply a contradiction between (38) and (39).

(d) Suppose that (39) holds whenever $\Delta\dot{\boldsymbol{\tau}} = 0$. Then when $\dot{\mathbf{e}}$ is assigned, any solution $\dot{\boldsymbol{\tau}}(\dot{\mathbf{e}})$ will be unique – because $\Delta\dot{\mathbf{e}} = 0$ and $\Delta\dot{\boldsymbol{\tau}} \neq 0$ imply a contradiction between (38) and (39).

Evidently (36) or (37) with (39) are enough to establish some quite definite uniqueness results concerning expression of the constitutive equations, which hold whether or not any generating functionals exist. Each of the six sets of hypotheses and conclusion in (a)–(d) may be regarded as holding separately, or some or all of them may be taken together. Results like those in (a) and (b) justify our viewpoint of regarding μ and γ merely as intermediate parameters convenient for analytical reasons. Existence theorems would be even stronger justification if these could be established.

When one or more generating functional *does* exist, the strict inequality (39) hypothesized in (a), (b), (c) or (d) may be interpreted in terms of the convexity or concavity indicated in Fig. 7, but now considered to be strict. For example, when $X[\dot{\boldsymbol{\tau}}, \gamma]$ exists such that (35) holds, hypothesis (39a) expresses its saddle property as being strictly concave in γ, and at least weakly convex in $\dot{\boldsymbol{\tau}}$; whereas hypothesis (39d) expresses strict convexity in $\dot{\boldsymbol{\tau}}$ and at least weak concavity in γ. Again, when

$Y[\dot{e}, \gamma]$ exists such that (cf. (9))

$$\dot{\tau} = \frac{\partial Y}{\partial \dot{e}}, \quad \mu = -\frac{\partial Y}{\partial \gamma}, \tag{40}$$

its joint convexity expressible by (39) is actually strict in γ under hypothesis (39a), and is strict in $\dot{e}$ under (39c). If single-valuedness of the gradient mappings like those in (35) and (40) is employed the uniqueness results can each be reached by more than one route. For example, the result of (36c) also follows by direct substitution of (36ai) into (35), whenever the latter mapping is single-valued.

(vi) A particular class of generating functionals

Here we examine consequences of a rather specific postulate which has the effect of associating the element γ of F, still for arbitrary F, with the "plastic part" of the strain-rate. We assume that

$$\dot{e} = M\dot{\tau} + \gamma\nu. \tag{41}$$

Here M symbolizes an assigned tensor M_{ijkl} of "instantaneous elastic moduli", symmetric with respect to the interchanges $i \leftrightarrow j$, $k \leftrightarrow l$, and $ij \leftrightarrow kl$, and $M\dot{\tau}$ denotes the symmetric second-order tensor $M_{ijkl}\dot{\tau}_{kl}$. The product $\gamma\nu$ is to be regarded as a symmetric second-order tensor of F-inner products between γ and the components ν_{ij} of ν, an assigned member of E. For example, in F_m there will be m assigned symmetric tensors ν_α and by (28) $\gamma\nu \equiv \gamma_\alpha \nu_\alpha$, a sum of m terms. Again in F_∞, there will be an assigned symmetric tensor function $\nu(\alpha)$ on the interval $[0, 1]$, and $\gamma\nu \equiv \int_0^1 \gamma(\alpha)\nu(\alpha)\, d\alpha$ in accord with (29).

Of course (41) is a generalization of one of the most elementary postulates of plasticity theory. Our approach allows it to be expressed entirely in terms of any one of the four generating functionals displayed in Fig. 7. We get (writing $K \equiv M^{-1}$ when M is invertible)

$$\frac{\partial X}{\partial \dot{\tau}} = M\dot{\tau} + \gamma\nu, \tag{42}$$

$$\frac{\partial Y}{\partial \dot{e}} = K(\dot{e} - \gamma\nu), \tag{43}$$

$$-\frac{\partial Z}{\partial \dot{\tau}} = M\dot{\tau} + \frac{\partial Z}{\partial \mu}\nu, \tag{44}$$

$$-\frac{\partial W}{\partial \dot{e}} = K\dot{e} + \frac{\partial W}{\partial \mu}K\nu. \tag{45}$$

Any one of these four equations may be regarded as an alternative statement of (41). Moreover, each contains only the two variables occurring in the associated functional, and by direct integration gives the form implied for that functional, regardless of whether the other three functionals exist or not. We find from (42)–(45) respectively, that

$$X[\dot{\tau}, \gamma] = \tfrac{1}{2}\dot{\tau}M\dot{\tau} + \gamma\nu\dot{\tau} + B[\gamma], \tag{46}$$

$$Y[\dot{e}, \gamma] = \tfrac{1}{2}\dot{e}K\dot{e} - \gamma\nu K\dot{e} + C[\gamma], \tag{47}$$

$$Z[\dot{\boldsymbol{\tau}}, \mu] = -\tfrac{1}{2}\dot{\boldsymbol{\tau}}M\dot{\boldsymbol{\tau}} + D[\mu - \boldsymbol{\nu}\dot{\boldsymbol{\tau}}], \tag{48}$$

$$W[\dot{\mathrm{e}}, \mu] = -\tfrac{1}{2}\dot{\mathrm{e}}K\dot{\mathrm{e}} + A[\mu - \boldsymbol{\nu}K\dot{\mathrm{e}}]. \tag{49}$$

Here *B, C, D, A* denote arbitrary functionals of the indicated arguments. They will be related in pairs whenever a corresponding Legendre transformation in Fig. 7 exists, and they will be restricted by whatever convexity/concavity requirements listed there are imposed.

For example, (1) in the form $X + Y = \dot{\boldsymbol{\tau}}\dot{\mathrm{e}} = -(W + Z)$, together with (41), implies

$$\begin{aligned} B + C &= \tfrac{1}{2}\gamma\boldsymbol{\nu}K\boldsymbol{\nu}\gamma \text{ in the notation for general } F \\ &= \tfrac{1}{2}\gamma_\alpha\boldsymbol{\nu}_\alpha K\boldsymbol{\nu}_\beta\gamma_\beta \text{ in the case of } F_m \\ &= \tfrac{1}{2}\int_0^1\int_0^1 \gamma(\alpha)\boldsymbol{\nu}(\alpha)K\boldsymbol{\nu}(\beta)\gamma(\beta)\,d\alpha\,d\beta \text{ for } F_\infty \\ &= D + A. \end{aligned} \tag{50}$$

Also, *B* and *D* are the duals of each other in a Legendre transformation. So also are *C* and *A*, according to the scheme of Fig. 9.

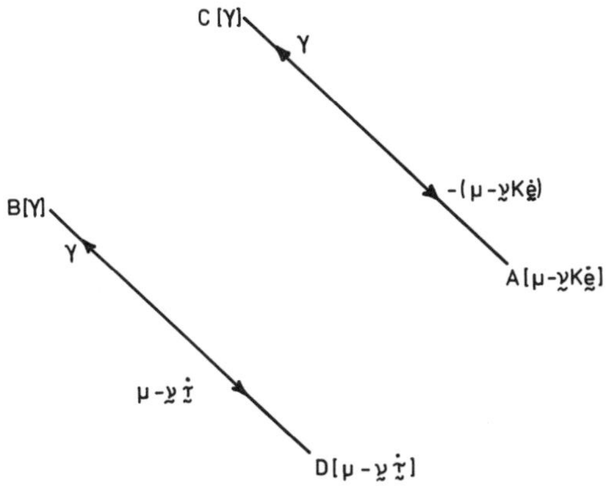

Fig. 9. Legendre-type transformations relating (46)–(49).

To treat the elastic/plastic case we assume that *M*, and therefore *K*, is finite and positive definite. In the rigid/plastic case $M = 0$ and *K* does not exist. In either case it follows that (46) is a saddle functional if *B* is concave, either strict or weak as required. A bi-linear function like $\gamma\boldsymbol{\nu}\dot{\boldsymbol{\tau}}$ is trivially a weak saddle, and so does not contradict even a strict saddle property so obtained. It also follows that (48) is a jointly concave functional if *D* is concave. When both *B* and *D* exist the concavity of one implies that of the other. With *K* positive definite, (50) will be at least semi-definite and therefore at least weakly convex. Then (47) will be jointly convex if *C* is more convex than (50). This can be expressed by adapting (7) into the form

$$0 \leqslant \Delta Y - \left.\frac{\partial Y}{\partial \dot{\mathrm{e}}}\right|_{-} \Delta\dot{\mathrm{e}} - \left.\frac{\partial Y}{\partial \gamma}\right|_{-} \Delta\gamma$$

$$= \tfrac{1}{2}\Delta(\dot{\mathbf{e}} - \gamma\boldsymbol{\nu})K\Delta(\dot{\mathbf{e}} - \gamma\boldsymbol{\nu})$$

$$+ \Delta[C - \tfrac{1}{2}\gamma\boldsymbol{\nu}K\boldsymbol{\nu}\gamma] - \left.\frac{\partial[C - \tfrac{1}{2}\gamma\boldsymbol{\nu}K\boldsymbol{\nu}\gamma]}{\partial\gamma}\right|_{-} \Delta\gamma. \tag{51}$$

In particular it will hold if B also exists and is concave. Again with K positive definite, (49) will be a saddle functional if A is more convex than (50). By this we mean that

$$0 \leqslant \Delta W - \left.\frac{\partial W}{\partial \dot{\mathbf{e}}}\right|_{+} \Delta\dot{\mathbf{e}} - \left.\frac{\partial W}{\partial \mu}\right|_{-} \Delta\mu$$

$$= \tfrac{1}{2}\Delta(\dot{\mathbf{e}} - \gamma\boldsymbol{\nu})K\Delta(\dot{\mathbf{e}} - \gamma\boldsymbol{\nu})$$

$$+ \Delta A - \left.\frac{\partial A}{\partial[\mu - \boldsymbol{\nu}K\dot{\mathbf{e}}]}\right|_{-} \Delta[\mu - \boldsymbol{\nu}K\dot{\mathbf{e}}] - \tfrac{1}{2}\Delta\gamma\boldsymbol{\nu}K\boldsymbol{\nu}\Delta\gamma. \tag{52}$$

This will hold in particular if D also exists and is concave.

When we wish to generate the constitutive equations of time-independent plasticity, we add to (41) the requirement that the generating functionals (46)–(49) be homogeneous of degree two in their arguments. By Euler's theorem it follows that their values will satisfy

$$B = D = \tfrac{1}{2}\gamma(\mu - \boldsymbol{\nu}\dot{\boldsymbol{\tau}}), \tag{53}$$

$$C = A = -\tfrac{1}{2}\gamma(\mu - \boldsymbol{\nu}K\dot{\mathbf{e}}). \tag{54}$$

(vii) Explicit examples of generating functions

In the previous subsection we did not need to specify the space F, or the actual form of the functionals A, B, C, D, and we did not need to employ either of the conditions (36) or (37). Here we give an explicit example of the constitutive equations of time-independent plasticity by choosing F to be the space F_m, by choosing homogeneous quadratics for A, B, C, D, and by adjoining the flow conditions (36).

For the purpose of this illustration we shall return to the use of Greek suffixes and the summation convention for them, for example to indicate inner products in F_m; but we shall retain the symbolic notation in E according to which, for example, $\tfrac{1}{2}\dot{\boldsymbol{\tau}}M\dot{\boldsymbol{\tau}} \equiv \tfrac{1}{2}\dot{\tau}_{ij}M_{ijkl}\dot{\tau}_{kl}$ and $\gamma_\alpha\dot{\boldsymbol{\nu}}_\alpha\dot{\boldsymbol{\tau}} \equiv \gamma_\alpha\nu_{ij\alpha}\dot{\tau}_{ij}$. The last "value" form given for each generating functional below would hold for homogeneous second degree functions (whether quadratic or not), and of course is simplified as soon as $(36)_3$ is actually incorporated.

We begin by letting

$$X[\dot{\boldsymbol{\tau}}, \gamma_\alpha] = \tfrac{1}{2}\dot{\boldsymbol{\tau}}M\dot{\boldsymbol{\tau}} + \gamma_\alpha\boldsymbol{\nu}_\alpha\dot{\boldsymbol{\tau}} - \tfrac{1}{2}h_{\alpha\beta}\gamma_\alpha\gamma_\beta \tag{55}$$

$$= \tfrac{1}{2}(\dot{\boldsymbol{\tau}}\dot{\mathbf{e}} + \mu_\alpha\gamma_\alpha)$$

in which $h_{\alpha\beta}$ is an assigned symmetric $m \times m$ matrix illustrated for $m = 2$ in [18]. According to (35) this generates the equations

$$\dot{\mathbf{e}} = M\dot{\boldsymbol{\tau}} + \gamma_\alpha\boldsymbol{\nu}_\alpha, \qquad \mu_\alpha = \boldsymbol{\nu}_\alpha\dot{\boldsymbol{\tau}} - h_{\alpha\beta}\gamma_\beta. \tag{56}$$

When the flow conditions (36) are imposed, in this example they take the form

$$\boldsymbol{\nu}_\alpha\dot{\boldsymbol{\tau}} \leqslant h_{\alpha\beta}\gamma_\beta, \quad \gamma_\alpha \geqslant 0, \quad \gamma_\alpha(\boldsymbol{\nu}_\alpha\dot{\boldsymbol{\tau}} - h_{\alpha\beta}\gamma_\beta) = 0. \tag{57}$$

Whenever the determinant $|h_{\alpha\beta}| \neq 0, \infty$ we can invert $(56)_2$ and so construct by a Legendre transformation the function

$$\begin{aligned} Z[\dot{\boldsymbol{\tau}}, \mu_\alpha] &= -\tfrac{1}{2}\dot{\boldsymbol{\tau}}M\dot{\boldsymbol{\tau}} - \tfrac{1}{2}h^{-1}_{\alpha\beta}(\mu_\alpha - \boldsymbol{\nu}_\alpha\dot{\boldsymbol{\tau}})(\mu_\beta - \boldsymbol{\nu}_\beta\dot{\boldsymbol{\tau}}) \\ &= \tfrac{1}{2}(-\dot{\boldsymbol{\tau}}\dot{\mathbf{e}} + \mu_\alpha\gamma_\alpha). \end{aligned} \tag{58}$$

Here $h^{-1}_{\alpha\beta}$ is the matrix inverse of $h_{\alpha\beta}$. From Fig. 7 and the convention associated with Fig. 1, this $Z[\dot{\boldsymbol{\tau}}, \mu_\alpha]$ generates the equations

$$\dot{\mathbf{e}} = M\dot{\boldsymbol{\tau}} + h^{-1}_{\alpha\beta}\boldsymbol{\nu}_\alpha(\boldsymbol{\nu}_\beta\dot{\boldsymbol{\tau}} - \mu_\beta), \quad \gamma_\alpha = h^{-1}_{\alpha\beta}(\boldsymbol{\nu}_\beta\dot{\boldsymbol{\tau}} - \mu_\beta). \tag{59}$$

In these variables the flow conditions (36) are

$$\mu_\alpha \leqslant 0, \quad h^{-1}_{\alpha\beta}(\boldsymbol{\nu}_\beta\dot{\boldsymbol{\tau}} - \mu_\beta) \geqslant 0, \quad h^{-1}_{\alpha\beta}\mu_\alpha(\boldsymbol{\nu}_\beta\dot{\boldsymbol{\tau}} - \mu_\beta) = 0. \tag{60}$$

Whenever the determinant $|M| \neq 0, \infty$, the inverse $K = M^{-1}$ exists, and we introduce the notation

$$g_{\alpha\beta} \equiv h_{\alpha\beta} + \boldsymbol{\nu}_\alpha K\boldsymbol{\nu}_\beta. \tag{61}$$

We can then express the inverse of (56), via a Legendre transformation beginning with X, in terms of the generating function

$$\begin{aligned} Y[\dot{\mathbf{e}}, \gamma_\alpha] &= \tfrac{1}{2}\dot{\mathbf{e}}K\dot{\mathbf{e}} - \gamma_\alpha\boldsymbol{\nu}_\alpha K\dot{\mathbf{e}} + \tfrac{1}{2}g_{\alpha\beta}\gamma_\alpha\gamma_\beta \\ &= \tfrac{1}{2}(\dot{\mathbf{e}} - \gamma_\alpha\boldsymbol{\nu}_\alpha)K(\dot{\mathbf{e}} - \gamma_\beta\boldsymbol{\nu}_\beta) + \tfrac{1}{2}h_{\alpha\beta}\gamma_\alpha\gamma_\beta \\ &= \tfrac{1}{2}(\dot{\boldsymbol{\tau}}\dot{\mathbf{e}} - \mu_\alpha\gamma_\alpha). \end{aligned} \tag{62}$$

According to (40) this generates the equations

$$\dot{\boldsymbol{\tau}} = K(\dot{\mathbf{e}} - \gamma_\alpha\boldsymbol{\nu}_\alpha), \quad \mu_\alpha = \boldsymbol{\nu}_\alpha K\dot{\mathbf{e}} - g_{\alpha\beta}\gamma_\beta. \tag{63}$$

In these variables the flow conditions (36) become

$$\boldsymbol{\nu}_\alpha K\dot{\mathbf{e}} \leqslant g_{\alpha\beta}\gamma_\beta, \quad \gamma_\alpha \geqslant 0, \quad \gamma_\alpha(\boldsymbol{\nu}_\alpha K\dot{\mathbf{e}} - g_{\alpha\beta}\gamma_\beta) = 0. \tag{64}$$

Finally, whenever the determinant $|g_{\alpha\beta}| \neq 0, \infty$ we can invert $(63)_2$ and so from Y construct the Legendre transformation indicated in Fig. 7 leading to the function

$$\begin{aligned} W[\dot{\mathbf{e}}, \mu_\alpha] &= -\tfrac{1}{2}\dot{\mathbf{e}}K\dot{\mathbf{e}} + \tfrac{1}{2}g^{-1}_{\alpha\beta}(\boldsymbol{\nu}_\alpha K\dot{\mathbf{e}} - \mu_\alpha)(\boldsymbol{\nu}_\beta K\dot{\mathbf{e}} - \mu_\beta) \\ &= -\tfrac{1}{2}(\dot{\boldsymbol{\tau}}\dot{\mathbf{e}} + \mu_\alpha\gamma_\alpha). \end{aligned} \tag{65}$$

According to the convention we have associated with Fig. 7 this function generates the equations

$$\dot{\boldsymbol{\tau}} = K\dot{\mathbf{e}} - g^{-1}_{\alpha\beta}(\boldsymbol{\nu}_\alpha K\dot{\mathbf{e}} - \mu_\alpha)(\boldsymbol{\nu}_\beta K), \quad \gamma_\alpha = g^{-1}_{\alpha\beta}(\boldsymbol{\nu}_\beta K\dot{\mathbf{e}} - \mu_\beta). \tag{66}$$

The flow conditions (36) this time become

$$\mu_\alpha \leqslant 0, \quad g^{-1}_{\alpha\beta}(\boldsymbol{\nu}_\beta K\dot{\mathbf{e}} - \mu_\beta) \geqslant 0, \quad g^{-1}_{\alpha\beta}\mu_\alpha(\boldsymbol{\nu}_\beta K\dot{\mathbf{e}} - \mu_\beta) = 0. \tag{67}$$

When *both* the matrices M and $h_{\alpha\beta}$ are positive definite, this is sufficient to allow all three functions Z, Y and W to be obtained from the X of (55) in the way described; in addition these two conditions are sufficient for X to be strictly convex in $\dot{\boldsymbol{\tau}}$ and strictly concave in the γ_α. It also follows then that the Y of (62) is strictly jointly convex in $\dot{\mathbf{e}}$ and the γ_α, from the fact that the transliterated version of (7) is positive; and by expressing (7) in terms of Z and W we see that the particular examples (58) and (65) are strictly jointly concave and strictly saddle respectively when M and $h_{\alpha\beta}$ are positive definite, all in accordance with Fig. 7. These arguments are specific illustrations of those more general ones in the latter half of the preceding subsection.

The illustration we have given in (55)–(67) is the phenomenological analogue of the Mandel–Hill equations for incipient time-independent elastic/plastic distortion of metal single crystals by multislip on m glide systems (e.g. see Hill, [7]). The m assigned symmetric tensors $\boldsymbol{\nu}_\alpha$ are individually normal to the m facets of an envisaged locally pyramidal yield surface vertex (smooth if $m = 1$), and $h_{\alpha\beta}$ is a hardening matrix. At the single crystal level the m γ_α represent engineering shear strain-rates in the glide directions. Noble and Sewell [14, Section 12(iii)] showed that the Mandel–Hill equations could be generated from the single saddle function (55). Elimination of the intermediate parameters μ_α and γ_α via the flow conditions can be carried out in a unique way when M and $h_{\alpha\beta}$ are positive definite, in accordance with the uniqueness theorems established after (39). The resulting continuous functions are composed by joining different branches appropriate to different domains of $\dot{\mathbf{e}}$ and $\dot{\boldsymbol{\tau}}$ space. This particular type of complication is not present at the *outset* of the present approach, since we begin with the straightforward *single* generating saddle function $X[\dot{\boldsymbol{\tau}}, \gamma_\alpha]$ of (55), which is defined without restriction on $\dot{\boldsymbol{\tau}}$ or the γ_α. The branched constitutive equations then emerge as a consequence of adjoining (35) and (36).

These latter conditions pick out certain curves on the generating function, when this is plotted as a surface. In the case of the space F_1, i.e. for $m = 1$, there are just two branches of the constitutive equations, one of loading ($\gamma > 0$, $\mu = 0$) and one of unloading ($\mu < 0$, $\gamma = 0$). We can illustrate these in three dimensions as curves on the surface

$$A[\mu - \boldsymbol{\nu}K\dot{\mathbf{e}}] = \frac{1}{2g}(\boldsymbol{\nu}K\dot{\mathbf{e}} - \mu)^2 \tag{68}$$

obtained from (49) and (65) (the subscript 1 has been dropped). This part of the generating function $W[\dot{\mathbf{e}}, \mu] = -\frac{1}{2}\,\dot{\mathbf{e}}K\dot{\mathbf{e}} + A$ is a parabolic cylinder (Fig. 10). Then (35) and (36) pick out the unloading branch as the lowest semi-generator $\boldsymbol{\nu}K\dot{\mathbf{e}} = \mu < 0$, and the loading branch as the half-cross-section $\boldsymbol{\nu}K\dot{\mathbf{e}} > \mu = 0$ inclined at 135° to the unloading axis.

Degenerate or limiting cases can be accounted for by relaxing the strict monotonicity requirements on X, and accepting that Y, Z or W may then no longer exist. For example, the equations generated by X describe the rigid/plastic solid ($M = 0$, as we said), the non-hardening solid ($h_{\alpha\beta} = 0$) and the perfectly plastic solid ($M = h_{\alpha\beta} = 0$). For purposes other than time-independent plasticity it may be more appropriate to set down *a priori* some version of Y, Z or W instead of X as the starting point, and then to allow degeneracies in *that* function; and also to use (37) instead of (36).

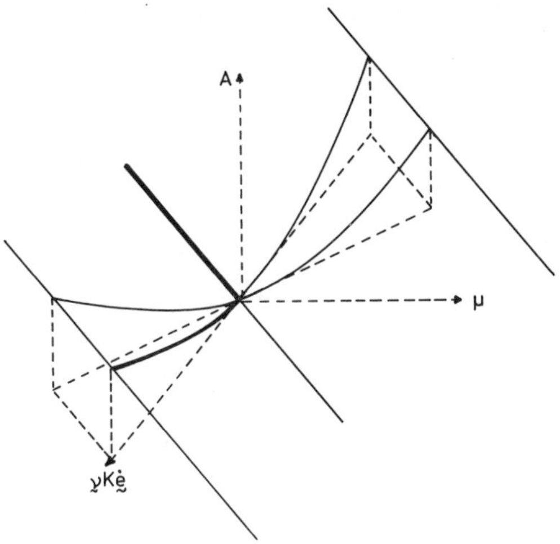

Fig. 10. Loading and unloading branches on the surface $A[\mu - \boldsymbol{\nu} K\dot{\mathbf{e}}]$.

(viii) Other possible generating functionals

The flexibility permitted by the present approach in the choice of constitutive rate equations to describe plasticity may be precisely described – it is twofold, involving freedom of choice of the space F, and then of the functionals A, B, C or D in (46)–(49). For the time-independent case the latter need to be homogeneous of degree two, and the particular spaces F_1, F_m and F_∞ are associated with yield surfaces which are locally smooth, pyramidal or conical. With F_m the quadratic functions of the last sub-section might be replaced by homogeneous degree-two functions or functionals subject to the requirements laid down in the latter half of subsection (vi) (but for F_1 a homogeneous degree-two function of *one* variable is necessarily quadratic).

In the space F_∞ we might consider beginning with the functional

$$X[\dot{\boldsymbol{\tau}}, \gamma] = \tfrac{1}{2}\dot{\boldsymbol{\tau}} M \dot{\boldsymbol{\tau}} + \dot{\boldsymbol{\tau}} \int_0^1 \boldsymbol{\nu}(\alpha)\, \gamma(\alpha)\, d\alpha - \tfrac{1}{2} \int_0^1 \int_0^1 h(\alpha, \beta)\, \gamma(\alpha)\, \gamma(\beta)\, d\alpha\, d\beta. \tag{69}$$

Here $\boldsymbol{\nu}(\alpha)$ is the previous assigned symmetric tensor function on the interval $[0, 1]$, and the assigned symmetric function $h(\alpha, \beta)$ is a "positive kernel" on $[0, 1] \times [0, 1]$, so that the Hilbert double integral in (69) is non-negative (see Tricomi, [20] Sections 3.11 and 3.12). Similar reasoning applies in (50) on the assumption that $\boldsymbol{\nu}(\alpha) K \boldsymbol{\nu}(\beta)$ is a positive kernel, i.e. with all its eigenvalues positive). To deal with a conical yield surface these quantities will need to have equal end-values in the sense that $\boldsymbol{\nu}(0) = \boldsymbol{\nu}(1)$, at least $h(0, 0) = h(1, 1)$, and the elements of F_∞ will be subject to the restriction $\gamma(0) = \gamma(1)$. It follows from $(35)_2$, (29) and (6) that

$$\mu(\alpha) = \dot{\boldsymbol{\tau}} \boldsymbol{\nu}(\alpha) - \int_0^1 h(\alpha, \beta)\, \gamma(\beta)\, d\beta. \tag{70}$$

According to (36) with (29) the loading assumption $\gamma(\alpha) > 0$ then implies an integral equation $\mu(\alpha) = 0$ to be solved in order to express $\gamma(\alpha)$ in terms of $\dot{\boldsymbol{\tau}}$. Since $h(\alpha, \beta)$ is known methods are available in the literature for solving such an equation. In particular consider the case in which

$$h(\alpha, \beta) = \tfrac{1}{2}\,\bar{h}(\tfrac{1}{2}(\alpha+\beta))\,\delta(\tfrac{1}{2}(\alpha-\beta)) + k(\alpha, \beta)\sqrt{\bar{h}(\alpha)\bar{h}(\beta)}. \tag{71}$$

Here the Dirac delta function of $\frac{1}{2}(\alpha - \beta)$ is multiplied by a known positive function $\bar{h}$ of $\frac{1}{2}(\alpha+\beta)$ with the property $\bar{h}(0) = \bar{h}(1)$, and the "remainder" kernel $k(\alpha, \beta)$ is symmetric. Then Hilbert–Schmidt theory (Tricomi, [20], Section 3.10, eqn. (121)) can be used* to express the solution $\gamma(\alpha)$ of the integral equation $\mu(\alpha) = 0$ in terms of the eigenfunctions and eigenvalues of $k(\alpha, \beta)$ if no such eigenvalue is -1. When $k \equiv 0$ the integral disappears from (70), and $\mu(\alpha) = 0$ can be inverted directly to give $\gamma(\alpha) = \dfrac{\dot{\boldsymbol{\tau}}\boldsymbol{\nu}(\alpha)}{\bar{h}(\alpha)}$. From this and $(35)_1$ applied to (69) we obtain

$$\dot{\mathrm{e}} = M\dot{\boldsymbol{\tau}} + \int_0^1 \boldsymbol{\nu}(\alpha)\left\{\frac{\dot{\boldsymbol{\tau}}\boldsymbol{\nu}(\alpha)}{\bar{h}(\alpha)}\right\} d\alpha \tag{72}$$

for the constitutive equations of loading. This establishes contact with the theory of Hill [8, eqns. (20)] who arrives at these equations (72) by a quite different approach. We have shown that the solution for $\dot{\mathrm{e}}(\dot{\boldsymbol{\tau}})$ in (72) can be generalized to handle non-zero $k(\alpha, \beta)$ in (71) if experiment subsequently warrants it.

(ix) An inequality in bifurcation theory

The quadratic function $W[\dot{\mathrm{e}}, \mu_\alpha]$ on $E \times F_m$ defined by (65) has the property that, for the differences of quantities associated with any two "points" of its argument,

$$\begin{aligned}
\Delta\dot{\boldsymbol{\tau}}\Delta\dot{\mathrm{e}} - \Delta\mu\Delta\gamma &= \Delta\mu\Delta\left\{\frac{\partial W}{\partial\mu}\right\} - \Delta\dot{\mathrm{e}}\Delta\left\{\frac{\partial W}{\partial\dot{\mathrm{e}}}\right\} \\
&= \frac{\partial^2 W}{\partial\mu^2}(\Delta\mu)^2 - \frac{\partial^2 W}{\partial\dot{\mathrm{e}}^2}(\Delta\dot{\mathrm{e}})^2 \\
&\equiv \frac{\partial^2 W}{\partial\mu_\alpha\partial\mu_\beta}\Delta\mu_\alpha\Delta\mu_\beta - \frac{\partial^2 W}{\partial\dot{e}_{ij}\partial\dot{e}_{kl}}\Delta\dot{e}_{ij}\Delta\dot{e}_{kl}.
\end{aligned} \tag{73}$$

The symbolic notation in the second line is explained via the summation convention in the third line. Write

$$\dot{\boldsymbol{\tau}}^L \equiv -\frac{\partial^2 W}{\partial\dot{\mathrm{e}}^2}\dot{\mathrm{e}} = \frac{\partial W_L}{\partial\dot{\mathrm{e}}} \tag{74}$$

where

$$W_L(\dot{\mathrm{e}}) \equiv W(\dot{\mathrm{e}}, 0) \tag{75}$$

is a function which generates a hypothetical "loading solid" in which all $\mu_\alpha = 0$.

*I acknowledge some discussion on this point with my colleague Dr. D. Porter.

Then (73) may be written

$$\Delta\dot{\tau}\Delta\dot{e} - \Delta\dot{\tau}^L \Delta\dot{e} = \frac{\partial^2 W}{\partial\mu^2}(\Delta\mu)^2 + \Delta\mu\Delta\gamma. \tag{76}$$

Since $W[\dot{e}, \mu_\alpha]$ is convex in the μ_α this Hessian $\partial^2 W/\partial\mu_\alpha\partial\mu_\beta$ will be positive definite, and the first term on the right of (76) will be positive or (if all $\Delta\mu_\alpha = 0$) zero. If we consider not all pairs of points of $E \times F_m$, but only those generated via the flow conditions (36) by different strain-rates $\dot{e}$ (so implying $\mu(\dot{e})$ according to (39(bii)), we see that $\Delta\mu\Delta\gamma \geqslant 0$ by (38). It follows that for any two strain-rates in the elastic/plastic solid generated by (65) (and therefore by (55) with positive definite M and $h_{\alpha\beta}$)

$$\Delta\dot{\tau}\Delta\dot{e} \geqslant \Delta\dot{\tau}^L \Delta\dot{e} \tag{77}$$

with equality for all pairs which make $\Delta\mu = 0$, and in particular for all strain-rates implying loading (all $\mu_\alpha = 0$). This is an alternative proof of a result of Sewell [17, eqn (3.65)] for general m which generalized one of Hill [6, lemma I] for $m = 1$. The two terms on the right of (76) may be identified with those given away in the earlier direct computation (Sewell, *op. cit.* Section 3(iv)).

The result is important in plastic bifurcation theory because it ensures that unloading will not begin *at* bifurcation, and so allows bifurcation loads to be calculated *as if* the solid had the single-valued quadratic potential $W_L(\dot{e})$, i.e. by replacing the actual solid by a hypothetical linear comparison solid (74) with strain-rate potential $W(\dot{e}, 0)$. An application of this method to calculate the bifurcation load of a plate when the stress is at a yield surface vertex with $m = 2$ has recently been made (Sewell, [18], with a sequel for $m = 3$ in 1974 in the same Journal).

The previous proofs of (77) mentioned above have seemed to require a somewhat *ad hoc* calculation. The present proof indicates "what kind" of inequality is involved – i.e. a combination of the saddle property of the generating functional $W[\dot{e}, \mu]$ and the property (38) of the flow conditions. It also suggests in what direction to seek generalizations. For example, if W were homogeneous of degree two but not quadratic (73) would still have the same general form, but the second derivatives would be evaluated at certain "mean values" and would no longer be constants. Also, if W were a functional rather than a function, and perhaps over $E \times F_\infty$ instead of $E \times F_m$, an appropriate definition of second derivatives would have to be introduced.

In these more general circumstances it is still an open question whether a useful "linear comparison solid" or a "loading comparison solid" (no longer necessarily the same thing) can usefully be defined. The present viewpoint suggests, however, that what might be required is an adaptation of (73) and a modification of (74), perhaps with something given away from the last term in (73) if that were possible. It would be of particular interest to know whether, for example, the constitutive equations calculated by Hutchinson [12] could be cast into the form implied here.

(x) Drucker's inequality

Finally we return to the strict form (39) of the monotonicity postulate (34) to derive from it an inequality of the type suggested by Drucker (see Hill, [9], for a statement using a precise stress-rate). If we consider in (39) not all pairs of points, but only those related by (36) (or by (37)), then (38) applies and we get

$$\Delta\dot{\boldsymbol{\tau}}\Delta\dot{\mathbf{e}} > 0 \tag{78}$$

for all such pairs. If we add the extra assumption that one such point has the property $\dot{\boldsymbol{\tau}} = \dot{\mathbf{e}} = 0$, and consider only those pairs consisting of this and another point, we get

$$\dot{\boldsymbol{\tau}}\dot{\mathbf{e}} > 0 \tag{79}$$

as a property of all the other points. This last constitutive inequality is of the Drucker type.

REFERENCES

1. Arthurs, A. M. *Complementary Variational Principles,* Clarendon Press, Oxford, 1970.
2. Callen, H. B. *Thermodynamics,* Wiley, New York, 1960, p. 119.
3. Coleman, B. D. and Noll, W. "On the Thermostatics of Continuous Media". *Arch. Rational Mech. Anal. 4,* 1959, 97–128.
4. Flower, J. O. and Evans, F. J. "Irreversible Thermodynamics and Stability Considerations of Electrical Networks". *J. Franklin Inst. 291*, 1971, 121–136.
5. Hill, R. "New Horizons in Mechanics of Solids". *J. Mech. Phys. Solids, 5,* 1956, 66–74.
6. Hill, R. "A General Theory of Uniqueness and Stability in Elastic/Plastic Solids". *J. Mech. Phys. Solids, 6,* 1958, 236–249.
7. Hill, R. "Generalized Constitutive Relations for Incremental Deformation of Metal Crystals by Multislip". *J. Mech. Phys. Solids, 14,* 1966, 95–102.
8. Hill, R. "The Essential Structure of Constitutive Laws for Metal Composites and Polycrystals". *J. Mech. Phys. Solids, 15,* 1967, 79–95.
9. Hill, R. "On Constitutive Inequalities for Simple Materials–I". *J. Mech. Phys. Solids, 16,* 1968, 229–242.
10. Hill, R. "On Constitutive Inequalities for Simple Materials–II". *J. Mech. Phys. Solids, 16,* 1968, 315–322.
11. Hill, R. "On Constitutive Macro-Variables for Heterogeneous Solids at Finite Strain". *Proc. Roy. Soc. London, A326,* 1972, 131–147.
12. Hutchinson, J. W. "Elastic-Plastic Behaviour of Polycrystalline Metals and Composites". *Proc. Roy. Soc. London, A319*, 1970, 247–272.
13. Noble, B. "Complementary Variational Principles for Boundary Value Problems I: Basic Principles with an Application to Ordinary Differential Equations". Mathematics Research Center Report 473, University of Wisconsin, 1964, 30pp.
14. Noble, B. and Sewell, M. J. "On Dual Extremum Principles in Applied Mathematics". Mathematics Research Center Report 1119, University of Wisconsin, 1971, 121pp. (Available in *J. Inst. Maths. Applics. 9,* 1972, 123–193).

15. Rice, J. R. "Inelastic Constitutive Relations for Solids: An Internal-Variable Theory and its Application to Metal Plasticity". *J. Mech. Phys. Solids, 19,* 1971, 433–455.
16. Sewell, M. J. "On Dual Approximation Principles and Optimization in Continuum Mechanics". *Phil. Trans. Roy. Soc. London, A265,* 1969, 319–351.
17. Sewell, M. J. "A Survey of Plastic Buckling". Mathematics Research Center Report 1137, University of Wisconsin (1971). (Available in H. Leipholz (ed.) *Stability*, pp. 85–198. University of Waterloo, Ontario, April 1972.)
18. Sewell, M. J. "A Yield Surface Corner lowers the Buckling Stress of an Elastic/Plastic Compressed Plate". Mathematics Research Center Report 1226, University of Wisconsin, 1972, 56pp. (Available in *J. Mech. Phys. Solids, 21,* 1973, 19–45.)
19. Sewell, M. J. "The Governing Equations and Extremum Principles of Elasticity and Plasticity generated from a Single Functional". Mathematics Research Center Report 1227, University of Wisconsin, 1972, 86pp. (Available in *J. Structural Mechanics, 2,* 1973, 1–32 and 135–158.)
20. Tricomi, F. G. *Integral Equations.* Interscience, New York, 1957.
21. Truesdell, C. *The Elements of Continuum Mechanics.* Springer, Berlin, 1966.
22. Truesdell, C. and Noll, W. "The Non-Linear Field Theories of Mechanics", in S. Flugge (ed.) Handbuch der Physik III/3 1–579, Springer, Berlin, 1965.

NOMENCLATURE

E, F	inner product spaces	c_p, c_v	specific heats
x, y	elements of E, having inner product (x, y)	c	sound velocity
u, v	elements of F, having inner product $\langle u, v\rangle$	$\mathbf{a}, \mathbf{b}$	second order tensors in space E
X, Y, W, Z	given functionals on $E \times F$	μ, γ	matrix or scalar elements of F_m, F or F_∞
T, T^*	mutually adjoint linear operators	$\dot{\boldsymbol{\tau}}$	stress-rate
		$\dot{\mathbf{e}}$	strain-rate
ϵ	internal energy	M, K	elastic moduli
ψ	free energy	$\boldsymbol{\nu}, \boldsymbol{\nu}_\alpha, \boldsymbol{\nu}(\alpha)$	plastic strain-rate directions
χ	enthalpy		
ζ	free enthalpy	A, B, C, D	given functionals
v	volume/unit mass	$\mu_\alpha, \gamma_\alpha$	matrix elements of F_m, the space of m-tuples
η	entropy		
θ	temperature	$h_{\alpha\beta}, g_{\alpha\beta}$	typical components of given $m \times m$ matrices
p	pressure		

BASIC CONCEPTS IN ENGINEERING AND ECONOMICS

Ole Immanuel Franksen
Electric Power Engineering Department
The Technical University of Denmark
Lyngby, Denmark

SUMMARY

The purpose of this paper is to establish a fundamental analogy between concepts and laws in electrical network theory and in economics. From an intuitive foundation, made up of an electrical analogy from 1935 of a simplified production system, the fundamental properties of measurements and scales come into focus. On this basis, it is straight-forward to establish an analogy between electrical currents and commodity flows and to transfer Kirchhoff's node law to economics. The analogy between voltages and economic prices, on the other hand, is far more difficult to substantiate. Confining the discussion to a barter economy, the main problem seems to be that the linear graph of price measurements in a closed economy forms a Lagrangian tree. The root of this tree is identified, by a generalization of the economic definition of a free commodity, with a mathematically empty set of goods. With this analogy established Kirchhoff's mesh may be introduced into economics. Next in line to Kirchhoff's laws are the economic analogies of the first and second laws of thermodynamics which pervade all of physics. The economic counterpart of the first law or the principle of conservation of energy or power is an equally important invariance principle known as Walras' law. On the other hand, a direct analogy of the second law or the principle of irreversibility is not traditionally formulated in micro-economic theory. It turns out, however, that the economic concept of rational behaviour may be recast, apart from a change of sign, in a similar form as the second law. Ohm's law occupies a very central position in the theory of electrical networks. The analogous concept, the so-called point elasticity measure, occupies an equally important position in micro-economic theory. Similarly, if we conceive voltage and current sources or mutual inductances as special cases of Ohm's law, perfectly elastic and inelastic curves or cross elasticities, are the corresponding special cases of the elasticity concept. Thus, it is demonstrated that, within a rather broad spectrum of theoretical endeavour, economic concepts and laws do not differ basically from those of physics.

1. INTRODUCTION

The paradox of modern expansion, or rather explosion, of information is the fact that, individually, we tend to become less and less knowledgeable. The significance of this paradox may be illustrated by considering two basic trends in our university curricula of the past decade. In the engineering schools, to a large degree, the emphasis has been on training specialists, each confined to an ever-narrowing subject. In the business and economics schools, on the other hand, the emphasis has been primarily on a general education for the man who wants to know something about an ever-increasing number of subjects. Thus, in an exaggerated future we may visualize two extremes: the true specialist who will know everything about nothing and the man of general education who will know nothing about everything.

From an educational viewpoint systems science may be regarded as an endeavour to harmonize teaching with expanding knowledge. Scientifically, systems science is an interdisciplinary approach which by generalization aims at reconciliation of the dilemma of the increasing number of specialized subjects. Generalization requires simplification. Therefore, in systems science we strive to find out what simple regularities or uniformities characterize natural or man-made systems. Also, we attempt to discover the simplest consistent set of basic concepts and abstract models in terms of which we can talk about and understand such systems in our environment.

A basic hypothesis in systems science is that, from a general and interdisciplinary viewpoint, the number of basically different concepts and models is extremely limited. The argument behind this assumption being that our concepts and our models, possibly, are more typical of our way of thinking than of anything else. This, however, endows systems science with the additional quality that it becomes useful to discuss philosophically oriented questions about the meaning of scientific concepts and explanations. Accordingly, systems science may be said to have its historical origin in the classical discipline of natural philosophy. The new aspect brought into systems science, is that we expand the field of interest to also include man-made phenomena such as economic.

The purpose of this paper is to illustrate in terms of the network approach advocated in this volume the fundamental line of thought in the discipline of systems science. To make it simple we shall focus exclusively on seemingly trivial concepts in well-known engineering equilibrium models. To indicate the power and the possibilities of this unifying approach we shall expose the meaning of these concepts when related to neo-classical equilibrium models of economics. A more exhaustive treatment of the analogy between engineering and economics has been given elsewhere (Franksen, 1969 [1]).

2. P.O. PEDERSEN'S ELECTRICAL ANALOGUE

Often, the intuitive establishment of an analogy is the first step in the formulation of a basic set of concepts and of a corresponding, abstract structure common to two, previously separated disciplines. Historically, Pedersen's electrical analogue of a simplified economy may have been such a first step in the unification of physical and economic equilibrium models (Pedersen, 1935 [2]). The remarkable feature of

Pedersen's contribution, was that he envisaged his model as just a simple illustration of a new and powerful, scientific approach. For this very reason it will be of interest briefly to discuss the implications of Pedersen's analogue.

The Pedersen model is a rather unsophisticated economy in which commodities are manufactured in a production system for either consumption or inventory. Two dual types of variables characterize this economy. On the one hand, we have flows of commodities and orders and, on the other, we have propensities to produce, to consume, or to store goods. These propensities are related to money values in some undefined way which is not included in the model.

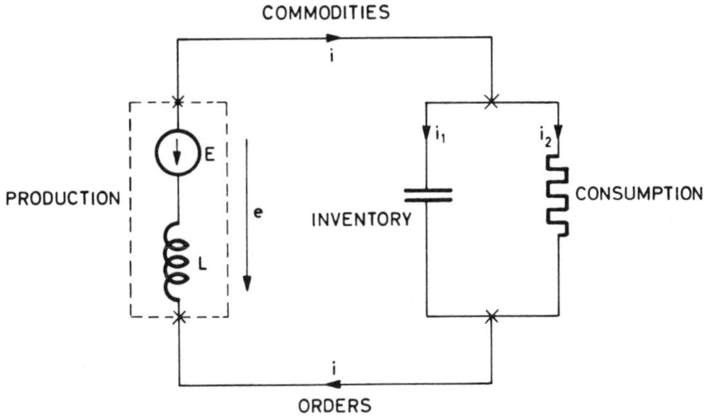

Fig. 1. P.O. Pedersen's electrical analogue of a simplified economy.

In order to study the dynamic behaviour of his economic model Pedersen established the electrical network analogy in Fig. 1. In this network the production system is depicted by a voltage source E in series with an inductance L; the consumers are represented by a conductance G; and the inventory is described by a capacitance C. The flow of commodities and orders are reproduced by electric currents i, whereas the propensities are portrayed as electric potential-differences e driving these currents.

Since electric current i is the time-derivative of dielectric charge q:

$$i = \frac{dq}{dt} \tag{1}$$

the chosen analogy has the advantage that dielectric charge q appears automatically as a unit amount of the commodity or as a single order.

Basically, the flow of commodities originates in the production system while the flow of orders originates in the inventory or with the consumers. Kirchhoff's node law for the currents:

$$i = i_1 + i_2 \tag{2}$$

depicts the simple fact that all the commodities produced are used either for inventory or for consumption. Respectively, we may also say that the total number of orders for production is the sum of orders from inventory and consumption.

In the production system the voltage source E is the propensity to produce, while the inductance L is the resistance to changes in the rate of flow of production. Since by definition the voltage drop over the inductance is:

$$L\frac{di}{dt}$$

the resulting increase in voltage over the production system, i.e. the resulting propensity to produce, is:

$$e = E - L\frac{di}{dt} \tag{3}$$

By Kirchhoff's mesh law the corresponding voltage drops over the capacitance or the conductance are also e. Thus, as we shall see, the propensities to store or to consume are directly related to the price of production underlying the propensity to produce.

In the inventory system by eqn 1 the amount stored of flow i_1 of commodities in time period t_1 to t_2 is depicted by the dielectric charge:

$$q_1 = \int_{t_1}^{t_2} i_1\, dt \tag{4}$$

By definition, from electric network theory, the capacitance C relates the stored amount, expressed by dielectric charge q_1, to the propensity to produce, represented by the voltage drop e, by:

$$q_1 = Ce \tag{5}$$

The capacitance C, in other words, expresses the propensity to store or perhaps rather the inventory capacity as some constant multiple of the propensity to produce.

Finally, in the consumer system, P.O. Pedersen assumes by Ohm's law:

$$i_2 = Ge \tag{6}$$

that the conductance G expresses the propensity to consume as another constant multiple of the propensity to produce. It is easy to visualize, as a first approximation, that the conductance G is directly proportional with the number of consumers.

After thus having formulated the electrical analogue of his economic model, Pedersen established and solved its differential equation. The immediate result of his investigation was the intuitively satisfactory observation that the system is stable, if consumption follows production, but unstable in the opposite case. More important is it, however, that he presents his analogue as an illustration of a much more fundamental approach. Basically, Pedersen says, the use of analogies offers a two-fold, interdisciplinary reward: the creation of powerful concepts and the transfer of highly-developed, computational models.

3. A REMARK ON SCIENTIFIC EXPLANATION

P. W. Bridgman, the American physicist, once said that the essence of scientific explanation consists in reducing a situation to elements with which we are so familiar that we accept them as a matter of course, so that our curiosity rests. Theories and models are compressions and reorganizations of empirical data for use. Thus, theoretical results are not conclusions deduced logically from some philosophical first principles, but conclusions drawn from centuries of experiments. However, to facilitate our handling of the growing mass of disparate, observed facts we map the empirical domains into logico-mathematical constructs. In these constructs "reducing a situation to elements" means discovering for the data, representing the measurements on the system in question, familiar relationships of which the situation is composed.

Concepts and laws are the generalized formulations of empirically found relationships. To be accepted as familiar, i.e. useful, they must be simple, precise, and explicit. Often, as Pedersen has remarked, the most successful concepts and laws may appear both awkward and secluded. Thus, for example, the Newtonian law that a particle under the influence of no force travels with constant velocity, is a complete fiction which never has been experienced in practice. To be sure, the usefulness of theoretical concepts and laws need not be imperilled by the difficulty of making decisions in borderline cases. It is useful to distinguish between day and night despite the penumbral passage through twilight.

Because of the difficulty or impossibility of making laboratory experiments in economics, there seems to be a general belief that economic concepts or laws differ basically from those of physics. The viewpoint adopted here, is that the difference is not in principle, but rather in accuracy of prediction. Either discipline is a practical study of what can be observed, and the prediction from that of what will be observed.

The establishment of an analogy between two disciplines reveals that the set of relationships, of which the situation is composed in either discipline, can be depicted by exactly the same abstract structure. Thus, it is the empirical substrate in the original data or measurements that gives each of the models its specific meaning. For this reason it seems natural to further decompose the process of explanation in an attempt to understand the meaning of concepts representing measurements like price or electrical current.

At this point, it is important to realize that it is outside the scope of science to explain "what price is" or "what electrical current is". Of course everyone is at complete liberty to put forward his opinion which by necessity will depend upon the experience, age, maturity and philosophical outlook of the individual. Clearly, we find it neither helpful nor meaningful to explain electrical current as the pushing around of electrons by blue fairies with red noses who know exactly what they are doing, only it happens that every time we look in their direction these fairies make themselves invisible. Still, it should be understood that also less naive answers are equally useless, since all they state are subjective ideas.

To paraphrase Poincaré, the important thing is not to know what electrical current is, but how to measure it (Poincaré, 1905 [3]). That is, it is necessary by instrumentation of one kind or another to relate the physical observations to the real-number system. Hence, the elements to which the situation can be reduced are associated with the characteristic properties of the measurement process itself and of its real-number result. Since measurements serve as the only link between theoretical equations and physical observations, the organized establishment of analogies must begin with some structural characteristics common to two or more, basic sets of measurements. Consequently, we shall find it useful to discuss how meaning can be given to measurements.

4. THE MEANING OF MEASUREMENTS

Formally, measurement has been defined as the assignment of numbers to observable qualities of objects or events according to rule (Campbell, 1921 [4], Stevens, 1946 [5]). It should be realized that the acceptance of this definition, makes it possible to ascribe meaning to measurements from at least two different viewpoints.

The viewpoint of *operationalism* focuses on the words: "according to rule" in the sense that meaning is given to the concepts of measurements by relating them to a well-defined activity on the part of the observer (Bridgman, 1927 and 1936 [11, 12]). That is, the measurements are qualitatively classified by an interpretation in terms of the set of operations making up the procedure of measurement in question. Basically, dependent upon the manner in which, at least conceptually, the measuring device is used, the kinds of measurements under discussion here can be divided into two distinct classes (Firestone, 1933 and 1938 [6, 7]; Trent, 1955 [8]; Koenig and Blackwell, 1961 [9]).

A *transvariable** (or across-variable) is measured by simultaneously attaching the instrument to two actual connection points of the system without cutting the interconnections in the latter. Examples of transvariables are voltage, displacement, rotation, and temperature. Note, that the one connection point can be a fixed point of reference such as is exemplified in the measurements of nodal voltages, displacement of masses in a gravitational field, and, in economics, prices.

An *intervariable** (or through-variable), on the other hand, is measured by cutting the system at an actual connection point and inserting the instrument at the point of cut. Examples of intervariables are electric current, mechanical force, torque, and heat transfer. In economics, flows of resources and commodities must be characterized as intervariables.

The viewpoint of *symbolism* takes as its basis the words "the assignment of

*The author prefers the terminology: *transvariables* and *intervariables* to across- and through-variables. The latter, original terms of Firestone (1933 [6]), are impossible to transfer, in a meaningful and yet mnemotechnically striking manner, into other languages like Scandinavian, for example. To remedy this situation he introduced about 1962 the former terminology in terms of Latin prefixes, which any European language can incorporate. As a matter of fact, even in England some authors of electrical engineering textbooks have discarded the original terminology in favour of terms like transvariables and pervariables. The latter word, though a more correct translation of through-variables, does not really catch however. It is as if we are more accustomed to words like, say, *intercontinental* and *transatlantic.* Hence, the author's conclusion is that he would like to maintain his terminology concerning these two concepts. Besides, since this volume wishes also to convey its ideas to readers who do not have English as their native language, this might give them some idea on how to introduce this terminology in their native languages.

numbers" in the sense that observed relations among empirical entities are mapped upon someone or more properties of the number system. The image in the number system is called a scale. Distinct combinations of properties of the number system, the so-called organized sets, are termed scale forms. Thus, for example, measurements of temperature in degree Fahrenheit and in degree Celsius are commonly known to exhibit the same scale form, since transformation from one scale to the other preserve the empirical information. Mathematically, we characterize scales of measurements in terms of the group of transformations that leaves the scale form invariant (Stevens, 1946 and 1959 [5, 10]).

In mathematics sets of real numbers can be organized by any combination of binary relations like "equal-to" or "greater-than" and of binary operations like addition and multiplication. Two scale forms represented by such distinct sets of properties will be of interest in the present context. Both scale forms satisfy a binary relation of quasi-serial order like "less-than-or-equal-to". The differences between the two scale forms of concern here is related to the kind of binary operations that are permissible.

The *interval* scale is determined by the fact that successive intervals of differences are of equal size at the same time as it is meaningless to talk about ratios between two measurements. Temperature, measured in degree Celsius or Fahrenheit, dates on a calendar, and electric voltage potentials are typical examples. In general, interval scales are invariant under any affine transformation:

$$x' = ax + b \tag{7}$$

where $a > 0$ defines the unit and b the arbitrarily selected zero point. In other words, the scale form is not affected by the addition of a constant. It should be noted, though, that differences between measurements on an interval scale, i.e. total lengths of intervals like periods of time or electric voltage drops, may correctly be defined as double or some proportion of each other. Thus, lengths of intervals can be measured on ratio scales.

The *ratio* scale is defined as an interval scale augmented with a determination of the equality of ratios. Hence, measurements on a ratio scale infer that all the features of the real number system carry the imprint of meaningful properties in the empirical domain. An absolute zero point is always implied by virtue of the fact that the scale form remains invariant when each value is multiplied by a constant. That is, ratio scales are invariant under any similarity transformation:

$$x' = ax \tag{8}$$

where $a > 0$ defines the unit. A large number of measurements such as voltage drops, electrical currents, geometrical lengths, and mechanical forces are measured on ratio scales.

The fact that we believe that some transformation on the numbers in the real number system will leave intact the information gained by the measurements, is tantamount to saying that, from the viewpoint of symbolism, theoretical concepts are given meaning by being derived from certain postulates, the satisfaction of which cannot be proven, but which experience indicates are in agreement with observations. The type of postulates, in which measurements on interval and ratio scales may be said to originate, can be exemplified by the following two.

The first of these two postulates is introduced in order to permit establishment of unified concepts that express the *sameness* of operationally different observables (Bridgman, 1927 and 1936 [11, 12]). The viewpoint of symbolism does not recognize that, as the physical range increases, the fundamental observables cease to exist and therefore must be replaced by other observables which are operationally quite different. An example of the point in question is the concept of length which is used to describe astronomic length, engineering length, as well as atomic length (Seeger, 1956 [13]; Brillouin, 1964 [14]).

The second postulate concerns the introduction of *continuity* or, at least, piecewise continuity. Continuity is a mathematical idealization the invocation of which measurement, by its very nature, cannot verify. The justification for introducing continuity lies in the operational discovery of conclusive evidence as to the possibility of postulating the existence of a *binary operation of addition*. Thus, it is possible to represent the theoretical concepts by symbolic entities which, at least within certain ranges, submit to the applicability of calculus.

Evaluation of operationally defined measurements, transvariables and intervariables, from the viewpoint of their interpretation as symbolic entities measured on interval or ratio scales, yields the rather significant result that it is possible to *unify* the operational and the symbolic viewpoints into a general *systems viewpoint*.

Experience indicates, namely, that direct measurement of an intervariable, for example an electrical current, a mechanical force, or a commodity flow, is always based on a ratio scale. On the other hand, direct measurement of a transvariable is normally related to the use of an interval scale. Thus, the zero or reference point in measuring electric nodal voltages or economic prices is a matter of convention or convenience. From the systems viewpoint, therefore, we relate intervariables to direct measurements on ratio scales and transvariables to direct measurements on interval scales. Of course, the computational manipulations of transvariables in the formal models are based on differences that are measured on ratio scales. For example, the voltage drop over an electric element, determined as the difference between two nodal voltages, or the increase in economic value along a production activity, determined as the difference between two prices, are such derived transvariables measured on ratio scales. In physics, transvariables measured on interval scales are generally conceived as scalar potentials, whereas intervariables, being measured on ratio scales, are apprehended as vectors.

5. ON ECONOMIC MEASUREMENTS

Probably, one of the most characteristic features of our conception of physical systems is the fact that all these systems can be described in terms of only two physical types of measurements, one of which invariably will be an intervariable measured on a ratio scale, while the other correspondingly is a transvariable measured on an interval scale. For example, in mechanics the two types of measurements will be forces and coordinates or velocities, whereas in electrical network theory it will be currents and voltage potentials. Similarly, in neo-classical economics which is the analytical approach to economic systems in equilibrium, we also confine ourselves to these two types of measurements that here are known as respectively commodity flows and prices.

From psychology and sociology it is a well-known fact that human tastes and desires exhibit a complete lack of constancy in comparison with the phenomena of physics. For this reason, the data of economics are taken to refer to possible events at just one moment in time. That is, we conceive economic measurements as the observation of instantaneous values of the variables. For instance, the quantity q demanded of a certain commodity is always denoted q/t on the corresponding demand curve (relating consumer demand of commodity to alternative prices) to emphasize the fact that the curve depicts the situation at a single point in time t. Clearly, in the limiting case the entity q/t can be conceived as a commodity flow:

$$i = \frac{dq}{dt} \tag{9}$$

in complete analogy with eqn 1 for the electrical network current. In the following, to stress this temporal quality of quantities or amounts of commodities, we shall consistently speak, in contradistinction to the usual economic practice, of flows of commodities, goods, or products. Of course, the flows in economics need not all be flows of physical entities. However, it is a straight-forward abstraction to extend this concept to also include services of different kinds such as labour or the use of a hotel room.

While the meaning of economic flows as intervariables, measured on a ratio scale, is evident from our experiences in and understanding of the physical world, the meaning of price measurements seems to elude a scientific explanation in the sense of Bridgman. To be sure, we all have an intuitive understanding of the price concept in its common sense notion as the exchange value, expressed in money terms like kroner or dollars, of a single unit of a good. Yet, many properties ascribed to the price concept appear to be both confusing and contradictory. Among other things, if we consider the monetary unit, it does seem strange that both price (i.e. exchange value for one unit good) and total value (i.e. price times quantity of the good) have the same unit: kroner or dollars. After all, we never find in physics that, for example, both displacement and displacement times force (i.e. mechanical work) are measured in the same unit.

Market economy implies a self-regulating system of markets in which all production is for sale at the same time as all income derives from such sales. That is, we consider only economic barter-type systems directed by market prices. This implies that, although money can be introduced as a standard of value for the measurement of price, it does not serve as a store of value, since it is desired only as a productive factor or a consumable commodity on the same instantaneous basis as all other goods. Therefore, the theory of money is not touched upon in this discussion. Rather, we will say that any good may be selected as a standard of value and serve as money in the sense that all prices are expressed in terms its unit. The good, arbitrarily selected as standard of value, is termed the *numéraire.*

In order to reveal the meaning of price measurements in a market economy it will be advantageous first to consider the concept from the operational viewpoint. Conceptually, it may be said that we determine price by a two-terminal measurement as a potential-like difference in the same manner as we determine position or voltage with respect to an arbitrarily selected reference. From the symbolic viewpoint, two things of importance are involved in this statement which in essence defines price as a transvariable. One of these is that the reference or zero point can in fact be chosen arbitrarily. The other is that a difference and not a ratio is the basic operation.

In modern economics a commodity is said to be *scarce, free,* or *noxious* depending upon whether the price is positive, zero, or negative (Debreu, 1959 [15]). The designation a noxious commodity or a *discommodity* refers to the costly disposal of products like industrial waste. Thus, prices are not confined to non-negative values. The fact that the reference or zero price also can be selected at will, can be demonstrated by considering problems like the transportation problem. Mathematically, this problem is a linear programming problem in which we try to minimize the cost of transportation along possible routes from a set of factories to a set of destinations. Dependent upon which location among the factories or destinations that we choose as price reference for our goods to be shipped we will end up with positive or negative prices of the goods at the remaining locations.

By virtue of the fact that the zero or reference price is arbitrary prices are measured on an interval scale. Therefore, determination of differences is the basic operation. Measurements on interval scales are characterized by the affine transformation of eqn 7. Basically, this transformation permits us to change either the zero point or the unit or both. Change of the zero point or reference alone implies in eqn 7 that $a = 1$ while b adopts the negative of the value, say k, of the new zero point measured on the old scale:

$$x' = x - k \tag{10}$$

On the other hand, change of the unit alone implies in eqn 7 that $b = 0$ while a adopts the reciprocal of the value, say m, of the new unit-commodity or numéraire measured on the old scale:

$$x' = \frac{x}{m} \tag{11}$$

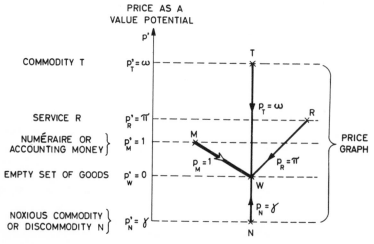

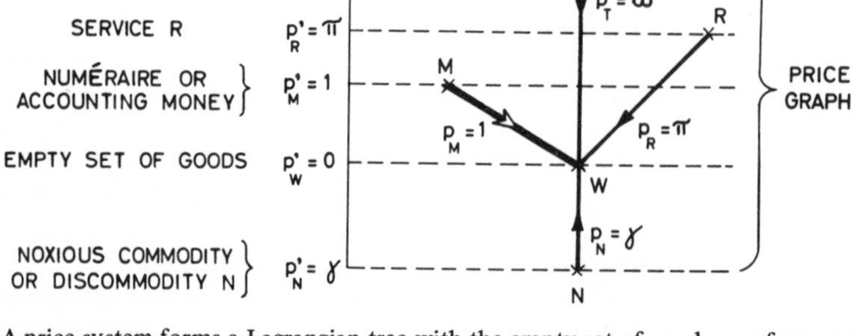

Fig. 2. A price system forms a Lagrangian tree with the empty set of goods as reference and the numéraire as unit.

That is, change of zero point or reference is based on differences, whereas change of unit or numéraire is based on ratios. Clearly, when we speak of price in the sense of exchange value, we are actually referring to the latter exchange ratio with respect to some arbitrary unit or numéraire.

However, price measurement has an additional property which usually does not characterize measurements of physical transvariables. This property is that, although the zero point or reference can be chosen arbitrarily, we all intuitively tend to select the same reference point which may be termed, by a loan from mathematics, *the empty set of goods*. Economics is the science of scarce commodities which we compete to acquire and discommodities which we compete to get rid of. All commodities or services that are plentiful or free are of no interest in the economic sense and we therefore define their price to be zero. The result of this choice is that, topologically, a price system will form a so-called Lagrangian tree, an example of which is shown in Fig. 2. The characteristic property of a Lagrangian tree is the fact that it is not possible to distinguish between price conceived as a potential measured on an interval scale and price conceived as a length or difference measured on a ratio scale. The reason being, of course, that we only consider differences with respect to the common zero point or reference in a similar manner as Newton's second law only refers to movement of masses with respect to a common reference like the earth. Obviously, it is this last mentioned property that causes price to become such an elusive concept.

6. KIRCHHOFF'S LAWS

With prices and flows of commodities and services as the two, fundamental types of instantaneous variables for an economic system, the next problem to be examined is determination of the mathematical relationships which characterize the abstract structure of each of the two sets of variables when considered independent of each other.

For a physical system we formulate the structure of its set of intervariables in terms of a *postulate of continuity*. In electric network theory this postulate says that dielectric charges cannot be created, accumulated, or destroyed at a node. The consequence of this postulate is Kirchhoff's node law which tells us that an algebraic summation of electrical currents confluent in the same node must equal zero. For a mechanical particle in static equilibrium, by Newton's first law, the postulate expresses the fact that the vectorial summation of forces on the particle equals zero. For a mechanical particle in motion the dynamic equilibrium condition duplicates this formulation if, as a consequence of d'Alembert's principle, we add the force of inertia to the other forces. Similarly, in a barter economy we require in equilibrium that each market must be *cleared*, i.e. aggregate demand must equal aggregate supply.

In general, therefore, we can conceive *Kirchhoff's node law* as the resulting relation which for a physical or economic system describes the abstract structure of its set of intervariable measurements. Mathematically, this is tantamount to saying that Kirchhoff's node law organizes the set of intervariables.

The potential-like character of a set of transvariables implies that the value, found at some point or node of the system in question relative to that at another, is independent of the route chosen in traversing the system from one of these points to another. Consider, for example, the measurement of a typical transvariable like the altitude of a peak in a mountainous terrain. By the customary methods used in surveying we do this through successively measuring the differences in altitude between appropriately chosen intermediate points extending over some route from the arbitrarily selected reference, say sea level, to the mountain peak. With negligible errors in measurement, the net difference in altitude is expected to be independent of the route chosen. Now, instead suppose that our surveying trip starts from some arbitrary reference point, traverses all over the mountainous terrain, and finally returns to the same reference point. In this case, since the end points of our excursion are identical, we would expect to find that the net difference in altitude is zero.

A formal proof of this obvious fact was published by Kirchhoff for potential differences in electrical networks as early as 1847. It is interesting to know that the proof of this mathematical relation, the so-called Kirchhoff's mesh law, was based on the use of all those topological graph-theoretical concepts which only in the past two decades have found their way into modern engineering curricula. In this connection, of course, we have to except the use of matrix algebra since its invention by Cayley had to wait until 1858.

Since Kirchhoff's proof is valid for any set of transvariables, we can in general conceive *Kirchhoff's mesh law* as the resulting relation which for a physical or economic system conveys the abstract structure of its set of transvariables.

In summary, therefore, we have found that the structure of the set of intervariables and the set of transvariables, when considered independent of each other, can be described by respectively Kirchhoff's node law and Kirchhoff's mesh law for any physical or economic system characterized by only these two types of variables.

7. THE SECOND LAW OF THERMODYNAMICS

After thus having considered, for a physical or economic system, the structural relationships between the set of intervariables taken by themselves and the set of transvariables taken by themselves, it is of course of interest to see whether any relationship can be established between the two sets of measurements. In order to investigate the nature of such a relationship, it is necessary to adopt the arbitrary convention that, every time we perform an intervariable measurement between two nodes of the system in question, we will also perform a corresponding, transvariable measurement across the said two nodes. In other words, we endow the total set of measurements on the system with an arbitrary, but complete symmetry so that to each intervariable there corresponds a transvariable and vice versa.

Now, to get an idea of the kind of relationship we are looking for let us consider a waterfall which sets a waterwheel into rotation and by means of that furnishes mechanical work. For a given amount of water in the stream experience teaches us that the force upon the waterwheel, and thereby the amount of work to be gained, depends upon the height of the fall. In his theoretical study of steam engines it was exactly by this hydrodynamical analogy that Carnot concluded that the work furnished by a steam engine was generated not by the heat as such, but by its fall. Or, as Carnot stated his principle, work cannot be derived from heat unless this heat is allowed to fall from a higher to a lower temperature.

This principle of asymmetry or irreversibility connected with heat transfer, which is known as the *second law of thermodynamics*, was later given an alternative formulation by Clausius who for this purpose introduced entropy as a new concept. In this new formulation we say that the entropy of an isolated system tends to a maximum. Later again, it was shown in statistical mechanics that, from a microscopic viewpoint, the macroscopic formulation of Clausius simply said that the *direction of change* of an isolated system is from a less random to a more random situation. Thus, the second law is not absolutely certain, but only highly probable.

From a macroscopic, interdisciplinary systems viewpoint it will be most advantageous to maintain Carnot's formulation, which for an isolated, physical system can be stated in general terms as: *a flow will occur only from a higher to a lower potential*. Hence, in electrical network theory for example, the second law expresses the experimental fact that electrical currents will flow only from a higher to a lower voltage potential.

8. A LAW OF RATIONAL BEHAVIOUR

Also in economics do we have a relationship like the second law of thermodynamics between the set of transvariable prices and the set of intervariable flows of products and services. In this case, of course, we cannot speak of a statistical formulation of a natural phenomenon. Rather, we have to introduce the principle as a statistical distribution of the rational behaviour of producers and consumers. In neo-classical economics we conceive this distribution function as a macroscopic average expressing a constant rational behaviour. For example, in so-called perfect competition it is assumed that all firms seek to maximize profits while all consumers endeavour to maximize utility.

The general experience, resulting from macroscopic observations of closed economic system under well-defined influences, is that all changes which occur have a very definite direction. Thus, in an industrial production process, it is impossible to reverse the direction of the physical flow. From the economic point of view, the flows will always occur in the direction from the resources or production factors and to the demands. The use of economic terms, like input and output, illustrates the point. But more can be said. The fundamental problem in economics is the best allocation of limited means toward desired ends. The latter are demands as function of prices. From a macroscopic point of view, in a closed economy, it will not make sense to attempt a production if the corresponding flow of products has a higher price at the resources than at the demand.

With the transvariable prices conceived as economic potentials, therefore, it is possible to formulate the conception of rational behaviour as a statistical principle of product flow orientation which, apart from a change of sign for the flow direction, is completely analogous to Carnot's formulation of the second law of thermodynamics. In words, this law of rational behaviour for a closed economic system can be stated: *a flow will occur only from a lower to a higher price.*

From information theory it is a well-known fact that the irreversible increase of entropy in an isolated, physical system must be interpreted as nature's tendency towards the most random state of disorder. The purpose of an economic production system is to encounter this tendency. That is, the aim of a production system is to create order. It does make sense, therefore, that the economic analogy to the second law of thermodynamics depicts situations which physically may be illustrated by, say, the orientation of the current inside a voltage source.

9. SYMMETRY AS A PRINCIPLE OF FORMULATION

Symmetry has two meanings, an intuitive, particular, geometric meaning, and an abstract, general, algebraic meaning. Bilateral symmetry, the symmetry of left and right, is a familiar illustration of the concept in its first mentioned sense. In the last mentioned sense, a whole is termed symmetric if it has interchangeable parts. There are many kinds of symmetry in this abstract acceptance of the word. They differ in the number of interchangeable parts, and in the operations which exchange the parts. Algebraically, we describe these different kinds of symmetry in terms of structural properties which remain invariant under certain mappings of any set of elements. Thus, the abstract idea of symmetry is a refined, but general concept derived from the common sense notion of harmony of proportion.

The advantage of organizing physical observations into the rational patterns of modern algebra, is the fact that structural similarities rather than differences become emphasized in the formulation. Indeed, the recognition of a structural symmetry is of vital importance in any situation of *a priori* physical prediction. Archimedes, for example, concluded *a priori* that equal weights balance in scales of equal arms. Similarly, in statistical investigations, *a priori* probabilities are assigned as a consequence of the symmetry in the pertinent experimental situation. Just consider the age-old pastime of flipping a coin or rolling a die. Possibly, it might be said that all *a priori* statements in physics have their origin in symmetry considerations. As Weyl has expressed it: "If conditions which uniquely determine their effect possess certain symmetries, then the effect will exhibit the same symmetry" (Weyl, 1952 [16]). Obviously, observed deviations from a predicted symmetry are clues of paramount importance in the search for an improved physical insight and understanding.

From a symbolic viewpoint the theoretical formulation of fundamental relationships or laws in physics or economics is based on the discovery of structural properties that are invariant under certain well-defined changes. Mathematically, to bring out this invariancy of structure or form we aim at the largest possible amount of symmetry in the abstract formalism, completely disregarding the fundamental role played in experimental practice by the direct measurements making up the basis of the individual unit systems.

For example, when James Clerk Maxwell formulated his theory about 1860, the laws of electrodynamics admitted up to his time accounted for all known facts. Maxwell did not perform a new experiment which came to invalidate them. As a matter of fact, Maxwell's laws had to wait nearly thirty years for an experimental confirmation. How it is, then, that he was able to be so many years ahead of experiment. It was because he was profoundly steeped in the sense of mathematical symmetry. Looking at the laws of his time under a new bias, Maxwell saw that the equations became more symmetrical when a term was added, and besides, this term was too small to produce effects appreciable with the old methods. Thus, the symmetry in the formulation revealed to him the hidden harmony of the empirical world in making him see it in a new way.

The aim of the previous sections was to emphasize the symmetry in the abstract modelling of physical and economic systems. Clearly, for such a system, we can obtain a complete symmetry between the set of intervariable measurements and the set of transvariable measurements. In particular, we find that the sets of relationships on these two sets, i.e. Kirchhoff's laws, are symmetrical or, as the more technical term is, *complementary* to each other.

This complementarity or symmetrical relationship between the two sets of fundamental variables was emphasized qualitatively by the mere existence of either the second law of thermodynamics or the law of rational economic behaviour. The problem now, is to use this complementary relationship in the formulation of basic laws in the quantitative sense. In other words, we want to investigate which binary operations are meaningful on a pair of measurements, consisting of one transvariable and one intervariable.

Obviously, it does not make sense to add electric current to electric voltage for an electrical network, or to subtract the number of carrots from the price of carrots. Therefore, the only meaningful operations by means of which we can combine a transvariable from a system with an intervariable from the same system are those of multiplication and its converse operation: division. Actually, the theories of physics and neo-classical economics are nothing but a simple and symmetrical play upon these two possibilities.

Analytical mechanics or, in more modern terms, energy conversion and the state-function methods of control theory are based upon multiplication, since the *product of a transvariable and an intervariable* is energy or power, i.e. rate of change of energy. On the other hand, electric network theory and vectorial mechanics are based on division since the *ratio between a transvariable and an intervariable* is the essence of laws like Ohm's and Hooke's. In the following section we shall see that these two possibilities also make up the foundation of economic theory.

10. ELASTICITY OR OHM'S LAW IN ECONOMICS

The development of physics as a science, rather than as a taxonomy of empirical interdependencies was due to their abstract representation in terms of mathematical relations among mathematical elements, subject to manipulation by algebraic and differential processes. Therefore, the modern student, in a world well supplied with calibrated apparatus and recognized systems of measurement, naturally considers experimental research to involve the operations of making measurements on unknown quantities and then expressing their relations to known quantities by appropriate equations.

However, a perusal of the writings of the earlier classical workers in any field shows that their results were usually stated merely as constant ratios or, shall we call it instead, direct proportionalities. Thus, Newton wrote: "The alteration of motion is ever proportional to the motive force involved"; Ohm: "The voltage drop across a copper wire of given thickness and length is proportional to the current through it"; Hooke: "The amount by which one terminal of a spring is displaced with respect to the other is proportional to the applied force"; and, among a multitude of other examples, the definitions of inductance and capacitance introduced in the previous discussion of Pedersen's analogue.

Also in economics the endeavour has been to establish equally simple relationships between corresponding pairs of a transvariable and an intervariable parameter. It should be realized, however, that the observations in physics of direct proportionalities were extremely lucky coincidences without which, possibly, modern science would not have existed. In economics such simple, linear relationships between a transvariable price and an intervariable flow of commodity are very seldom observed in practice. Consider, for example, the demand for a good as a function of its associated prices at a given market at a particular point in time. In a linear diagram with price as ordinate and flow of commodity as abscissa such a demand schedule will usually be a downward sloping curve like a negative power function. Similarly, in the same kind of diagram the supply curve, depicting the relationship between different quantities of a commodity that sellers will offer at different prices at a given market and at a given unit of time, will most often be an upward sloping curve like a positive power function.

The fact that demand and supply curves look like power functions in a diagram with linear scales on both axes, makes it natural for us instead to try to map such curves on double logarithmic diagrams in our search for a relationship of the direct proportionality type. Thus, for example, the demand curve:

$$i = 10p^{-2} \tag{12}$$

shown in the linear diagram of Fig. 3A becomes the straight line with the gradient −2 shown in the double logarithmic diagram of Fig. 3B.

In economic theory the straight line representation on double logarithmic diagrams of relationships like demand and supply is the direct proportionality counterpart to physical laws like Ohm's and Hooke's. That is, just as the straight line relationship on a linear diagram, known as Ohm's law, tells us how sensitive the electric current through a resistor is to voltage changes across it, so does the economic straight line relationship on double logarithmic paper tell us how sensitive demand or supply in a market is to price changes.

In electrical network theory, if in accordance with the previously used terminology we designate by e the transvariable voltage drop across and by i the intervariable current through a two-terminal element, then by Ohm's law in its generalized form the gradient in the linear diagram:

$$Z = \frac{de}{di} \tag{13}$$

is called the *impedance*, whereas its reciprocal:

$$Y = \frac{di}{de} \tag{14}$$

is termed the *admittance*.

Although we do not use the designation "a law" to describe demand or supply in economic theory, it has been found equally useful also here to introduce a special

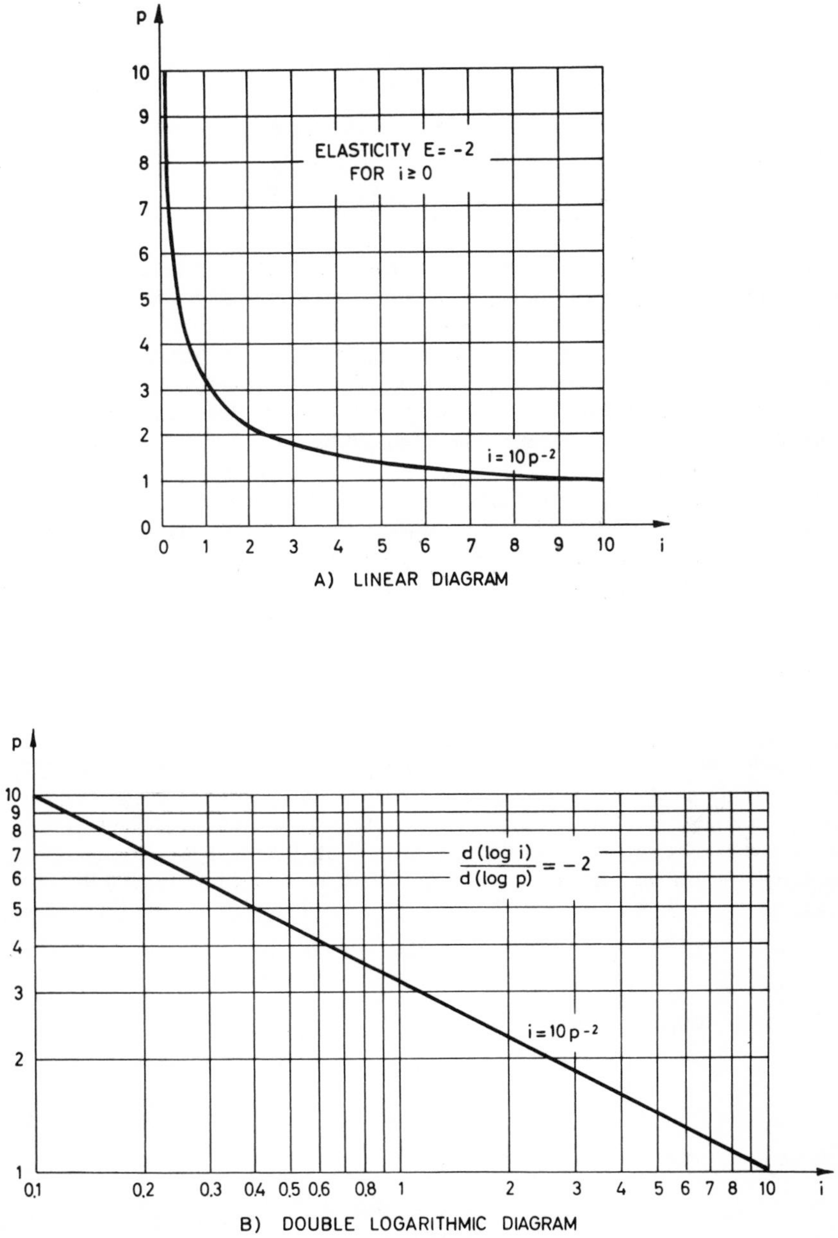

Fig. 3. Elasticity of the Falling Power Function: $i = 10p^{-2}$.

terminology to describe the gradient in a double logarithmic diagram of the relationship between the transvariable price p and the intervariable flow i of commodity. Thus, corresponding to the admittance concept we introduce the concept of *point elasticity*, or just *elasticity*, of price:

$$E = \frac{d\,(\log i)}{d\,(\log p)} \tag{15}$$

and, corresponding to the impedance concept, its reciprocal the so-called *flexibility* of price:

$$F = \frac{d\,(\log p)}{d\,(\log i)} \tag{16}$$

Evidently, as can be seen by comparison of eqns 14 and 15 respectively eqns 13 and 16, the difference between, say, Ohm's or Hooke's law of physics and the elasticity concept of economics is merely in the practical matter of providing a straight line diagram which exhibits the direct proportionality between the transvariable and intervariable involved. Of course, either of the definitions in eqns 13–16 can be used also in more generalized cases than the straight line relationships, but from the systems viewpoint that is considered simply as an additional refinement of the original core concept.

Thus, after having established that the conceptual foundation is the same in physics as well as in economics, let us consider the differences originating in the practical formulation for use in either discipline of the concept.

In electrical network theory, since the impedance Z is the gradient of a straight line, we usually just consider a single point of the straight line for its determination:

$$Z = \frac{e}{i} \tag{17}$$

and conversely for the admittance Y.

In economics, on the other hand, it is normally preferred to transfer the expressions for price elasticity and price flexibility from the mapping of the pertinent curve on a double logarithmic diagram into a mapping upon a linear diagram. For the sake of illustration let us consider the price elasticity concept in eqn 15. Clearly, this equation assumes that the flow i of commodities is a continuous and differentiable function:

$$i = i(p) \tag{18}$$

of the price p.

Now, if in eqn 15 we put:

$$v = \log i \tag{19}$$

$$u = \log p \tag{20}$$

then we can rewrite the point elasticity as:

$$\frac{d\,(\log i)}{d\,(\log p)} = \frac{dv}{du} \tag{21}$$

Obviously, eqn 19 tells us that v is a function of i which again, by eqn 18, is a function of p. Further, by eqn 20 u is also a function of p. Hence, we can rewrite the right-hand side of eqn 21 into:

$$\frac{dv}{du} = \left(\frac{dv}{di} \cdot \frac{di}{dp}\right) \bigg/ \left(\frac{du}{dp}\right) \tag{22}$$

From eqn 19 we find:

$$\frac{dv}{di} = \frac{1}{i} \tag{23}$$

whereas from eqn 20 we find:

$$\frac{du}{dp} = \frac{1}{p} \tag{24}$$

By inserting the results of eqns 23 and 24 into eqn 22 we see that *price elasticity* of a curve, transformed into a linear diagram, can also be written:

$$E = \left(\frac{di}{i}\right) \bigg/ \left(\frac{dp}{p}\right) \tag{25}$$

Similarly, *price flexibility* of a curve in a linear diagram can be written as the reciprocal of the elasticity:

$$F = \left(\frac{dp}{p}\right) \bigg/ \left(\frac{di}{i}\right) \tag{26}$$

At this point, it should be added, our systems oriented approach runs counter to the historical development. Actually, it was by adopting eqn 25 as a definition that the British economist, Alfred Marshall, introduced the elasticity concept into economics. However, in bridging the gap from one discipline to another more insight can be obtained by letting the concepts originate in eqns 15 and 16.

But, if Alfred Marshall did not visualize the straight line picture of the double logarithmic diagram, how was he inspired, then, to define the price elasticity concept? Clearly, he was looking for an analytical tool for calculating the degree of sensitivity of the demand or supply of a commodity to a change in its price. The first measure which nearly automatically comes into mind in such a situation, is the gradient of the demand or supply curve. In physics, as we have seen in the case of Ohm's law, this measure works quite well indeed. In economics, on the other hand, the outcome of using this measure is not nearly as satisfactory on account of two reasons.

First, the absolute value of the gradient is critically dependent upon the unit of measurements used on the axes. In physics, this problem is solved by introducing a standard unit like Ohm (i.e. Volts/Amperes) for the gradient. In economics, however, it has not been possible to introduce a similar standard unit for the gradient, since flows of commodities are measured in different units like whiskey in quarts and land in acres, even if we may agree upon a given currency, say dollars, as the standard measure of prices.

Secondly, the problem extends beyond the dissimilarity of units, because, also in the measurement of price changes, the magnitudes are not readily comparable. Consider, for example, a ten-kroner fall in price, respectively, of a bottle of schnapps and of a car. We would not be at all surprised to find the schnapps sale booming, but it is difficult to believe that it would have any influence what so ever on the car sales. Still, even if the absolute measure of the gradient would yield a much larger number in the case of schnapps than in that of cars, we would surely hesitate to conclude from this that the demand for schnapps was significantly more sensitive to price changes.

To by-pass these two problems Marshall concluded that an appropriate measure of responsiveness of demand (or supply) to price changes should employ relative (e.g. percentage) rather than absolute change figures. In retrospect, however, we realize that it was exactly those structural properties, which remain invariant in the transformation from the linear to the double logarithmic diagram, that guided Marshall to his definition of the elasticity concept.

Clearly, Marshall's definition of price elasticity as the percentage change in commodity flow divided by the percentage change in price, implies that elasticity E is a pure number. Because of the inherent inaccuracy in the establishment of demand and supply curves, economists usually prefer to compare price elasticities of different commodities on a more qualitative basis. That is, economists use the elasticity concept to distinguish between three cases of sensitivity to price changes:

1. *Inelastic* demand or supply occurs when a small relative price change is accompanied by a *proportionately smaller* relative change in quantity of the commodity taken or offered. That is, numerically, the elasticity is less than unity:

$$|E| < 1 \tag{27}$$

2. *Unitary elastic* demand or supply is the borderline case when a small relative price change is accompanied by a (proportionately) *equal* relative change in quantity of the commodity taken or offered. That is, numerically, the elasticity is equal to unity:

$$|E| = 1 \tag{28}$$

3. *Elastic* demand or supply is the remaining situation when a small relative price change is accompanied by a *proportionately larger* relative change in quantity of the commodity taken or offered. That is, numerically, the elasticity is greater than unity:

$$|E| > 1 \tag{29}$$

As an illustration of the use of this terminology let us consider the idealized, straight line demand curve shown in the linear diagram of Fig. 4A. Transformation of this curve into the double logarithmic diagram of Fig. 4B readily reveals that, above the midpoint price, demand is elastic; below it demand is inelastic; and at it demand is unitary. Thus, we cannot infer the degree of elasticity from inspection of the linear diagram alone since, usually, demand and supply curves change elasticity from point to point.

The use of numerical values for the elasticity measures in this classification scheme is one of those charming idiosyncrasies which for historical reasons characterizes economics as a science. The great classical economists abhored negative numbers which to them were tantamount to losses and thereby failure. Now, demand curves are usually negatively inclined, like the idealized one of Fig. 4A, with the result that demand elasticity becomes negative, since price change and

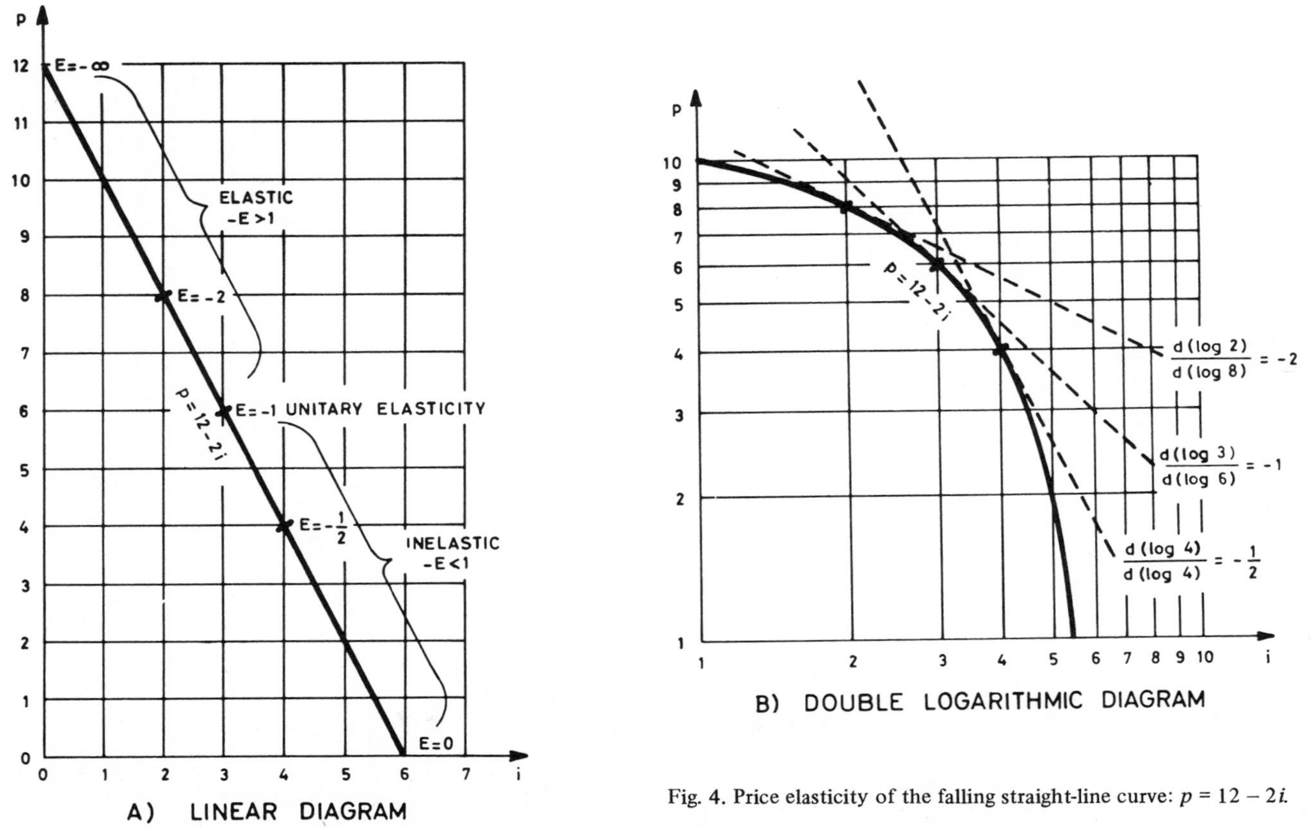

Fig. 4. Price elasticity of the falling straight-line curve: $p = 12 - 2i$.

quantity change take place in opposite directions. To overcome this emotional disturbance it became customary to ignore signs and refer only to the numerical values of the elasticity magnitudes. And that tradition has been kept up ever since in economic literature.

11. A REMARKABLE RESEMBLANCE

In the Introduction it was postulated that, above all, it was the build-in constraints on our way of thinking which determined the theoretical foundation of our concepts and models. This hypothesis, of course, cannot be proved although its likelihood may be demonstrated.

In computer programming it is a well-known fact that it is not enough to test a subprogram or a subroutine by some typical numerical examples. It is also necessary to verify that the program in question can handle the extreme cases. The topic of the previous section was the fundamental properties characterizing the analogy between Ohm's law and the price elasticity concept. Hence, let us now examine this analogy in the light of some extreme applications.

The linear diagram formulation of Ohm's law is one of those geometrical simplifications which, typically, our imagination squeezes for its innermost, conceptual structure by carrying its application into the extremes. Empirically, to being with, Ohm found an approximate linear relationship, a direct proportionality, between the current through and the voltage across a two-terminal, electrical network element. By mapping this relationship on a linear diagram with the current as abscissa and the voltage as ordinate a positively inclined, straight line characteristic through the origin appeared. In this characteristic, which may look like the graph of Fig. 5A, the positive gradient is known as the impedance. Later, theorists concluded by generalization that it also must be possible to introduce negative impedances corresponding to the graph of Fig. 5B. But the human mind is inquisitive. Would it not be possible to imagine characteristics parallel with one of the axes. And, thus, the concept of idealized sources was born as extreme cases of Ohm's law.

A characteristic, parallel with the current axis like the graph of Fig. 5C, is known as an *ideal voltage source*, since it fixes the voltage at a given value independent of the current. Evidently, by eqn 13 the impedance Z of a voltage source is zero whereas, by eqn 14, its admittance Y is infinite. Similarly, a character, parallel with the voltage axis like the graph of Fig. 5D, is known as an *ideal current source*, since it fixes the current at a given value independent of the voltage. Clearly, by eqns 13 and 14, we find for a current source that its impedance is infinite and its admittance zero. The theoretical power of these idealized concepts is based on the fact that any other straight line characteristic, like the graphs of Figs 5E and F, can be established by superposition of the characteristics of these four simple, but extremely idealized concepts.

It is quite remarkable that exactly the same idealized characteristics, independently have gained merit as fundamental concepts in economics too. To see this let us again consult Fig. 5 which, apart from the different graphs, also contain calculations of the elasticities and the flexibilities by, respectively, eqns 25 and 26.

It has been demonstrated earlier that, normally, the elasticity (or flexibility) is

	GRAPH	FUNCTION	di/dp	p/i	ELASTICITY E	FLEXIBILITY $F = 1/E$	UNITARY ELASTICITY
A	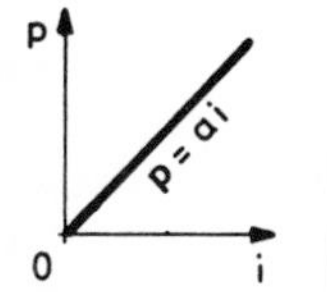	$p = ai$ OR $i = p/a$	$1/a$	a	$+1$	$+1$	$E = +1$ FOR $i \geq 0$
B	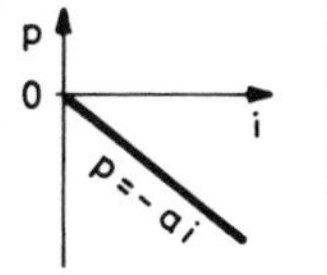	$p = -ai$ OR $i = -p/a$	$-1/a$	$-a$	$+1$	$+1$	$E = +1$ FOR $i \geq 0$
C	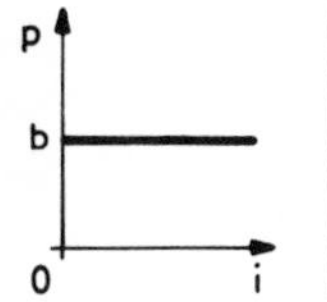	$p = b$	$+\infty$	—	$+\infty$	0	PERFECTLY ELASTIC
D	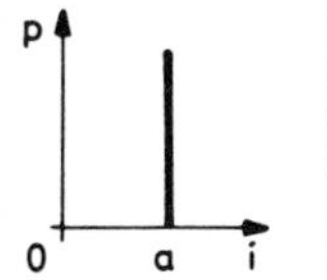	$i = a$	0	—	0	$+\infty$	PERFECTLY INELASTIC
E	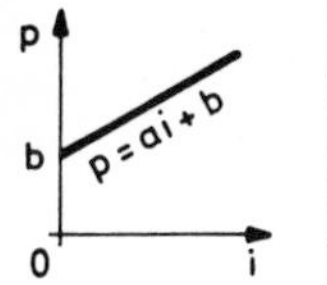	$p = ai + b$ OR $i = (p/a) - (b/a)$	$1/a$	$\frac{ai+b}{i}$	$\frac{ai+b}{ai}$	$\frac{ai}{ai+b}$	$E = -1$ FOR $i = -b/2a$
F	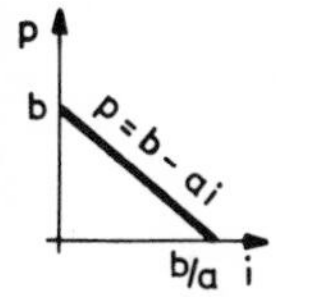	$p = b - ai$ OR $i = (b/a) - (p/a)$	$-1/a$	$\frac{b-ai}{i}$	$\frac{ai-b}{ai}$	$\frac{ai}{ai-b}$	$E = -1$ FOR $i = b/2a$

Fig. 5. Price elasticity and flexibility of straight-line curves.

not constant along a straight line curve. Therefore, the constancy of elasticity along the straight line curves through origin in Fig. 5A and B, emphasizes the special nature of these characteristics. And indeed, it does it in such a manner as to make us visualize the elasticity (or flexibility) concept as a further generalization of Ohm's law.

Before we introduce the exceptional cases of Figs 5C and D it is worthwhile to remember, as mentioned earlier, that elasticity corresponds to admittance whereas flexibility corresponds to impedance. The horizontal curve in the graph of Fig. 5C is *perfectly elastic* in complete accordance with the fact that an ideal voltage source has infinite admittance. Similarly, the vertical curve in the graph of Fig. 5D is designated *perfectly inelastic* in complete agreement with the fact that an ideal current source has zero admittance. In economics these two counterparts of the idealized sources of electrical network theory play equally important roles as fundamental concepts as do the idealized sources in physics. Thus, for example, the linear programming technique is based on an economic model of production in which the supply curves are perfectly inelastic, corresponding to limited production factors, whereas the demand curves are perfectly elastic, corresponding to the demand faced by a single producer in a perfectly competitive market.

Figures 5E and F yield us the algebraic expressions for the elasticity and the flexibility of respectively, a supply curve, corresponding to the use of a factor with a certain "entrance fee', and a demand curve of a type which has already been discussed in connection with Fig. 4. A complete summary of the terminology, introduced so far in connection with the price elasticity concept, is given in Fig. 6.

After this discussion of the similarity between, on the one hand, ideal voltage

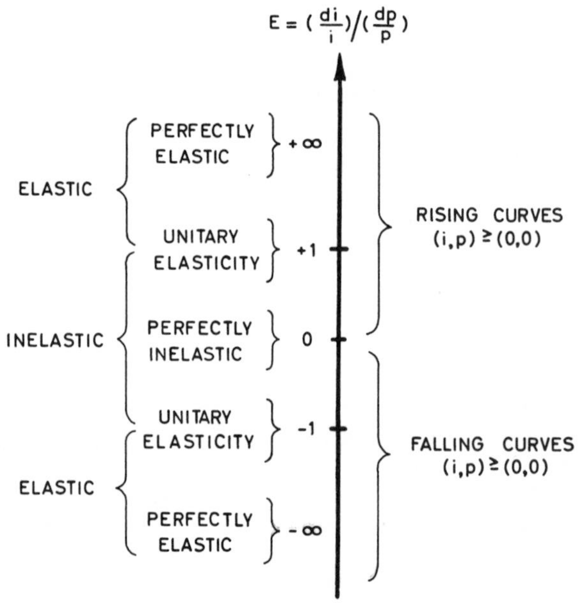

Fig. 6. Terminology of the price elasticity concept.

and current sources in electrical network theory and, on the other, perfectly elastic and inelastic demand and supply curves in economics, let us instead focus the interest on another extreme application of the analogy between Ohm's law and the price elasticity concept.

In 1820 Professor H. C. Ørsted, founder and first president of the Technical University of Denmark, discovered an interesting phenomenon while demonstrating Volta's new battery for a group of students at the University of Copenhagen. In a hurry he took a note on his observation on the back of an old envelope. Fortunately, he did not mislay this note in a mood of professorial absent-mindedness, but remembered it and wrote a paper about it so that today, from all textbooks on electricity and magnetism, we can learn that an electrical current set up a magnetic field about the wire in which it is carried. The time was ripe, then, for the French physicist, André-Marie Ampère, to show that two parallel wires carrying currents behave like two magnets by virtue of the interaction between their respective magnetic fields. It remained, however, for a self-educated, ex-bookbinder's apprentice, Michael Faraday, who was working in England, and a schoolmaster, Joseph Henry, of the Albany Academy of New York, to discover the remaining essential link between electricity and magnetism, namely, that an electrical current will be induced in a wire if its surrounding magnetic field is varied.

The essence of the effort of these eminent scientists can best be illustrated, perhaps, by considering a conventional two-winding transformer. With an a.c. current through the one winding an a.c. voltage will be produced on the terminals of other winding. That is, in the transformer a direct proportionality will be established between the current through the one winding and the voltage across the other. Let us assume that both windings are placed in identical manners with respect to their orientation relative to the magnetic core. Then, if both windings are right-hand screws or both are left-hand screws, fields and currents will have the same positive orientation in either of them with the result that the constant of proportionality is positive. Alternatively, if one winding is a right-hand screw and the other is a left-hand screw, the constant of proportionality is negative. The thus introduced constant, which may be called M, is termed the *mutual inductance* in order to emphasize that it is a generalization of the self-inductance L introduced in connection with eqn 3.

The important thing to note here, is that the mutual inductance is the resistance to changes in the current of a winding B as perceived from the terminals of another winding A:

$$e_A = M_{AB} \frac{di_B}{dt} \tag{30}$$

where the positive or negative sign of the constant M tells us whether the windings work together or oppose each other. Definitely, from this point of view, the concept of mutual inductance is another extreme generalization of Ohm's law in its impedance formulation.

The economic analogy to this utmost application of Ohm's law is the concept of *cross elasticity* of demand and supply. Evidently, since elasticity is an admittance analogy, the cross elasticity establishes the converse relationship of that of the mutual inductance. Hence, while the mutual inductance is used to establish the change in the transvariable, electric voltage across winding A as a function in the change of intervariable, electric current through winding B, we find that the cross elasticity concept yields us the relationship between the relative change in the intervariable flow of a commodity A in response to the relative change in the trans-variable price of another commodity B. Basically, therefore, cross elasticities are used when the price of one commodity effects the quantity demanded of another commodity.

Let us assume that the demand for any of a set of goods, A, B, C, etc., is a function of all market prices. Thus, for good A the demand can be stated:

$$i_A = f(p_A, p_B, p_C, \ldots) \tag{31}$$

From this function we find that the demand of, say, A in response to price changes of, say, B with other prices remaining constant, defines the cross elasticity:

$$E_{AB} = \frac{\partial(\log i_A)}{\partial(\log p_B)} = \left(\frac{\partial i_A}{i_A}\right) \Big/ \left(\frac{\partial p_B}{p_B}\right) \tag{32}$$

From a purely logical viewpoint and completely disregarding the assumption of eqn 31, it may be mentioned in passing that goods A and B are said to be *independent*, if E_{AB} is zero. More interesting than this special situation is it, however, to consider the two cases where, in complete analogy with the two possibilities for the mutual inductance, the cross elasticity is either positive or negative.

Goods A and B are said to be *substitutional* or *competitive*, if the cross elasticity E_{AB} is *positive*. For example, an increase in the price of pork would bring with it an increase in the quantity demanded of beef. Alternatively, a decrease in price would be accompanied by a decrease in quantity demanded. Pork and beek, therefore, are substitutes.

Goods A and B are said to be *complementary*, if the cross elasticity is *negative*. For example, a rise in the price of coffee would bring with it a decrease in the quantity demanded of cream. Thus, coffee and cream are complements. Another illustration originates in the Danish drinking habits of a non-alcoholic beer in the early 1920s (Fog and Rasmussen, 1964 [17]). The habit was to drink this light ale together with schnapps. The price of the last mentioned, however, had some tremendous increases during the years 1917–20 as a result of a temperance legislation which, in the politically most pleasant manner, clothed financial needs with moral necessities. The consequences of this legislation were not only a drastic decrease in the sale of schnapps, but also that the demand for this type of beer was more than halved.

In summary, investigation of extreme applications of the analogy between Ohm's law and the price elasticity concept revealed the following two further resemblances between physics and economics. First, we found that ideal voltage and current sources correspond to, respectively perfectly elastic and perfectly inelastic demand and supply curves. Secondly, we discovered that the converse of the mutual inductance concept, corresponded to the cross elasticity concept in such a manner that a positive mutual inductance matched substitutional goods and a negative mutual inductance conformed with complementary goods.

12. HOW TO USE THE ELASTICITY CONCEPT

To most engineers it is a well-known fact that Kirchhoff's mesh and node laws together with Ohm's law make up the complete mathematical model of an electrical network. That is, the solution to the system of equations, depicting these three laws, yields us the state of the network in question.

Evidently, since elasticity (or flexibility) is only a relative and dimensionless measure we cannot expect it to furnish us with any absolute state of the economic system under consideration. After all, the structural refinement from the linear diagram of Ohm's law to the double logarithmic diagram of the elasticity concept, must definitely constrain our possibilities of application of the last mentioned concept in relation to the first mentioned. But, if the analogy between Ohm's law and the elasticity concept ceases to work at this most important point what, then, is the use of the elasticity concept?

To see that the elasticity concept is still useful, albeit in a more narrow sense than Ohm's law, let us consider how a producer can utilize the elasticity of demand concept. Clearly, his aim is to make more profit. The problem is how to do it. For example, shall he cut his price on the goods produced in order to sell more, so as to make more profit. Or, will a price cut result in a decrease of his total revenue. Basically, the correct answer to problems like these involves application of the elasticity concept.

Let us assume the functional expression of the pertinent demand curve to be single-valued, continuous, and monotonic decreasing so that, instead of eqn 18, we can write it conversely as:

$$p = p(i) \tag{33}$$

By definition, at a given unit of time, *total revenue TR* is price times quantity:

$$TR = p \cdot i \tag{34}$$

The gradient of total revenue with respect to changes in the flow i of commodities is known as the *marginal revenue MR:*

$$MR = \frac{d(TR)}{di} = \frac{d(p \cdot i)}{di} = p + i \cdot \frac{dp}{di} \tag{35}$$

Introducing in this expression for MR the price elasticity E from eqn 25 yields:

$$MR = p \cdot \left(1 + \frac{1}{E}\right) \tag{36}$$

This is an important result, because it tells our businessman how his total revenue changes as a result of his price cut from a given working point W of his demand curve:

1. It demand is *inelastic* $(0 > E > -1)$ at point W, he finds:

$$MR_W < 0$$

   which indicates that total revenue TR decreases as a result of the price cut.
2. If demand is *unitary elastic* $(E = -1)$ at point W, he finds:

$$MR_W = 0$$

which indicates that total revenue *TR* has a stationary (usually a maximum) value.

3. If demand is *elastic* ($E < -1$) at point *W*, he finds:

$$MR_W > 0$$

which indicates that total revenue *TR* increases as a result of the price cut.

Thus, to maximize his total revenue, our friend the businessman should raise his price if demand is inelastic, but cut it if demand is elastic.

13. THE FIRST LAW OR THE LAW OF CONSERVATION

Previously, in our discussion of symmetry as a principle of formulation, it was mentioned that analytical mechanics and the state-function methods of modern control theory are based upon *inverse proportionality relationships* between a corresponding pair of a transvariable and an intervariable measurement. The result of such a two-factor product is always an additive scalar of the dimension energy or power.

It is a well-known fact that the so-called *first law of thermodynamics* or, equivalently, the law of *energy or power conservation* of analytical mechanics is based exclusively on summation of scalars of this origin. The idea that, in an isolated universe, there must be an invariant entity, changeable in form but indestructable, was first recognized in 1693 by Leibnitz in the study of simple mass points in a terrestrial gravitational field. As additional physical types of systems were considered, the established form of the conservation principle repeatedly failed. In each case, however, it was found possible to revive it by the addition of "a new kind of energy", which was perceived as the contribution of the remaining and, up to then, unknown part of the universe.

From a systems point of view the principle of conservation is conceived as our only way of expressing the *constraint equation among sub-systems of different physical kinds.* From this point of view an electric motor, for example, consists of three different subsystems: an electrical, a mechanical, and a thermodynamic. These subsystems are constrained together by the conservation principle saying that, per unit time, the input of electrical energy equals the output of heat, due to friction and losses, and mechanical work.

The usefulness of the law of conservation originates in the fact that, by suitable differentiation or integration of this law, we can establish state-functions which, like the virtual work or the Lagrangian function, together with constraint equations of the Kirchhoff's mesh law type yields us the state of the system. Thus, by this approach we preserve constraint equations like Kirchhoff's laws, but substitute Ohm's law by a suitable state-function derived from the principle of conservation.

It is interesting to note that this alternative to the formulation in terms of Ohm's law is equally valid in economics. As a matter of fact, the whole concept of mathematical programming which originates in economics, is nothing but an organized extension of the principle of virtual work (Franksen, 1969 [1]). The purpose here, however, is only to indicate that also in economics do we have a fundamental principle of conservation which in all respects is analogous to that of physics.

To see this, we must first of all realize that the product of an intervariable flow and a transvariable price is an additive scalar like the total revenue *TR* of eqn 34. Secondly, we find that an isolated economic system, a so-called *closed economy*, submits to a conservation principle which appear as a constraint equation in terms of such additive scalars. As an illustration let us formulate this principle for a market in perfect competition.

Considering the perfectly competitive market earlier economists discovered an important identity which since has come to be called *Walras' law* (Dorfman *et al.*, 1958 [18]; Baumol, 1965 [19]). Any person who demands a commodity is, by definition, prepared to supply, in exchange, an amount of money or other commodities of equal value. Similarly, anyone who supplies some amount of goods on the market demands, in exchange, its value equivalent in money or other commodities. Thus, each person's demands and supplies are subject to a budget constraint which says that outlays on goods equals income from factor services. Since this is true for each individual separately, it is true for the aggregate. In words, therefore, Walras' law can be stated to say that, for a closed economy, the total money value of all items supplied must equal the total money value of all items demanded. According to Baumol (Baumol, 1965 [19]), this relationship, which is "little more than an accounting relationship", must hold independently of the state of the system – "whether it is in equilibrium of disequilibrium".

In economic theory, assuming constant returns to scale, Walras' law is often deduced as an equilibrium condition from the Walrasian system of equation, applied to a perfectly competitive market in equilibrium. The point of view taken here is that *Walras' law expresses a principle of conservation for a closed economy.* In the words of Baumol: "It is difficult to imagine an economy in which this law does not hold." Hence, we can consider Walras' law as the economic analogy to the first law of thermodynamics or, equivalently, the energy conservation principle.

14. CONCLUSION

The aim of this paper has been to elucidate the general philosophy underlying the interdisciplinary approach of modern systems science. A basic assumption of systems science is the idea that, at the abstract core-content level, the number of fundamentally different concepts and models is rather limited. Thus, even if we in this paper have touched upon a diversity of fields: thermodynamics, classical mechanics, electrical network theory and neo-classical economics, we found that, conceived as an abstract pattern, a single equilibrium model sufficed to describe an amazingly wide span of the theoretical endeavour in each of these fields.

The establishment of this interdisciplinary model originates in the fact that our universe of discourse can be partitioned into two complementary sets of measurements: intervariables like mechanical forces, electrical currents, and flows of heat or commodities; and transvariables like displacements, electrical voltages, temperatures, and prices. Basically, the idea that a universe of discourse consists of something, which we arbitrarily call a *system*, and the remaining part, which we consequently call its *environment*, is founded on this complementary bipartitioning of the original set of measurements. Thus, for example, in scheduling or information retrieval problems where we do not have such two complementary sets of measurements, it is not useful to talk about a system and its environment.

The next step in the establishment of the model is to determine the relationships which characterize, respectively, the set of intervariable measurements and the set of transvariable measurements, when each of these two sets of the given universe are considered independent of each other. The intervariables of the one set are all measured on a ratio scale; and the relationship among them is based upon a postulate which may be called a generalized form of Kirchhoff's node law. The transvariables of the complementary set, on the other hand, are all measured on an interval scale with the result that it is possible to prove, as Kirchhoff first did it, that the relationship among them is a generalized form of his mesh law.

In the third step we establish a qualitative relationship between the two complementary sets of measurements in the given universe of discourse. If the latter is of a physical nature, this relationship, with a loan from thermodynamics, is termed the second law. Alternatively, if we are considering an economic universe, the relationship is a law of rational behaviour which, apart from a change in sign, is completely analogous to the second law.

In the fourth and final step we establish a quantitative relationship between the two complementary sets of measurements in the given universe of discourse. Here, two possibilities are left open to us. Either, we can introduce products of an intervariable and a transvariable like energy, power, or total revenue. Or, we can introduce ratios between an intervariable and a transvariable like Ohm's and Hooke's laws or like the elasticity and flexibility concepts. If the first mentioned approach in terms of products is selected, we end up with the energy conversion methods of classical mechanics, the state-function methods of modern control theory, or the mathematical programming techniques of econometrics. If the last mentioned approach in terms of ratios is adopted, the end product is vectorial mechanics, electrical network theory, or the so-called partial equilibrium analysis of economics which deals with demand and supply of individual markets.

To illustrate into depth the validity and limitation of this interdisciplinary system structure we discussed, in some detail, the analogy between Ohm's law of electrical network theory and the elasticity concept of economics. To confirm the validity of this analogy we carried it into its extreme applications in either discipline. Thus, we saw that ideal voltage and current sources had their counterparts in perfectly elastic and inelastic demand (or supply) curves, whereas, the concept of mutual inductances corresponded, albeit in a converse manner, to that of cross elasticities. To exhibit the limitations of this analogy we showed that, where Ohm's law provided us with a quantitative answer regarding the state of the electrical network, the logarithmic refinement inherent in the elasticity concept constrained us to a qualitative answer concerning the state of the economic system.

REFERENCES

1. Franksen, O. I. "Mathematical Programming in Economics by Physical Analogies. Part I: The Analogy between Engineering and Economics". *Simulation*, no. 6, June 1969, 297–314. "Part II: The Economic Network Concept". *Simulation*, no. 1, July 1969, 25–42. "Part III: System Equilibrium and Mathematical Programming". *Simulation*, no. 2, August 1969, 63–87.

2. Pedersen, P. O. "Et produktionsdynamisk problem (A Dynamic Problem of Production)". *Nordisk Tidsskrift for Teknisk Økonomi*, Serial no. 1, Sept. 1935, 28–48.
3. Poincaré, H. *Science and Hypothesis*. 1905. (Reprinted by Dover Publications, New York, 1952.)
4. Campbell, N. *What is Science?* 1921. (Reprinted by Dover Publications, New York, 1952.)
5. Stevens, S. S. "On the Theory of Scales Measurement". *Science, 103,* no. 2684, 7 June 1946, 677–680.
6. Firestone, F. A. "A New Analogy Between Mechanical and Electrical Systems". *J. Acoust. Soc. A. 4,* 1933, 249–267.
7. Firestone, F. A. "The Mobility Method of Computing the Vibration of Linear Mechanical and Acoustical Systems: Mechanical–Electrical Analogies". *J. Appl. Phys. 9*, June 1938, 373–387.
8. Trent, H. M. "Isomorphisms between Oriented Linear Graphs and Lumped Physical Systems". *J. Acoust. Soc. A. 27,* no. 3, May 1955, 500–527.
9. Koenig, E. and Blackwell, W. A. *Electromechanical System Theory.* McGraw-Hill, New York, 1961.
10. Stevens, S. S. "Measurement, Psychophysics, and Utility", in Churchman, C, West and Ratoosh, P. (eds.). *Measurement: Definitions and Theories*. Wiley, New York, 1959, pp. 18–63.
11. Bridgman, P. W. *The Logic of Modern Physics*. 1927. (Reprinted by MacMillan, New York, 1960.)
12. Bridgman, P. W. *The Nature of Physical Theory.* 1936. (Reprinted by Dover Publications, New York.)
13. Seeger, R. J. "On Teaching Physical Mechanics". *J. Eng. Education,* June 1956, 842–854.
14. Brillouin, L. *Scientific Uncertainty, and Information.* Academic Press, New York, 1964.
15. Debreu, G. *Theory of Value – An Axiomatic Analysis of Economic Equilibrium.* Wiley, New York, 1959.
16. Weyl, H. *Symmetry.* Princeton University Press, N.J., 1952.
17. Fog, B. and Rasmussen, R. *Driftsøkonomi* (*Micro-Economics*), Vol. I. Einer Harcks Forlag, Copenhagen, 1964.
18. Dorfman, R., Samuelson, P., and Solow, R. *Linear Programming and Economic Analysis.* McGraw-Hill, New York, 1958.
19. Baumol, W. J. *Economic Theory and Operations Analysis*. Prentice-Hall, Englewood Cliff, N.J., 1965.

NOMENCLATURE

Electrical quantities

E	voltage source
e	voltage drop (potential difference)
q	dielectric charge
i	current
Z	impedance
Y	admittance
G	conductance
C	capacitance
L	inductance
M	mutual inductance
t	time

Economic quantities

q	commodity amount
i	commodity flow
p	price
TR	total revenue
MR	marginal revenue
E	point elasticity
F	point flexibility
t	time

Mathematical quantities

a, b, k, m, x, u, w	constants
x, x', v, u	variables

ECONOMIC THEORY—NEW PERSPECTIVES

J. M. English
Engineering Systems Department
School of Engineering and Applied Science
University of California, Los Angeles
U.S.A.

SUMMARY

Some concepts are developed that are new, at least to one whose formal background has been in engineering. Also, they are believed to be either new or at least different from the point of view of the economist. In many ways they are speculative and in no way represent fully developed economic principles. They are submitted in this chapter for the purpose of generating discussion and with the hope that they will spark further research in quite new directions.

The approach is to extend into the realm of economics, fundamental concepts that have been fruitful in developing and helpful in understanding engineering. The concepts of inertia and storage, essential elements for treating the dynamics of physical systems, are employed in the economic system and are shown to have their equivalent effects in *capital*. The entropy concept, important in all systems, is extended to the economic system and is demonstrated to exist in the information system known as money. A final extension of these concepts suggests an explanation of inflation and its relationship to unemployment. In all, it is an ambitious undertaking and is recognized all too clearly by the author to just scratch the surface of what is felt to open up the possibility for new insight into economic theory.

1. ECONOMICS AND PHYSICS

1.1. Introduction

In the physical sciences certain causality relationships have been found to be invariant and as a consequence were formulated into laws that always hold true under certain well defined conditions. Very often the underlying principles on which the laws hold were not understood until long after the law was formulated. In the meantime the usefulness of the relationships was valid in that they satisfactorily explained observed phenomena. An example might be the gas laws of Boyle and Charles without an understanding of the atomistic system of molecules on which they are based.

From time to time some new principle was discovered which at first seemed to violate a previously established law. In every case the new principle was found to represent a greater generalization under which the old law could be subsumed as a special case. For instance, Newtonian mechanics is a special case of the more general Einsteinian mechanics. While controversies have occurred in physics they have lasted only so long as it took to show that the physical phenomena could be explained in alternative ways provided that the constraining conditions are adequately described. The fundamental condition is that the phenomena are always predictable within limits of statistical variations associated with their measurement.

In economics no such happy state of affairs seems to exist as yet. In almost every aspect of economic theory one finds strong differences of opinion on the validity of opposing hypotheses. These develop emotional adherents to one "school" or another.

For example, consider just a few such situations:

One of the most basic concepts in economics is that of *value*, while most economists disavow the labour theory of value there are many others as strongly committed to it.

While most economists subscribe to the principle that the market is the ideal mechanism for resource allocation, we seem to move further and further away from it in practice. Thus we observe, that the strongest adherents of economic policy based on dependence on the market for establishing prices, have become the agents for administered pricing.

In the realm of monetary policy, controversy is particularly evident. Some contend that money is the key variable for economic stabilization policy while others feel just as strongly that it is of secondary importance. On international exchange rates, arguments become extremely heated.

On the question of capital theory there still are complex views ranging from the Marxian exploitation theory through the savings–investment concepts of Keynes.

The very fact that controversy can exist at all is evidence of a lack of valid theory. As soon as it is possible to demonstrate a single exception to a theory, that theory is by definition invalidated. Thus a valid economic theory is incontrovertible and not subject to debate.

1.2. The Walrasian law of equilibrium in exchange

When we search for economic laws, we do find one that is valid – the Walrasian law of exchange under conditions of equilibrium. This law simply states that all goods offered in exchange must equal all goods demanded. In other words, nothing is left over when the operation has been completed. However, the time to reach the equilibrium state is never considered. The Walrasian law is usually identified with a market process, but this is not essential. The central concept is that of static equilibrium. It is no different from the Kirckhoff node law in electrical networks or the static equilibrium law in mechanics – the sum of the forces must equal zero ($\Sigma F = 0$). In economics the sum of the values, considering sign, equate to zero.* At

*Franksen [2, 3, 4] has utilized these same laws, along with the power conservation principle, to describe the production process in an economic system where the technical coefficients are constant. The difference in point of view will be discussed later.

any one time when a set of goods and services supplied to the market equal the set demanded, then there is a unique set of relative prices establishing the equality condition.

$$\Sigma x_i p_i = \Sigma y_i q_i \tag{1}$$

where X and Y are the quantities and P and Q the prices. At the next point in time a new set of supplies and demands will exist and correspondingly a new set of prices will hold. It takes time for all economic agents to know what the new set should be in order for the market to be cleared. Therefore, there is an implicit relaxation time for the equilibrium state at t_1 to come to the new equilibrium state at t_2.

It will be noted that when a physical quantity is multiplied by a price the product becomes a measure of value, v. Therefore when sign is considered as being positive into the exchange and negative out of it (1) can be written as

$$\Sigma v_i = 0 \tag{2}$$

A second point to be emphasized is that the Walrasian equations represent exchanges of values based on supplies on the one hand and demands on the other. These exchanges are the result of the perceived collective desires of all economic units participating in the exchange process. It is applicable to a set of already produced goods being exchanged between individuals desiring to obtain a better mix of what they hold. Only implicitly does the relationship subsume goods produced on one hand and goods consumed on the other. While the exchanges take place between economic units, there is nothing that requires these to flow exclusively from producing units to consuming units.

Furthermore the exchange process is essentially continuous in time. The exchange process represents a flow of values. Equation (1) would be just as valid if sign considered, it were to be expressed as

$$\Sigma \dot{v} = 0 \tag{3}$$

where
$$\dot{v} = \frac{d(x_i p_i)}{dt} = \frac{d(y_i q_i)}{dt}.$$

It is to be noted that the equation, while implicitly including time is still one of static equilibrium as is Kirchhoff's node law that

$$\Sigma i = \Sigma \dot{q} = 0 \tag{4}$$

where $\dot{q}$ is the rate of flow of charge that we identify as a current.

The same applies in mechanics.

$$\Sigma F = \Sigma \dot{\phi} = 0 \tag{5}$$

where $\dot{\phi}$ is rate of change of a mass potential that we call momentum.

It will be observed that both of the two physical laws, the Kirchhoff node law and the Newtonian static equilibrium law, while extremely important, are highly specialized cases. No dynamic term is included. Both in mechanics and electrical networks the truly useful extension comes with the introduction of a term to allow for change. In both cases a term to take account of inertia or storage capacity is introduced. In mechanics for example, the d'Alembert principle, that of introducing a force that depends on a mass acceleration, extends the generality of the equation of statics to encompass the dynamics of the system. In economics there is no satisfactory equivalent of the d'Alembert principle. There have been attempts to find one (for example Grubbström, [7]) but they are not entirely satisfactory.

One may be led to question why we seem to have gone so much further in the understanding of physics than of economics. I suggest that the answer may be in the inherent limitations of our human perceptions. In attempting to explain our perceptions we form concepts or mind-constructs of the physical world. These lead us to formulate hypotheses which we then test by searching for further perception. In the physical world we always started by considering something external to us. Our attempts at explanation have, over time, led to a conceptualization of an atomistic system, the average effects of which, is what we perceive. Thus temperature is a perception of an average of the activities of millions of molecules. We have a concept of the motion of a molecule and how it interacts with other molecules to produce the perception of hotness that we measure by temperature.

In economics we necessarily must proceed in the other direction. Our perception, in an economic sense, is the effect on our individual welfare of an exchange with other individuals. *We are the molecules.* From this perceptual level it is understandable that economic theory started with models dealing with individual interactions (micro theory). The explanation of economic systems or macro economics was then largely constructed by some method of aggregation. We have tended to add effects rather than to conceptualize a macro system independently of the individual actions.

In order to examine how we may take dynamics into account it will be necessary to consider some basic concepts in human interactions that lead to the exchange process.

1.3. Organization

Economic activity comes about because of the nature of the organization of human affairs. The concept of organization is elusive. In a general sense any arrangement of elements is a system. To the extent that these elements are arranged according to some recognized purpose, it may be said to be organized. Correspondingly arrangements that go counter to the purpose represent degrees of disorganization.

In turn the condition that the system be arranged according to purpose implies an intelligence or knowledge on the part of an individual (or collective) who is capable of providing the signals needed to effect the organization. In other words organization is a function of information or lack of randomness, and disorganization that of randomness.

A system may be organized as a collection of sub-systems, each of which may be organized for an original purpose unrelated to that of the system itself. A very simple example will make this concept clear. A pencil is an organization comprising a writing element and a wooden holder. The holder is a collection of molecules arranged according to some biological purpose that exhibit the characteristics of wood. This sub-level organization is of no concern to the designer of the pencil. He is only interested in the net effective properties of wood.

A further characteristic of organization to be kept in mind is that the purpose arises from the viewpoint of an intelligence external to the system. In other words the system may only be considered as an isolated universe disjointed from its environment, and its performance characteristics perceived from a point of view and purpose arising from outside of it.

In physical systems it is the motion of atomistic elements that results in a manifestation of the system phenomena, e.g. temperature. (For the present the nature of this motion in economic systems will be postponed.) The activity of the elemental units in any system may, when acted on by external influences, result in organization or conversely when uncoordinated, degeneration of the organization results. A system left to itself tends to disorder. The latter notion is associated with the term *entropy*. The external influence may be termed a signal and the amount of its effect on the system as a whole is identified by the increase in the organizational level or *neg-entropy* of the system.

From the foregoing definitions the system may be examined from the point of view either of its entropy as a measure of its lack of order at any time or of the change of its entropy caused by the information content of the external influence, i.e. the received signal. In other words the concept of entropy in thermodynamics and information theory are essentially the same. Concept of entropy has also been used in economics from other perspectives, for example Theil [9], Georgescu-Roegen.**

2. THE BASIS FOR ECONOMIC ACTIVITY

All biological processes are organizing processes even where, in the end, the ultimate result represents a greater degree of disorganization or increase in the entropy of the universe. The individual animal, as a highly organized biological system in itself, is contributing to the organizing function of its species. Thus man in response to some cosmic purpose is caught up in the process of decreasing the entropy of the earth's resources by organizing them in a more complex way. For example, he mines copper from an already partially organized state in nature, he processes it into mechanical devices, and then uses these for some purpose that he perceives to meet his objective. Ultimately, he consumes it by degrading it again and then discards it as waste. In the end the entropy of the copper has increased in that it is more diffused than it was originally as an ore body. In economics we may categorize the organizing or entropy decreasing processes as *production* and the disorganizing or entropy increasing processes as *consumption*.

On the basis of the cosmic influences acting on mankind the conclusion from this avenue of reasoning would have to be that the primary motivation of the species, man, is production and that consumption is at least in part incidental to the production process.* (Galbraith, [5], English, [1].)

On the other hand, the individual does not perceive the cosmic picture. He is concerned with maintaining himself as an individual living organism. To do this he

*From both a different point of view and line of reasoning, this is consistent with the conclusions of Galbraith (1967). Galbraith contends that the consumer responds to the demands induced by large corporations.

**This paper was written before the author had encountered the thought-provoking work of N. Georgescu-Roegen [11]. Georgescu-Roegen contends that economics is essentially entropic and that economic theory has gone astray by reliance on mechanistic models.

perceives his motivations as satisfying a consumption need. Nevertheless, the viability of the species depends on the net averaged effect of all activity resulting in a decrease of entropy internal to the species. In other words, in some incomprehensible way the individual must be motivated to produce on balance more than he consumes. He perceives this as a need to ensure his survival by providing reserves against unforeseen needs. However, in this regard it will be observed that even when all possible future needs are provided the individual man will still tend to produce. At the same time as his productivity is increasing his real needs may also be growing. This may be a result of the changing complexity (organization) of society that dictates an increasing consumption level as a concomitant of organizational growth.

An analogy might be that of a heat engine. The production of useful work must be accompanied by a waste (consumption) of some heat. The need for greater output requires matching increased output of waste heat. In any economy the effect is evidenced by the increased dependence on new technologies that have evolved to the point that societal consumption patterns make them necessities. For example, not many years ago an automobile was not a necessity, but now it is.

2.1. Economic activity

All individuals are constantly active. This activity may be dichotomized as production and consumption both of which are processes that transform material resources and energy from one form to another. As stated earlier, production is a process of transformation from a less organized to a more organized form and consumption is the reverse or the transformation into waste.

The purpose of both is the satisfaction of individual needs. To the individual the primitive need is survival and this requires consumption. However, once his basic consumption needs are satiated in the context of his particular environmental demands, he may only increase his satisfaction by directing his surplus energies to production. This may lead to outputs that become available for consumption by others whose needs exceed their productivity, by increasing societal consumption levels, or by the net increase in organized resources that remain unconsumed.

2.2. Some important economic concepts

It may be well at this point to recall certain terminology used in economics. In the first place, economics or economic activity is concerned only with some form of signal transmission characterized by a value exchange between individuals. This requires that we differentiate between two kinds of human activity:

Self-satisfying activity representing production-consumption cycles, that does not involve an exchange between individuals.

Organized activity that represents transfers from one individual to another.

Economic activity is represented only by the latter. Each individual within limits may vary the amount of self-satisfying activity by choice. When he selects more of it, he regards it as *leisure* activity. A man may work in his garden and gain satisfaction both from the activity of gardening and the subsequent enjoyment (consumption) of the flowers he has produced but no economic activity is identified with this process. On the other hand, if he employs a gardener to do the work, the exchange of the gardener's services for wages represents an economic value. While it may be presumed that the man, employing the gardener, does so because he prefers to spend his own time on other activities, it does not follow that the net value to him equals the wages of the gardener. Rather it is the difference between the values of enjoyment in the work of gardening and the alternative activity that he substitutes for gardening. Economics does not measure this effect but only the amount of the exchange between the man and his gardener.

Up to this point we have considered only individuals. However, the individual may be considered as an economic unit compromising a collection of individual persons. Therefore, we may properly identify economic activity by exchanges between economic units. These economic units, as collective individuals, will divide their activities between the self-satisfying kind – no economic effect – and the interactive kind or in other words, economic activity.

Now we may also differentiate between economic units by the essential purpose of the activity, whether of production or consumption. Since we are interested in collectives, we define production units as firms and consumption units as households. However, it should be emphasized that this differentiation represents what the net economic activity of each of economic unit is. Within each a great deal of activity is of the self-satisfying type, for example the work of the housewife in the family (household) is productive but this does not change the essential economic purpose of the household. It is still consumption.

Also within the firm a great number of related and complementary activities proceed concomitantly. These require transfers from department to department but because they are internal they are not counted as economic activity. The accounting system to keep track of such exchanges may use the same monetary units as for external transactions but this does not affect the overall economic measures. These are the net value flows *into* and *out* of the firm as a single economic unit.

2.3. Macro economic measures

The sum total of all transactions, excluding certain transfer payments and imputed value exchanges that are clearly recognized as having economic significance, constitutes *gross national product* (GNP). Thus it may be observed that all human institutions that contribute to increasing organizational complexity and specialization of production, contribute to increasing GNP. When the housewife goes out to work and at the same time hires a maid, there may be only a slight increase in utility to each but both now contribute an increment to GNP. In effect the GNP, as a measure of economic well being, is in reality a measure of increased organization, that is, decreased entropy of the economy.

Not all decisions to change organizational form necessarily reduce economic entropy. When one firm merges with another to become a single economic unit, it is, often with the intent of reducing costs. If the effect hoped for does not materialize then entropy will have increased. Even where the merger is successful, some of the economic exchange that had taken place between the two units before the merger become internal operations after the merger and so are excluded from the economic accounting. This would imply an increase in entropy but on the other hand the internal entropy of the reorganized production operation should have decreased and the total entropy of the economy and the submerged firm may have in fact decreased.

2.4. Money in the economic activity

We have restricted economic activities to those that require transfers between economic units. The individual has a choice between self-satisfying activity and entering into the organization game that we have described as economic activity. This means that to the extent that an individual is motivated to engage in the latter he does so by means of transmitting a signal that elicits a response activity on the part of other units. In turn his activity is increased by received signals. For the production unit, this signal is most visible in the actual transfer of influence (or control) over use of physical resources. That is, the output of production is passed on to a receiving unit that is either a final consumer or another producer at a consecutive stage in the production process. While the actual effects may be discrete and repetitive it is reasonable to conceptualize these transfers as flows.

There are other effects that do not represent physical flows but must be considered in the same sense because they are economic influences of just as much concern as the physical flows. These fall into the broad classification of services. The effect is the same, in that the activity of the service-producing unit influences the receiver. In the case of a physical product, the activity of production is a transformation process acting on the physical goods (including moving them) thereby affecting the activity of the receiver of these goods. In the case of service the same sort of thing occurs without a visible intermediary set of goods; the initiating activity directly influences the receiving activity.

Such induced activity occurs because of transmission of a signal. This signal is the flow phenomenon that contains all the information of the economic activity and as such is what we must examine in greater depth.

It should be further observed that as soon as we can divorce our thinking from the physical flows, they become incidental to the process we wish to describe in economic terms. This is equivalent to a mental construct we use for understanding the nature of perceived phenomena in ways that we can measure and predict cause-effect relationships. This concept is no different in principle to the mind constructs we use to explain physical phenomena. For example, we are perfectly content to explain electrical phenomena by means of a current flowing through a conductor. The current is conceptually a signal that carries information. It in no way describes the actual effects at an atomistic level. These are complex interactions of orbiting electrons. If a true flow occurs, it must be in the opposite direction from what was originally imagined but this anomaly makes no difference at all to our ability to predict effects exactly as if current flowed as we conceptualize. The actual activity is of a micro nature but the predicted effects are macro phenomena. The important thing for the understanding of the process is not the physical flows but the information contained in the signal that induces them.

2.5. Value and its measurement

Since we are interested in the response stimulating signals, it will be necessary to establish some way to measure both the intensity of the signal and the quantity of the response. This can only be done by observing under what circumstances an individual responds statistically to the signal. No measure is afforded by a non-response. This may simply be interpreted as not being of sufficient intensity to reach the threshold stimulus for an exchange to take place.

The degree of satisfaction or utility deriving from any activity, whether production or consumption, is not directly measurable. Only when the individual reaches his threshold, the point where he is willing to trade rights to engage in a particular activity at the margin of his utility, does an opportunity arise to measure it. The rights to a consumption activity are represented by goods offered in exchange and the rights to a production activity are represented by labour. Under a condition of static equilibrium all transactions are presumed to obey the Walrasian law.

One difficulty with (1) is that it is indeterminate. It cannot be solved for any absolute scale of price but only a relative one based on selecting either a single good or an index based on some arbitrary set of goods as a reference or *numéraire.*

The individual is interested in maximizing his present utility but encompassed in this is a concern for his future well being. Thus included in the trades he is prepared to make, is the foregoing of some present value in expectation of a later satisfaction. Therefore, contracts for future delivery form part of the set of trades. A second way to ensure future satisfaction is to buy and store goods that may be expected to have enduring value and which accordingly may be re-traded at a later date. This represents hoarding. Finally, if for any reason, it is expected that the demand for the stored good will increase then the future price will increase. Current prices will reflect this by speculation. These expectations all influence the equilibrium price set in the present. In particular the general acceptance of one commodity as the *numéraire* will unduly influence its relative price by creating an added demand for the hoarding of it. Gold is the classical example of such a situation.

Relative prices will change over time as a result of two effects: (1) the increasing organization of activity will affect the production outputs of different goods as differing rates of technological change, and (2) the different supplies of goods as well as different perceptions of individual utilities will alter. For these reasons a price set can only be described probabilistically and cannot be related to any fixed reference value scale.

One further observation is important. No self-satisfying activity is ever measured; only actual exchanges are of interest. Thus all measures or attempted measures of economic activity are concerned only with value as determined by exchange.

2.6. Nature of the exchange mechanism

Exchanges except for the insignificant element of barter are accomplished by means of the individual either surrendering goods or services first for a promise of receiving a later equivalent value, or alternatively of consuming first and promising to repay later. In other words, transactions represent a time separation of production and consumption and so require the creation of debt. The process requires that each individual responds to a message by deciding to surrender something of value. This message carries the information required to organize the economic system. Stated another way, all economic activity is a result of organization of many individuals into production and consumption activities. Since all organization is accomplished by means of communication, that is by an information system, this information system is a fundamental characteristic of economics.

The ways in which the exchange-effecting messages are transmitted are numerous. In an auction the sign of a finger or a particular coded sound may suffice. A message on the telephone, a letter, or a conventional set of symbols on a piece of paper all represent ways of transmitting signals. However, in all of these the information is later stored in some form of record. For each individual the summation of all his transactions are made by various categories. Some represent positive and some negative balances. By various other signals he may be free to change one category to another. For example, he may purchase shares of stock in some enterprise thus reducing his credit balance in the bank and increasing the implied credit balance with his broker.

It might be argued that his ownership of stock does not represent a debt on the part of the company. However, this is a legal technicality. In an economic sense his ownership of stock is a way to obtain a prospective future value. It is probabilistic but so is every other prospective value. The differences are only those related to differences in the probability density functions. These differences derive from accountability or rules categorizing legal rights but do not affect the principle of futurity of values and this is why in an economic sense they may be viewed as debt or its equivalent, *money*.

If each message transmitted perfect information, then the values surrendered would be replaced eventually by equivalent values. However, this implies perfect knowledge of both all amounts and all relative prices at the time of the exchange. Because this is impossible the transmission is made with some entropy. The message is subject to *equivocation* or loss of information.

Money is the conceptual device that we have invented to accomplish the information transmission. In its first characteristic, we denominate it as an index or *numéraire* to relate relative values of different goods. However, more importantly the quantities of money transferred represent the accounting for values of goods transferred at the time of their surrender. The recording of these amounts either in a formal record keeping sense or in men's minds store the information of a future promise. Although there is nothing more than this as a concept, nevertheless it leads to the notion of a second property of money, that of a *store of value*.

The significant thing to remember is that at no time is this store of value real, but is storage of a signal that has the potential for initiating transfer of goods. Only goods available at any time can be delivered at that time. Therefore the (money) information of stored value is subject to entropy.

2.7. Analogous physical systems

In the physical world all measurement is based on comparison of one system chosen as a standard and separated from the system to be measured. Franksen [3, 4] has neatly classified physical measurement and extended this to economic measurement. He points out persuasively that our approach to measure is dictated more by limitations of the human thought process than by the physical nature of the phenomenon to be measured. For this reason economic phenomena can be developed on the same basis as mechanics, electrical networks, and thermodynamics.

Fundamentally there are two classes of variables – *across* or *trans* variables and *through* or *inter* variables. In the first we measure differences in potential to produce an effect between two points of reference. In the second we measure the effect, interpreted as a flow from the point of higher to the point of lower potential. In mechanics the potential may be referred to geometric location within a gravitational field and the flow as force. In the electrical sphere it is electrical potential (volts) and current. At any point in either the physical or economic system in static equilibrium, the rather uninteresting condition must hold that the sum of the flows (sign considered) must equate to zero. In economics as will be recalled this is in exact correspondence with the Walrasian law of exchange.

There are two principles corresponding to the conceptualization of the universe that enable us to use these measures to predict the action of any system, disjoint from the environment in which it is embedded. The first is that of conservation of something across the system boundaries and which may always be shown to be the product of a transvariable and an intervariable. The second is the principle of flow orientation that requires the flow effect to move in the direction from the higher to lower potential.

2.8. The principles in economics

Franksen has argued that these principles apply to economics in the context of the Walrasian model. He takes prices as potential and the physical flows of goods to satisfy the Walrasian equation. A possible flaw in his argument is the dimensionality of goods which requires that price be measured as a reciprocal of a commodity characteristic, e.g. dollars per pound of aluminium. Interestingly it required him to define his prices as neg-prices in order to satisfy the second law. In spite of this minor shortcoming his contribution represented a major advance in conceptualizing economics from a fresh perspective and has contributed significantly the ideas presented in this chapter.

One major problem in economics has been the lack of a storage mechanism. In physics it is represented by capacitance or inertia. The development of the d'Alembert principle in mechanics enabled the simple static condition of equilibrium to be extended by introduction of an inertia force to the solution of the dynamic problem. It is hypothesized that its equivalent in economics is capital, but, while it will be shown how this may come into a dynamic economy, the question of how to make it fully operational remains open.

2.9. Economic dynamics

The conservation principle is an analytical device that enables us to transform the potential for producing an effect into predicted effects of one disjoint system on another. Two things are needed to accomplish this: A conceptual model of how the effect is explained and a perceptual scale for measuring the potential. It is not essential that we know precisely how the effect is produced. All that is required is that cause and effect are invariant within the defined conditions of the model. At the same time the scale of measurement is necessarily separate from the system to be measured.

In mechanics, we observe that separate bodies located in a Cartesian coordinate system have a potential to change position in some mysterious way. We find it necessary to introduce the concept of force associated with a change of relative position of the bodies and reason that for any static equilibrium condition corresponding to a fixed set of positions, all forces must be mutually cancelling. In other words $\Sigma F = 0$.

We then proceed conceptually to separate out any body we choose and treat it in isolation (free body). Now consider a change in position of the body. In the first place change of position takes time and in the second place the new position will represent a change in the potential. This requires that the potential-induced effect must equate with the rate of change of potential or

$$F = -\frac{d\Phi}{dt}. \tag{6}$$

Conceptually, an equivalent relationship is:

$$F = \frac{d(mv)}{dt}, \tag{7}$$

the rate of change of momentum, where m is a characteristic of the body conceptualized as mass and v is relative rate of change of position in the coordinate system. The introduction of this term, as demonstrated by d'Alembert enables us to consider the dynamics of the body as an equivalent static equilibrium case.

Now let us translate these principles into economics. We recall that the economic system is described completely by the effects of each unit on all others with which it interacts. The potential for exchanges, i.e. transmission of signals, between economic units arises from some indefinable influence that these economic units perceive will increase their utilities. In order to make this system tractable, we need some reference scale for measuring effects, and along with this a scale for the potential.

Unfortunately there is, as yet, no way to define a meaningful separable scale for a macro-economy. We simply cannot get outside of it in order to observe it. However, on an abstract sub-system level we may be able to do something. For example, consider a single firm. The value flows into and out of it may be measured as monetary flows. When these do not sum to zero, the firm is not in static equilibrium. A net accumulation (or dispersal) of money values is occurring. In an analogous physical sense we may introduce the equilibrating force term.

$$K \,.\, \ddot{v} = -\frac{d\Phi}{dt} \tag{8}$$

We must now question the interpretation of these terms.

It would seen evident that the economic potential, Φ, must be a function of the organizational effectiveness of the people comprising the firm. This would be represented in the complex of motivations of the individuals comprising the team, the way in which they can work together in concert, and the like. In other words in some not readily understood way the potential is the organizational inertia of the economic unit.

Now consider the left-hand side of (8). An unbalanced flow of value into the unit represents a buildup of capital in the accounting sense. If for the time being we were to accept this as a true change in capital in monetary terms, then it is readily seen that K is the capital value of the firm. It remains only to call attention to the need for a reference acceleration required to make the equation dimensionally consistent. In the mechanical system, the mass acceleration is referenced to a coordinate system represented by the earth as the convenient system. Thus the mass constant includes the earth's gravitational acceleration. Although in economics we do not have a convenient external reference system, in the case we have chosen, that of book capital equating to real capital value, the reference acceleration reduces to unity.

This idealization is of course unrealistic. Firms that exhibit strong growth-characteristics, in general have capital values far in excess of book values. Whether it might be possible to find a representative firm as reference, or have to resort to some averaging process in order to establish a suitable index, is worthy of further exploration.

Another approach for the determination of K might be to consider the right-hand member of (8). This would require a measure for economic potential. In this connection the very interesting work of Grubbström [6, 7] should be mentioned. Grubbström has also been searching for a satisfactory explanation of economic dynamics. His interesting discussion of the work of other economists clearly calls attention to a distinction between dynamics and kinematics. In most economic literature, attempts at a dynamic analysis have been oriented towards examination of the change-of-state from one time-period to another. This, as he points out, is *economic kinematics.* Only when a causal relationship is introduced explicitly, can an explanation for a change-of-state be found. In another sense, the change-of-state from one time-period to another is predicted on an implicit assumption that causitive influences were introduced and the system then had sufficient time to relax to a new equilibrium state. Thus, such economic analyses are properly called comparative statics. Only with the introduction of a storage or inertia term can a system dynamics be treated.

Grubbström recognizes the need for the acceleration term and describes this as

$$\alpha y = -\frac{du}{dx} \tag{9}$$

which, except for symbols, is identical to (8). However, Grubbström approaches the problem from the micro-economic viewpoint of market exchange and defines the economic potential as the individual's utility, u. The difficulty here is that it is not easy to conceptualize a storage- or inertia-mechanism that may be associated with utility, and the approach seemingly did not lead him to recognize the essential capital characteristic of α. At the same time it must be recognized that motivational influences on individuals are manifested by utility, even though utility is a non-operational concept.

2.10. The exchange process

The foregoing discussion avoided consideration of the market in the Walrasian sense where, by means of the *tâtonnement* process all buyers and sellers come to know the required relative prices necessary to clear the market. The individual offering his services or demanding something from the market receives or pays money that contains imperfect information of the values that he can later demand or be called upon to supply.

He is in a position to originate a signal, i.e. pay money, in proportion to assets he holds, and that others recognize to be convertible to money. The ability to generate money with which to buy depends only on the degree of belief of others. Whether this belief originates because of real physical assets such as gold, a piece of real estate or the like, or whether it comes from a belief that there will be a subsequent money flow to the buyer is quite irrelevant. The important thing is that the individual is capable of transmitting a signal that produces a response.

Consider the hypothetical case of an economy in which at a particular point in time there is a certain set of goods and services to be exchanged. If the signal system contained perfect information of all demands and supplies, then the market would be cleared. Ostray [8], has discussed conditions of market information under which exchanges do take place and extended this to the resulting effect on the market. Market clearing simply follows from the definition of static equilibrium. Regardless of the time and iterative procedure needed, the exchange will have been completed only when perfect information is reached.

Because we are concerned with a stochastic dynamic model in the real world, new supplies and changing demands are continuously flowing in the process and perfect information under these conditions is even a less realistic ideal. It should be noted, however, that if the demand and supply flows reach a steady state with only random oscillations about mean values, the information of the money flows to facilitate the exchange would reach a point where only the random noise of these oscillations would remain in the signals. This is analogous to thermal noise. In effect it represents an irreducible *entropy of money* in an exchange process in static equilibrium.

Now if the economy is dynamic and the way in which it is changing is not known precisely but is learned after time lags, then the information of money flows can be expected to be even less perfect, that is, the entropy will increase.

From this reasoning it should be quite evident that imperfect information in the signals will result in unsatisfied demands and excess supplies. It is not so clear that to overcome the difficulty all that is needed is to increase money flows in such a way as to create general inflation of prices. A simplified model of a two-person economy may show what actually happens.

3. A TWO-PERSON ECONOMY

The way in which the economy works may be visualized by means of a highly idealized model, from which most of the real world complexities have been removed.* We will postulate two economic agents only, (A and B). In each time period each agent produces one good in excess supply above his own inelastic demand for it. We will also postulate that these two goods are perishable so that any quantity left over at the end of the period cannot be carried forward to the next period. Each agent will maximize his utility for the period if he is in a position to obtain the other's excess supply. Under these circumstances, utility will be maximized for the economy under the condition that A will trade all of his excess supply q_1 for all of B's q_2. Clearly if they know what their respective quantities are, they will agree on a relative price in the ratio of q_1/q_2. However, at this point we will introduce an essential constraint, entirely artificial for this idealized model, that neither A nor B is permitted to know the other's supply. They must agree on a price in the absence of that information but with knowledge of past experience.

Now if the game has been played in accordance with these rules in previous exchanges. They both will have *a priori* probability functions of each other's outputs. The question is: what is their best strategy for agreeing on a relative price?

We will define p as the price index at which the market will be cleared. The probability density function will be

$$f(p) = f(q_1, q_2) \tag{10}$$

Now they will agree on a price p_e, that will prove to be different from the actual p (*a posteriori*). Depending on whether $p > p_e$ or $p < p_e$ either A or B will be left with an undemanded surplus. We can define this as an unemployed resource and for simplicity at present call it *unemployment.*

*The author has developed the subsequent model in a later paper [10].

The value of unemployment will be the product of the quantity remaining and the price as a relative value.

The unemployment ratio will be

$$U = |p - p_e|. \tag{11}$$

Of course the probability of $p_e = p$ is zero, except for the case of perfect information and so

$$U > 0$$

The minimum level of unemployment is

$$U_{\text{Min}} = \text{Min}\,(|p - p_e|)$$

which by definition is

$$E(|p - p_e|) = \sigma \tag{12}$$

for the condition that

$$p_e = E(p) \tag{13}$$

Any negotiated exchange price either greater or less than $E(p)$ will result in a higher expected unemployment.

3.1. The distribution functions

The distribution function $f(p)$, is a function of supplies that are available for exchange in any one period. This will be dependent on natural variabilities (crop yield, weather, etc.). If the agents could know with certainty, *ex post facto*, what the actual quantities were, then $f_0(p_1)$ Fig. 1, would be valid. Unfortunately, they

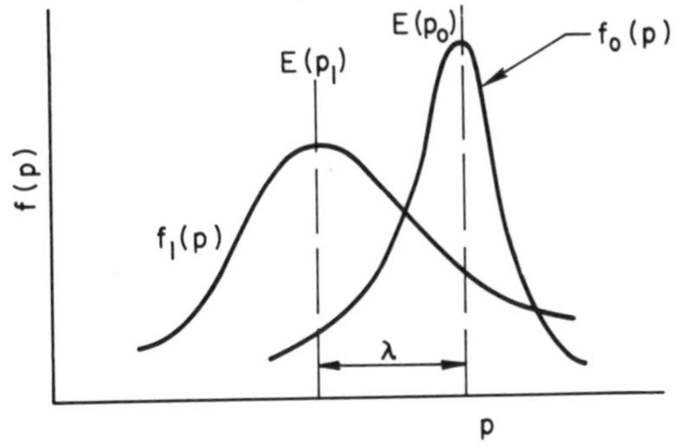

Fig. 1. Distribution functions for price.

do not have a perfect information system. There is loss in the (money) information system and accordingly the distribution function $f_1(p)$ will be the one perceived by the two agents. Under these circumstances, they may be expected to set the exchange price at $E(p_1)$. This requires that expected unemployment will be increased as a function of the shift of the expected exchange price from $E(p_0)$ to $E(p_1)$ or by assuming statistical independence.

$$U = \sqrt{U_0^2 + \lambda^2} \tag{14}$$

The relationship of (14) holds regardless of the position of f_1 relative to f_0. However, it may be reasonable to hypothesize that f_1 lies to the left of f_0 as depicted, rather than to the right. The rationale is that in an expanding economy there is a tendency to conservatism coupled with myopia. That tends to make the agents underestimate future quantities.

3.2. Inflation effect

The exchange price will be defined in respect to a reference money index value. The change of reference over time is defined as inflation. Thus the price distribution function for market clearance at time, t_0, will be simply shifted by the amount of the inflation to a new price scale at t_1, Fig. 2.

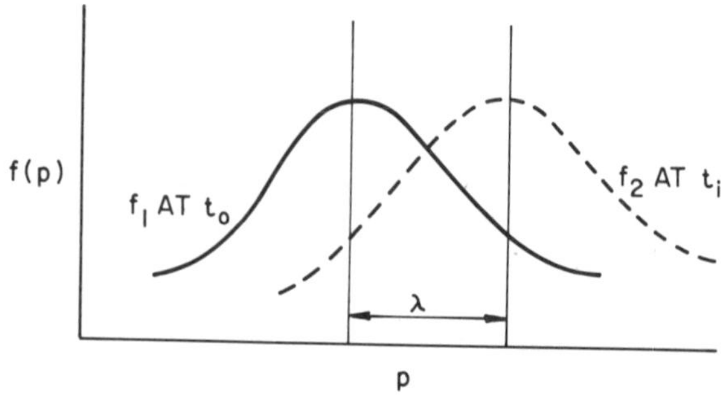

Fig. 2. Inflation effect.

If there were no inflation effects, the mutual expectations of the clearing price required to minimize unemployment would be $E(p_0)$. However, since each will surrender goods supplied for money at t_0 and which will be received later at t_1, as goods demanded, they will add their mutual inflation expectations to the expected clearing price in order to agree on an exchange price,

$$p_e = E(p_1).$$

A necessary condition for inflation to occur is this time lag between the supply-demand functions.

3.3. Stable inflation

It is apparent that where the entropy of money leads to the perceived distribution of the clearing price $f_1(p)$ to the left of $f_0(p)$, the function corresponding to zero money entropy, then any inflation policy will serve to reduce unemployment, provided only that $f_2(p)$ does not shift to the right of $f_0(p)$. In that case the inflation point of absolute minimum unemployment will have been passed and unemployment will be found to increase accompanied by increasing inflation.

3.4. Constraint on over-inflating price

The basis on which either agent agrees *a priori* on an exchange price is the money he has available to buy the other's outputs. This money is simply the credit deriving from assets they each hold. Each one is interested in obtaining all of the other's supplies and therefore would be inclined toward bidding high. If A overbids relative to B, he will obtain all of B's surplus but at the expense of a reduction in his asset values. This might represent a smaller loss of utility to him than failure to demand all available supply. For this reason alone he will be disposed to overbid.

In addition to this effect, a tendency for each to overbid will result in an inflation of the price index so that the subsequent money return will cancel the transaction with a concomitant inflation of asset values.

However, asset values ultimately depend on the capitalized value of future returns and so it will not pay either to overbid beyond the point that even with inflation they do not have sufficient credit to negotiate suitable exchanges in future periods.

3.5. Generality of the model

A question naturally arises as to whether this two-person economy can be generalized to a multi-person economy. While a formal proof is not offered here, it is suggested that each individual *A* faces an exchange situation with a multi-person composite individual *B*. In this case *B*, as a collective, does not know his excess supply but that does not alter the principle developed above. Thus the general economy is represented by a multi-variate distribution function of the price index.

The constraint imposed, that the goods in the two-person model be perishable, is not unrealistic for the general economy. Many goods can be inventoried and so carried forward from one period to another. This washes out the variation in production as a factor in unemployment. On the other hand, many things cannot be carried forward. In particular labour must be employed continuously in time and so fits the condition perfectly. As a result it tends to be the residual unemployment index.

3.6. Inferences drawn from the model

Some interesting inferences may be drawn from the two-person model. These lead to explanations of inflation phenomena that hitherto have been somewhat puzzling.

1. There can never be a known set of prices to clear the market and so there is an irreducible minimum level of unemployment that tends to the natural variance of supplies on the one hand and variance of peoples' values that translate into demands on the other.
2. Economic conditions will reflect a rapid change in technology or a rapid change in societal values by increasing inflation and/or unemployment. Rapidly developing economies must, therefore, expect higher inflation rates than mature economies.
3. There is a trade-off between inflation and unemployment but this will be modified as a result of (2) above.
4. Policies that drive unemployment down have the potential for over-shooting the optimal level and so result in an instability. This instability will be manifested by both inflation and unemployment growing concomitantly.

REFERENCES

1. English, J. M. "Economic Concepts to Disturb the Engineer". *The Engineering Economist, ASEE,* Vol. 20, no. 2, 1974.

2. Franksen, O. I. "A Physical Analogy of the Walrasian System". ch. 5 in J. Morley English (ed.) *Economics of Engineering and Social Systems.* J. Wiley, New York, 1972.
3. Franksen, O. I. "Mathematical Programming in Economics by Physical Analogies, Part I: The Analogy between Engineering and Economics". *Simulation*, no. 6, June 1969, 297–314. "Part II: The Economic Network Concept". *Simulation,* no. 1, July 1969, 25–42. "Part III: System Equilibrium and Mathematical Programming". *Simulation*, no. 2, August 1969, 63–87.
4. Franksen, O. I. "Basic concept in Engineering and Economics", in Dixhoorn, J. J. van, and Evans, F. J., (eds.) *Physical Structure in Systems Theory*, Academic Press, London, 1974 (this volume).
5. Galbraith, John K. *The New Industrial State.* Houghton Mifflin Co., Boston, Mass., 1967.
6. Grubbström, Robert W. *Market cybernetic processes.* Almquist & Wichsell, Stockholm, 1969.
7. Grubbström, R. W. *Economic Decisions in Space and Time*. Gothenburg Studies in Business Administration, Sweden, 1973.
8. Ostray, J. M. "Informational Efficiency of Monetary Exchange". *American Economic Review, LXIII,* no. 4, 1973.
9. Theil, H. *Economic and Information Theory.* North-Holland Publishing Co., Amsterdam, 1967.
10. English, J. M. "*Inflation: An Entropy Explanation*" (To be published).
11. Georgescu-Roegen, N. *The Entropy Law and Economic Processes.* Harvard University Press.

NOMENCLATURE

F	force	U	unemployment
i	electrical current	v	velocity
i	subscript number	V, v	value
K	constraint representing capital	$\dot{v}$	value flow (i.e. dollars per year)
m	mass	x	quantity of goods
P, p	price	y	quantity of goods
q	price	GNP	Gross National Product
q	charge	α	used in Grubbström equation as equivalent to K
q	time rate of change of charge	Φ	potential
t	time	λ	inflation effect
u	utility		

AUTHOR INDEX

SUBJECT INDEX